FOUNDATIONS OF NATURAL HISTORY

Foundations of Natural History is a series from the Johns Hopkins University Press for the republication of classic scientific writings that are of enduring importance for the study of origins, properties, and relationships in the natural world.

Published in the Series:

Materials for the Study of Variation: Treated with Especial Regard to Discontinuity in the Origin of Speices, by William Bateson, with a new introduction by Peter J. Bowler and an essay by Gerry Webster

Nicholas Copernicus: On the Revolutions, translated by Edward Rosen

Nicholas Copernicus: Minor Works, translated by Edward Rosen

Problems of Relative Growth, by Julian S. Huxley, with a new introduction by Frederick B. Churchill and an essay by Richard E. Strauss

Fishes, Crayfishes, and Crabs: Louis Renard's "Natural History of the Rarest Curiorisites of the Seas of the Indies," edited by Theordore W. Pietsch

Historical Portrait of the Progress of Ichthology, from Its Origins to Our Own Time, by Georges Cuvier, edited by Theordore W. Pietsch, translated by Abby J. Simpson

Cosmos: A Sketch of a Physical Description of the Universe, vol. 1, by Alexander von Humboldt, translated by E. C. Otté, with an introduction by Nicolaas A. Rupke

Cosmos: A Sketch of a Physical Description of the Universe, vol. 2, by Alexander von Humboldt, translated by E. C. Otté, with an introduction by Michael Dettelbach

COSMOS

A SKETCH

OF

A PHYSICAL DESCRIPTION OF THE UNIVERSE

BY

ALEXANDER VON HUMBOLDT

TRANSLATED FROM THE GERMAN
BY E. C. OTTÉ

Naturae vero rerum vis atque majestas in omnibus momentis fides caret, si quis modo
partes ejus ac non totam complectatur animo.—Plin., *Hist. Nat.*, lib. vii, c. 1.

VOLUME II

WITH AN INTRODUCTION
BY MICHAEL DETTELBACH

THE JOHNS HOPKINS UNIVERSITY PRESS
Baltimore and London

Introduction ©1997 The Johns Hopkins University Press
All rights reserved. Published 1997
Printed in the United States of America on acid-free paper
Reprinted from the English-language edition published by
Harper & Brothers, Publishers, New York, 1850
Johns Hopkins Paperbacks edition, 1997
06 05 04 03 02 01 00 99 98 97   5 4 3 2 1

The Johns Hopkins University Press
2715 North Charles Street
Baltimore, Maryland 21218-4319
The Johns Hopkins Press Ltd., London

Library of Congress Cataloging-in-Publication Data will be found at the
end of this book.

A catalog record for this book is available from the British Library.

ISBN 0-8018-5503-9

Frontispiece: Alexander von Humboldt, etching, Jacobs Pinx, Prudhomme.
From the 1858 Harper & Brothers edition of *Cosmos,* volume 1.

INTRODUCTION
TO THE 1997 EDITION
Michael Dettelbach

"You have traveled much and seen many ruins in your life, and now you have seen one more," Humboldt said, offering me his hand. "Not a ruin," was my immediate reply, "but a pyramid." I pressed the hand which had pressed those of Frederick the Great, Forster, the companion of Cook, Klopstock and Schiller, Pitt, Napoleon, Josephine, the marshals of the Empire, Jefferson, Hamilton, Wieland, Herder, Goethe, Cuvier, Laplace, Gay-Lussac, Beethoven, Walter Scott —in short, every great man which Europe had produced in three-quarters of a century. I peered into eyes which had not only observed the history of the present world pass before them, scene by scene, until the actors one by one disappeared from the stage, never to return, but which had also gazed upon the cataracts of Atures and the forests of the Cassiquiare, upon Chimborazo, the Amazon and Popocatepetl, the alps of the Siberian Altai, the steppes of Tartary and the Caspian Sea . . . Never have I looked upon so sublime a manifestation of venerable age, crowned with undying achievement, full of the greatest wealth of knowledge, animated and quickened by the most prized attributes of the heart. A ruin, truly? No, a human temple, as perfect as the Parthenon.
—Bayard Taylor, *New-York Tribune*, 1856

When the much-anticipated second volume of *Kosmos: Entwurf einer physischen Weltbeschreibung*, was published in November 1847, Prussia was experiencing a brief lull in the political excitement of the year. Prussia's first national representative assembly, a combined meeting of the provincial estates, had been convened by King Friedrich Wilhelm IV in April and dismissed in July, without winning recognition as a "national assembly," as liberal representatives had demanded, but also without acceding to the king's request for new loans for railroad construction. Within the first five months of 1848, protesters were cut down by royal troops in the streets of Berlin, an elected parliament was sitting in the National Theater, and a ministry including businessmen from the Rhineland was governing Prussia. A year later, Prussian soldiers had once again entered Berlin, and reaction held sway at court. The German public received the second volume of Humboldt's magisterial essay on the certain and progressive realization of universal natural order—the completion of the full outline of the doctrine of a "cosmos"—just when such progress in the political realm seemed anything but certain.

These events capped two decades of political and social transformation in Germany known as the Vormärz, the prelude to the March 1848 revolutions. The Restoration after 1815 had predicated both the international order and the internal stability of European states on the maintenance of a Europe composed of effectively absolute monarchies. This was above all true within the German Confederation, where Clemens von Metternich, the Austrian chancellor, had successfully and pragmatically organized reaction against the erosion of monarchical prerogative by constitutional and nationalist movements. But beginning with the repercussions of the July 1830 revolution in France, which finally overturned the Bourbon throne, the Vormärz threatened finally to dissolve the Restoration constitution of Germany. Prussia, a largely new and officially Lutheran state spanning northern Europe from the Rhine to Poland, which contained after 1815 some of the most Catholic and industrially advanced areas in Germany, especially seemed on the verge of change, as it had immediately after the defeat of Napoleon. After 1830, Prussia seemed ready once more to take up the constitutional promises of the Reform Era, and the enthronement of the intellectual and popular Friedrich Wilhelm IV in 1840 raised liberal hopes that Prussia might lead a renovation of German political life.[1]

While he was composing the first two volumes of *Cosmos* in the 1830s and 1840s, Humboldt had a keen appreciation of Prussia's precarious position between reaction, reform, and revolution. Before his move to Berlin in 1827, Humboldt had accompanied Friedrich Wilhelm III (1770–1840; r.1797–1840) to all the major summits of the great powers (at Aachen,

Troppau, Laibach, and Verona), by which the Vienna Congress of 1815 had attempted to maintain stability and monarchy in Europe, and for several summers Humboldt had accompanied the king to the royal spa at Teplitz ("the eternal springs," Humboldt complained), where Prince Metternich could often be found with the Prussian monarch's ear. In 1827, after having spent twenty-three years in Paris and finding himself in financial straits, Humboldt was forced to accept Friedrich Wilhelm III's request to return to Berlin, where, as royal chamberlain, he became a permanent fixture at the Prussian court. Humboldt had varied responsibilities at court. He was frequently present at the king's tables at Sans Souci and Charlottenburg, after which he was expected to hold his own in conversations with the king and his official ministers and read to the royal family. He accompanied the Crown Prince, later King Friedrich Wilhelm IV (1795–1861; r.1840–58), on diplomatic missions, attending the convocation of the Polish Diet in 1830 (after which Czar Nicholas I suppressed the body), the baptism of Prince Edward of Wales in 1842, or the dedication of the Beethoven Monument in Bonn in August 1847 (where Prince Albert of Saxe-Coburg, consort of Queen Victoria, told Humboldt that he felt sorry for the miserable rebels in Cracow so brutally treated by the Russians during the 1846 revolt—but really, the Poles deserved as little consideration as the Irish). Humboldt's French connections gave him an important role in keeping channels open between the July Monarchy and German courts, against pressures from Vienna and St. Petersburg that kept Germany on the verge of war with France throughout the 1840s.[2]

During the Vormärz, the very name *Humboldt* recalled Prussia's Reform Era, the years of economic and political reform when there was an attempt to build a national political life that followed the monarchy's humiliation by Napoleonic armies in 1806. Wilhelm von Humboldt (1767–1835), Alexander's elder brother, was at the center of the political, economic, and cultural reconstruction of the Prussian monarchy after 1808, and people recalled his angry departure from government as Minister for Constitutional Affairs in 1819, when Friedrich Wilhelm III decisively foreclosed the possibility of a written constitution for Prussia. Alexander, a living relic of the first French Revolution, had worked under the reform chancellor Hardenberg as a mining administrator and diplomat in the 1790s. In the growth of Prussian political consciousness in the 1830s and 1840s, Humboldt watched the lives and legacies of dead acquaintances become live issues again. Wilhelm's own death in 1835 earned him the same fate, much to Alexander's frustration.[3]

Although his nemeses at court considered him a "tricolor-waving Jacobin" and reactionary papers viewed him as a materialist demagogue, Alexander von Humboldt was no revolutionary.[4] He was certainly a "liberal," but before 1848 and the beginnings of parliamentary politics in Prussia, *liberal* denoted a broad sensibility, an intellectual direction rather than a commitment to a particular course of political action or set of political goals.[5] Until the end of his life, Humboldt maintained a politics grounded in the eighteenth-century Enlightenment. The economic, political, and intellectual emancipation of the individual were all inextricably interwoven; as the French Revolution had demonstrated, the violent introduction of republican institutions was no guarantee of individual freedom. Free (popular) political institutions were only one aspect of the individual freedom that was the "noble destiny of humankind," and Humboldt long lamented the eruption of the civil wars for independence in South and Central America. In his three "political essays"—on New Spain (Mexico, 1808), New Granada (Venezuela and Columbia, 1825), and Cuba (1826)—he used geographical and statistical tableaux to describe each colony's state of social development and the tremendous obstacles that stood in their way. But once independent republican governments had become established in the former Spanish colonies, Humboldt vigorously defended them against European intervention on the principle that "in all governments already established, whether republics or moderate monarchies, reforms must be gradual to be salutary."[6] Similarly, in order to maintain a westward openness in Prussian politics, he had to defend the July Revolution and the new regime at court while remaining privately skeptical of the depth of change and reform that political revolution could achieve. As one Humboldt scholar perceptively put it, "The oscillation between the two capitals Berlin and Paris [eight times, for as much as a year at a time, between 1830 and 1848] mirrored the tension between his republican convictions and the pragmatic political insights and practices evident in his American studies."[7]

Humboldt's ideal was reform introduced by an informed and liberal monarch who had the confidence of his subjects. Humboldt was quite sincere in his dedication of the *Essai politique sur la Nouvelle Espagne* to Spain's Charles IV, like that of *Cosmos* to Prussia's Friedrich Wilhelm IV. Humboldt especially enjoyed the trust of Friedrich Wilhelm IV, who admired the aged savant, appointed him to the Council of State in 1840, and established a royal order distinguishing achievement in the fine arts and sciences (but specifically excluding theology), with Humboldt as permanent chancellor. Friedrich Wilhelm IV liked to think of himself as a "progressive" and "modern" king, a new Frederick the Great, enthusiastically affirming his belief that church and state were both subject to inevitable improvement; that Prussia must advance intellectually, technologically, and culturally; and that government must become popular. (As Humboldt once said, though, one "progressed" with Friedrich Wilhelm IV as one might progress toward the North Pole on an errant ice floe or down a river with high, narrow banks: there was indeed progress, but it was impossible to tell if one were moving backward or forward.[8]) Humboldt returned his sovereign's good faith, if not his unalloyed trust. Even at low points for liberal ideals—when Humboldt narrowly dissuaded Friedrich Wilhelm IV from backing a series of new regulations against Jews (1842), or during the reaction of the 1850s—Humboldt blamed the king's romantic notions rather than his will and took every opportunity to preserve "the noble and youthful sensibilities" of the monarch against the partisan temper of the times.[9]

Humboldt's greatest fear during the Vormärz was politics—that the growth of political passions and the sharpening of ideologies would eventually consume Prussia in political intolerance and violence—and he placed great hope in "the noble and youthful character" of the king to move reform forward calmly, while maintaining the trust of all parties. An especially worrisome threat to reasonability, sociability, and tolerance came from the orthodox Lutheran party in Prussia. Conservatives in Prussia grew firmly convinced that maintaining social order depended on maintaining religious orthodoxy and uniting the people in subservience to king and God. In the minds of Prussian conservatives, unbelief and anarchy were linked, and an evangelical revival hostile to science and philosophy found support in the 1830s among Prussia's most important aristocratic families and officials. Friedrich Wilhelm III, who had first united the Lutheran and Calvinist churches into the Prussian Evangelical Church in 1822, remained aloof from the intensification of evangelical Lutheranism in the 1830s. Friedrich Wilhelm IV, however, was inclined to recall the inviolability of his sacred, paternal office against calls for constitutional reform, and to see the national mission as a Christian one. He actively sponsored orthodoxy at court and in the universities.[10] Friedrich Eichhorn, who was Friedrich Wilhelm IV's Minister of Religion and Education (*Kultus und Unterrichts*) imposed new regulations on lecturers and temporarily suspended both the historian Friedrich von Raumer and the philosopher Karl Michelet from their teaching posts in 1847.[11]

The paralyzing effect of the zealous atmosphere on Prussian intellectual life was a frequent subject of comment in the correspondence between Humboldt and August Boeckh, University Orator, permanent secretary of the Berlin Academy's historical-philological class, and one of Humboldt's chief authorities for classical civilization. In 1847, Boeckh himself threatened to require the academic senate to approve his orations before he would give them. There reigned "a madness for seeing political tracts and materialist catechisms [*Naturbibeln*] in every contribution to the world of letters," Humboldt wrote to Boeckh in 1842.[12] In 1846, Boeckh used the annual oration on the occasion of the king's birthday to contest the claim that philosophy ought to remain within limits dictated by theology by arguing for "the authority of the positive, which is essential to maintain the order of civil life and to protect that which has already been achieved": philosophy's limits and privileges were both dictated by its obligation "to investigate the nature of things in their positive character."[13] In the 1850s, Humboldt lamented the tendency of Prussia's most educated classes to indulge in mysticism and urged Cotta (who in addition to being Humboldt's publisher also published Germany's most respected newspaper, the *Allgemeine Zeitung*) to stop giving space to reports of table turning and spirit writing.[14] When in 1854, at the height of the reaction, Friedrich Wilhelm IV appointed Friedrich Julius Stahl, the most vocal ideologist of sacred monarchy, to the inner chamber of the Council of State, Humboldt in frustration finally resigned his own seat.[15]

Humboldt did his best to insulate *Cosmos* from political-religious polemics, while asserting the high, philosophical claims of physical science. Despite repeated mentions of "the creation" and "created things" (e.g., 2: 98),[16] Humboldt argues in *Cosmos* that one could understand the order of nature, "the plan of creation," without relying on religious faith or sentiment or resorting to natural theology; that one could grasp nature as a purely physical unity—that, conversely, this grasp is not synthetic or demonstrative, but aesthetic and premonitory—and that natural science steadily advances this understanding. Physical science would never succeed in reducing nature to rational unity. Humboldt's studious avoidance of theological concepts in his description of cosmic order was thus not just an attempt to insulate *Cosmos* from evangelical criticism but to affirm positively the secular character of scientific and social progress in the face of growing obscurantism and to show that theism had nothing to fear from natural science.

By closing volume 1 of *Cosmos* with a nod to the universal ethical significance of Christianity, taken from his brother Wilhelm's work on comparative linguistics, Alexander killed two birds with one stone: he elevated *Cosmos* above theological criticism and put the Humboldt name in a light sympathetic to Christianity.[17] Humboldt had hoped that in volume 2 of *Cosmos* the Church authorities would read as "edifying" his sympathetic treatment of the role of Christianity and the writings of the Church fathers in cultivating an interest in nature study (38–43, 199).[18] Humboldt actually added a passage to the English and French translations specifically to address fears that the doctrine of the "cosmos" led to materialism and to reassure theologians that a "physical description of the universe" that revealed nature as a physical whole posed no threat to "higher studies," borrowing the mantle of Immanuel Kant to divide physical science from theology (1: 30–31).[19]

Of course, such tactics risked displeasing all sides rather than reconciling them. The devout and aristocratic *Rhein- und Mosel-Zeitung* accused Humboldt of "Voltairianismus" (i.e., French materialism and freethinking) and ranked him with the heterodox Hegelians Feuerbach, Bauer, and Marheineke. The king (Humboldt made a point of telling Varnhagen) thought the accusation mere bluster.[20] On the other hand, Prince Metternich wrote to congratulate Humboldt for restoring the word *discipline* to a place of honor in the sciences, by showing the limits of natural philosophy: "God grant that the same idea assert its eternal rights in the civil sphere as well!" From the Left, the young physiologist Karl Vogt urged scientists to ignore the limits that *Cosmos* had set to physical research and through science to declare war on the immaterial soul.[21] A "Book of Nature" thus could not avoid being a political work in the Vormärz.

Cosmos clearly affirms Humboldt's liberalism, his faith in the inevitable progress of political and economic—hence moral—

HUMBOLDTIAN SCIENCE AND THE IDEA OF COSMOS

Cosmos was thus neither an encyclopedia of the sciences nor a popularization of the sciences but a construction of physical science itself. Physical cosmography was itself a science, a critical collection and meaningful ordering of empirical data; but it also stood above the various special sciences: it offered a scientific portrait of the current state of physical science itself, an evaluation of the significance of work in the particular sciences, as such work contributed to the progress of our comprehension of the world as a physical whole, a "cosmos." As a "physical cosmographer," Alexander von Humboldt occupied an awkward role that stood somewhere between philosopher and natural philosopher, critic and practitioner of science. As a scientific portrait of the sciences, taking for its object the physical unity of nature, Humboldt's "physical cosmography" had what we might call "philosophical" or "metascientific" ambitions.

To show that "work in the natural sciences can be conducted to a higher standpoint, from which all particular forms and forces are revealed to be one physical whole, animated by an inner impulse" (1: 36/39), Humboldt had to develop a model of science that rigorously separated the activity of fact-gathering from that of theorizing and speculating about causes; or rather, one that made facts incipient theories and made sensations incipient concepts, so that fact-gathering was always implicit theorizing and the office of nature-philosopher was made obsolete and given to history itself. Humboldt's picture of nature's unity—its reflection in the products of the artistic imagination and its gradual realization over the course of history—was to "strengthen the bond which, according to ancient laws governing the very core of the intellectual realm, ties the sensible world to the insensible" and to "stimulate the communication between that which the mind receives from the world, and that which, from its depths, the mind returns" (1: 63/82).

Humboldt modeled natural science on the researches that had occupied his own public life in the sciences. Indeed, *Cosmos,* which Humboldt described as "a work, whose undefined image has floated before my mind for almost half a century" (1: ix/v), has something of an autobiographical, apologetic character; it represents Humboldt's life's work as the characteristic expression of modern intellectual development. Most commentators, following Humboldt's lead, have traced the idea of a "physical cosmography" back to the 1790s, when Humboldt first announced that he had discerned the "bare outlines" of a new science, *physique du monde,* a "physics . . . of the sort Francis Bacon demanded," one that would unite all the branches of natural science.[26] His 1799–1804 expedition through the American tropics and his nine-month research trip through Central Asia in 1829 were conceived as exemplary endeavors in this new, comprehensive, Baconian "physics."[27] A "physics of the world," Humboldt told the Prussian Academy of Sciences

Figure 1. *Opposite page.*
Albert Berg (1825–1884), "Virgin forest, ca. 7000 ft.; in the background, Tolima," Plate 3 from *Études physiognomiques sur la végétation de l'Amérique tropicale; recueillis dans les forêts vierges sur la fleuve Magdalena et dans les Andes de la Nouvelle-Grenade* (London: Paul and Dominic Colnaghi, 1854). Courtesy of the Library of the Arnold Arboretum, Harvard University.

This lithograph is one of thirteen Berg did from sketches he made while duplicating Humboldt's 1801 journey up the Magdalena River to its headwaters in the Andes. Such "physiognomical studies" attempted to capture the associations of plant forms which Humboldt felt especially determined the particular aesthetic character of a landscape. Such studies did not merely aim for picturesque effects, however; in their ability to convey a physiognomy, they were also demonstrations of physical law and the "cooperation of forces." In a letter that Berg used as a preface, Humboldt praised Berg's depictions for their ability not only to excite the viewer's emotions with the grandeur and wild profusion of their objects, but also to raise the viewer's intellect to the concepts of physical geography. Here, in a clearing near the volcano Tolima, Berg portrayed dense vegetation that included *Heliconia* (lobster claw), the waxy trunks of the Andean palm *Ceroxylon andicola,* the aborescent fern (*Cyathea*), the young fronds of *Oreodoxa frigida,* framed by the larger trunk of a *Urostigma,* festooned with vines and epiphytes.

immediately after his return to Berlin in 1805, would "track the great and constant laws of nature manifested in the rapid flux of phenomena and trace the reciprocal interaction, the conflict, as it were, of the divided physical forces" by attending to all the forces and phenomena of nature as they varied over time and space.[28]

The temperature, pressure, humidity, electrical tension, optical and acoustical properties, and chemical composition of the atmosphere; the temperature and movements of the oceans; the intensity of the Earth's magnetism; the elevation and stratification of the Earth's crust; the development and character of the Earth's vegetation; every variable force or phenomenon drew the attention of the true physicist, especially insofar as those forces or phenomena were measurable with precise, reliable, and comparable instruments. By amassing a number of sufficiently sensitive and accurate observations over a large expanse of territory or time, and then comparing them—or better yet, mapping or graphing them—the lawfulness and regularity of a particular phenomenon, and thus its essentially physical character, as the effect of other physical forces, could be, if not mathematically expressed, at least seen and recognized. Simultaneous attention to the full range of regularly varying forces and phenomena thereby generated a "picture" or image of nature's self-determination, its physical unity. As Humboldt explained in his 1807 "physical portrait of the tropics," a summary of the results of his American expedition:

> In this great web of cause and effect, no fact can be considered in isolation. The general equilibrium which reigns amongst disturbances and apparent turmoil is the result of an infinity of mechanical forces and chemical attractions balancing each other out. Even if each series of facts must be considered separately to identify a particular law, the study of Nature, which is the greatest problem of general physics, requires the combination of all the forms of knowledge which deal with the modifications of matter.[29]

Putting precise, standardized instruments in the hands of travelers or extended networks of stationary observers to measure and map the Earth's surface (topographic maps), magnetic field (isodynamic maps), heat (isothermal maps), weather (isobaric maps), geology (stratigraphic atlases), and so on, was characteristic enough of research in the period and sufficiently identified by its practitioners with Alexander von Humboldt to prompt historians to invent the category of "Humboldtian" science.

"Physical cosmography" was clearly the last incarnation of Humboldt's novel science. The very word *cosmos* was resurrected as shorthand for the dynamic unity of nature addressed by Humboldt's Baconian physics: the "connection of phenomena," "the cooperation of forces," "the contemplation of corporeal things under the rubric of a physical whole, animated by indwelling forces" (1: 52; 1: ix–x/vi–vii)—formulas with which Humboldt explicated the "cosmic" character of nature—denote nothing other than the "general equilibrium" of forces that was the object of "world physics." To perceive the "connection of physical facts" or the "cooperation of forces in the universe" is to recognize the particular phenomena of nature as the product of dynamic physical causes (forces and specific matters) or laws and their mutual interaction (reinforcement, limitation, disturbance). Humboldt took his paradigmatic examples of such "connection" from the fields he pursued in the Americas and Asia: meteorology, climatology, and plant geography. The recognition of the connection of the distribution of climates to the distribution of heat and the shapes of continental landmasses, and then, of the geography of plants to the distribution of climates, typified the manner in which the physical unity of the entire creation might be apprehended (1: 28/29). Marshaled under the banner of a physical cosmography, the natural sciences strive toward a common end, the cognition of nature in its character of self-determination and physical necessity: "the discovery of physical laws, the establishment of a lawful and regular classification of natural forms, the recognition of the necessary connection of all phenomena [*Veränderungen*] in the universe."[30]

In a very real sense, Humboldt's formulation of nature's dynamic unity made the discovery of physical law and the progress of physical science into a geographical and historical process. His argument for the scientific character of plant geography, as for meteorology and terrestrial magnetism, relied on cartographic analogies and techniques. By the steady accumulation of precise observations, fixing the "numerical elements" of the phenomenon, its periodic regularity gradually emerged, just as the shape of a coastline or the course of a river emerges from patient and precise mapping. "Cartography was not abandoned when map-makers encountered the sinuousness of rivers and the irregularity of coastlines," Humboldt reminded those skeptical of numerical plant geography. "The laws of magnetism manifested themselves as soon as we began to trace lines of equal declination and inclination and to compare a great number of observations that at first appeared contradictory." This was "the path by which the physical sciences were progressively raised to certain results."[31] The graphic techniques that Humboldt developed to display the emergence of order from the combined observations—isolines, pasigraphies (or universal symbolic languages), directional arrows—were applied to all of Humboldt's studies, because each phenomenon revealed its regularity in the same graphic fashion. In the configuration of the Earth's surface, the geology of its interior, the laws of its magnetism, the patterns of its weather, the distribution of its organic creatures,

> as in all other phenomena of the physical universe . . . amidst the apparent disorder which seems to result from the influence of a multitude of local causes, the unchanging laws of nature become evident as soon as one surveys an extensive territory, or uses a mass of facts in which the partial disturbances compensate one another.[32]

Comprehension of the physical lawfulness of the universe advanced with the accumulation of secure facts under the organizing conception of the conflict, equilibrium, or combined action of forces. The security of facts was guaranteed by the accuracy and frequency of observation and measurement.

Cosmos also elaborates as a general epistemological principle a rule that Humboldt had previously developed only in the context of separate works and particular phenomena: to grasp nature in its physical unity is also to grasp it as the product of a physical history.[33] Leibniz's *Protogaea*, the fanciful cosmogony that marks the modernity of Humboldt's "seventh epoch" (2: 349), recalls the speculative cosmogony that Humboldt himself published in 1799 as an exercise in *physique du monde*.[34] The same emphasis on apprehending cartographic forms—lines, curves, and so forth—that characterizes Humboldt's physics also determines the relation between physical description of the universe and its history. The existing object is to be seen as the historical product of forces; a true history of nature would just be the cooperation of forces projected back in time. "In the material world the reflection of the past in the present is clearer, since [unlike languages] we see analogous products formed before our eyes. . . . Their form is their history. That which *is* is completely known, in its entire extent and inner existence, only when it is known as that which has *become*" (1: 54–55/63–64). To see form is not just to recognize nature's dynamic physical regularity but also to recognize its physical history. It is in this way that "general views" (*generelle Ansichten*) afford insight into Earth's history. Isothermal lines, for instance, record the nebular condensation and gradual cooling of the planet. The visible regularity and lawfulness of the universe, the manifestation of the contemporaneous combined action of forces, indicated its historicity. Such knowledge must always be speculative, however: cosmography was not cosmology, but only gestured toward it, as it gestured toward truly rational, causal laws.

Under the regime of *Cosmos*, physical science becomes an endeavor in natural history, composed of systematic descriptions of phenomena that together would lead gradually to a causal history of nature. In trying to define the manner in which the separate sciences were both "sciences" unto themselves and yet subordinate to a larger and truer science,

Humboldt resorted to the concept of contiguity, an intellectual geography that mapped directly onto terrestrial and celestial geography, the surface of the Earth and the visible sky. Defining scientific disciplines was literally a matter of natural-historical classification, that is, more or less arbitrarily classifying natural objects and forces by their "inner analogies." The separate sciences were related as the objects they treated were related in nature itself, "connected by the same ties which link together all the phenomena of nature."[35]

> The particular objects of nature-study are by their very essence, as though by an assimilative power, capable of mutual illumination. Descriptive botany . . . leads the observer, who wanders distant lands and lofty mountains, to the theory of the geographic distribution of plants over the surface of the earth, in proportion to their distance from the equator and their vertical elevation. In order now to determine the complex causes of this distribution, the laws of temperature variation according to climate, as well as those of the meteorological processes of the atmosphere, must be traced. Thus each class of phenomena leads the observer thirsty for knowledge to another class of phenomena, by which it is determined or which determines it. (1: ix–x/vi–vii)

Recognition of this mutual dependence of physical phenomena is the goal of natural knowledge, the very idea of a cosmos, the notional unity governing the diverse territories of natural knowledge. The separate disciplines were self-governing "provinces," making their own decisions about language and research, but each also relied on the concept of an overarching discipline, physical cosmography, to escape the arbitrariness and impermanence of mere accumulations of facts or distorting systems. The individual disciplines thus appeared as temporary and ultimately artificial groupings of analogous phenomena through which laws and regularities had been traced—this was true even of astronomy and mathematical physics (*Naturlehre*), where (thanks to the "new organ" of analytic calculus) the abstractness and simplicity of the particular phenomena under consideration permitted laws to be "constructed according to atomistic conceptions" and expressed as numerical relation-ships.[36] Like all classifications, those that determined the bound-aries between the separate disciplines were more or less phe-nomenal and provisional, just as the very multiplicity of phenomena themselves (light, electricity, chemistry, and mag-netism, for instance, which came together in the phenomena of photography and galvanoplasty) ultimately hid a deeper unity (see 2: 354/398). *Cosmos* insists just as strongly, however, on maintaining "the limits of the separate sciences." By strictly separating the activities of observation and theorization, the same "critical method" that identified modern science would prevent the premature or illegitimate unification of sciences through "delusive analogies" and vague, symbolic language, too generally applied (2: 355). The sciences must remain as sepa-rate as the natural phenomena they treat, and Humboldt was as convinced that the investigation of nature would multiply the number of specifically different forces and substances as that it would reduce long-studied phenomena to manifestations of a single force.

In taking for its dual object the physical unity of nature and the unification of the physical sciences in a comprehensive "Baconian" science of nature, Humboldt's "physical cosmogra-phy" had what we might call "philosophical" or "metascientific" ambitions, which went beyond "Humboldtian science." Hum-boldt conceived of his new science not just as a new science alongside the other sciences, but as the recovery of the Baconian project for a true physics: a discipline that united all the physical sciences and transformed them by defining their place in a larger project: a truly philosophical investigation of nature.

THE AESTHETIC EFFECT OF NATURE
AS AN INCITEMENT TO NATURE STUDY

The first section of volume 2, "The Incitements to the Study of Nature," seems perhaps the most puzzling and least durable part of *Cosmos*, fitting uncomfortably between the portrait of

nature and the historical development of that portrait. To Humboldt, however, it was one of the most important sections and, as noted below, required a great deal of new research. The immediate contact with nature produces in susceptible spirits a sense of grandeur, of universal connection and harmony. And modern civilization is blessed with increasingly perfect means and increasingly experienced and sensitive spirits to *reproduce* and *render* for others the effects of landscape, the "exalted grandeur of the creation," in the absence of nature itself. The arts are substitutes (*Ersätze*) for the immediate presence of nature itself. The capable artist has the ability to study the individual forms of nature and to so connect and combine them in a landscape as to endow them with ideal significance. These means—linguistic description, graphic representation, and actual re-creation of nature—constitute the chief incitements to the study of nature.

Academically, descriptive poetry, landscape painting, and landscape gardening have never ranked high among the arts, being considered merely imitative and ornamental, and they do not now seem capable of containing the elevated spiritual significance with which Humboldt wanted to invest them, of inspiring in us "a conception of natural unity and the feeling of harmonious accord pervading the universe . . . a knowledge of the works of creation and their exalted grandeur" (2: 98). This contention now seems embarrassingly sentimental, and though Humboldt insisted on it repeatedly in his work, commentators

Figure 2. *Opposite page.*
Karl Blechen, *Inneres des Palmenhauses*, 1834. From Michael Seiler, *Das Palmenhaus auf der Pfaueninsel* (Berlin, 1989).

In volume 2 of *Cosmos*, Humboldt cites the palm house on the Pfaueninsel (Peacock Island), a royal landscape park on the Havel River near Potsdam, as a preeminent example of the modern taste for characteristic impressions of nature and the ability of modern art to reproduce them. The building itself, inspired by a Hindu temple designed by K. F. Schinkel for an opera taken from Thomas Moore's collection of "Oriental tales" *Lalla Rookh* (1817), heightened the sense of local character, and Blechen here confirms its exoticism by peopling his Prussian greenhouse with harem figures.

have never been comfortable with it. Traditionally, Humboldt's "aestheticism," his "holism," is dismissed as an artifact of contemporary "Romanticism."[37] But Humboldt was well aware of the low status of the arts that he recruited to promote natural science, the tendency of nature poetry to become maudlin (2: 80–81), and the need for landscape painting to be idealizing or "heroic," not just imitative (2: 94–95).

Humboldt, on the contrary, was concerned precisely with the relative novelty and popularity of the cultural phenomena he described: these incitements only exercise their power where "modern cultivation" and a "special direction in the development of their mental activity" have made individuals "more susceptible" to the impressions of nature on the senses (2: 21). It is the section in which Humboldt most directly attached his enterprise to elements of the popular culture of the day: to Goethe and Chateaubriand, to Barker's panoramic buildings in Leicester Square and Daguerre's in Regent's Park, to photography, and to exotic gardens like the steam-powered palm houses and waterfalls on the Pfaueninsel in the Havel or Prince Pückler's landscaped park at Muskau, both popular for public promenading.[38] Rather, Humboldt wanted to show that the sciences have the relevance and impact of "popular" culture, to interpret the popularity of panoramas, nature poetry, palm houses, and even his own works and lectures, as manifestations of a particularly modern taste for *experiencing* the lawfulness and regularity of nature itself. The lesson of Humboldt's history of nature representation is that individuals of sufficiently open senses and elevated views had achieved strikingly vivid and true pictures of nature in past ages, but that this taste for nature had become the general character of the age. The modern perfection of these means of imaginative stimulation demonstrates "how, at periods characterized by general mental cultivation, the severer forms of science and the more delicate emanations of fancy have reciprocally striven to infuse their spirit into one another" (2: 20). A physical cosmography was to afford the true aspect of nature to the reader unable to witness nature itself.[39]

THE HISTORICAL DEVELOPMENT OF COSMIC VIEWS

Volume 2 was absolutely central to the exposition of the idea of a "cosmos," though Humboldt had already offered the definition and sketch of physical cosmography in volume 1. The equivalent of empiricizing nature, that is, making the unity of the physical creation the subject of empirical science, was not only historicizing nature but also historicizing natural science. Given that, within the "empirical path" Humboldt had chosen, the physical unity of nature could not be demonstrated but only contemplated, in volume 2, Humboldt turned to history in order to demonstrate that "though the unending endeavor to grasp totality remain incomplete, the *history of cosmic contemplation* . . . teaches us how, over the course of centuries, humankind has gradually achieved a *partial comprehension* of the relative mutual dependence of phenomena" (1: 64/81–82). The "History of Cosmic Views" that Humboldt composed for the second volume of *Cosmos* therefore serves didactic ends: to show how the idea of a *cosmos* silently governs the past path and future course of progress in the physical sciences.

> To those who know how to pursue the primitive origin of our knowledge through the deep strata of the past to its roots, history teaches how for millenia the human race has striven to discover the permanence of law amidst the incessant periodic flux of phenomena, and thus gradually, by the power of intelligence, to conquer the whole globe. To inquire of this past is to trace the secret course of ideas in which the same image, a harmoniously ordered whole, a *kosmos*, which, dimly shadowed forth to the human mind in the primitive ages of the world, is now revealed to the maturer intellect of mankind as the result of long and laboriously gathered experiences. (1: 2/4)

In this respect, Humboldt's *Cosmos* manifests the tendency of the age to replace theoretical with historical exposition; more specifically, *Cosmos* follows the historically oriented philosophies of science of Auguste Comte and William Whewell.[40]

Indeed, *Cosmos* makes natural science thoroughly historical, that is, an embodiment of development, progress, and evolution: a product of external and internal causes acting in a temporal theater, the fixed· points of which Humboldt has located with all the precision afforded by the current state of chronology and historical research. Like nature itself, science is always in process, never still, never fixed in a particular system or language, and yet always the same. Humboldt pointedly deprived apriorism and mysticism of their chief crutches: the beliefs in a primitive, now-lost science of nature and in a unique origin—science is ineluctably the product of positive historical events and its development can be pursued only through scientific history. Seen as the endeavor to comprehend nature's cosmic unity, the history of natural science (*Weltbeschreibung*) reads exactly like the history of humanity (*Weltgeschichte*). *Cosmos* is not only a snapshot of nature's unity: it is also a "portrait of an epoch in the intellectual development of mankind," a self-consciously modern epoch beyond artificial systems and following nature, "that which is ever growing, ever in the process of formation and evolution" (1: 21–22, quoting C. G. Carus's grand summary of idealist morphology, *Of the Fundamental Parts of the Skeleton and Skull* [1828]; see also 2: 354/398). That both nature and science had to be understood as the lawful products of ongoing development was the implication of the compliment Humboldt paid to Whewell upon reading the *Philosophy of the Inductive Sciences*: "What a delight it is, after following you in your ingenious researches on the movements of the ocean, to follow you across these fluctuations of the human mind, varying across space and time, but always subject to periodic returns."[41] As an exemplary branch of "modern physical research," Whewell's tidological science demonstrated not only nature's lawfulness, but that of the human intellect as well.

Both the history of cosmic views and the history (and geography) of artistic sensibility toward nature represent historical scholarship that is as scientific and critical as the portrait of nature in volume 1. Both sections of the second volume make a point of using the latest results of empirical research in political and social history, comparative philology, archaeology, and the history of art and literature. Humboldt incorporated his brother Wilhelm's (1767–1835) final work, *Über die Kawi-Sprache auf der Insel Java* (1836), on the theory of comparative linguistics and the filiations of Sanscrit. From Egypt, where he was leading a Prussian expedition in 1845–46, Richard Lepsius sent regular archaeological reports that Humboldt inserted into *Cosmos* as he received them (e.g., 2: 124). Gustav Friedrich Waagen, director of the Berlin Gallery, fed Humboldt the results of his surveys of English and Dutch landscape painting (e.g., 2: 86–87). August Boeckh kept *Cosmos* current on the interpretation of Greek and Roman literature. Much of the second volume of *Cosmos* relied on Humboldt's own researches into the history of the navigation and the geography of the Americas, some recently published in his *Examen critique de l'histoire de la géographie du Nouveau Continent et des progrès de l'astronomie nautique aux XVe et XVIe siècles* (1834), some published for the first time in *Cosmos* (e.g., 2: 256n). One is advised to read volume 2 armed with a chronology of Egyptian dynasties, the Punic Wars, and the Alexandrian campaigns, as well as with a detailed map and classical synonymy of the Old Continent from Gadeira (Cadiz) to Pentapotamia (Punjab).

Humboldt's history of cosmic views is, like the rest of the work, strangely decentering to the twentieth-century reader because it is essentially a history of Humboldtian science. It is not, Humboldt emphasized, a history of the sciences, yet it is a history of the intuition of nature as subject to physical law, which is at the foundation of all natural scientific investigation, and the progressive realization of this insight in hard-won observations. The epochs of this history are marked by the growing knowledge of the outlines of landmasses and seas, the directions of mountain chains, the distributions of plants, and by the invention of instruments like the barometer, hygrometer, thermometer, and eudiometer, as well as the telescope.

Humboldtian science is not the only subject of Humboldt's history of cosmic intuition. Humboldt's historical investigations also revealed that the mind advances toward a Humboldtian comprehension of the universe in a Humboldtian fashion, by a process of geographical expansion. The gradual realization of the universe as a cosmos is marked by epochal extensions of the range, quantity, and variety of our perceptions, by the "enlargement of the sphere of ideas" (2: 166), and the "generalization of views" (2: 106). Humboldt occasionally described the history of cosmic intuition as revelation, "the account of events and relationships that have revealed to humankind the existence of a cosmos, a well-ordered whole."[42] Navigation, military campaigns, and commercial expeditions—"external events"—mark the turning points of history and the chief forces in the generalization of ideas, until, endowed with the circumference of the globe and the heavens by the voyages of the Columbian Age, the intellect began generalizing and connecting of its own accord.[43]

The figure of Columbus is central to Humboldt's historical narrative, as is the whole Columbian Age that the Genoese admiral represents. The moral and intellectual centrality of the late-fifteenth- and early-sixteenth-century discoveries—its preparation in the medieval recovery of Greek and Roman science and its liberating ramification into the modern age—was the principal theme of Humboldt's *Examen critique de l'histoire de la géographie du Nouveau Continent* (1834). In the expanded historical scope of *Cosmos*, Columbus is foreshadowed by Alexander the Great, as Columbus prefigures Humboldt himself, in a line of sensitive spirits of sufficiently elevated views (2: 155–56 / 186–87).[44] Columbus landing in the New World sounds uncannily like an untutored and unequipped Humboldtian scientist, "without intellectual culture, and without any knowledge of physical and natural science," but endowed by nature with a marvelous ability "to seize and combine the phenomena of the external world": "On his arrival in a new world and under a new heaven, he examined with care the form of continental masses, the physiognomy of vegetation, the habits of animals, and the distribution of heat and the variations of terrestrial magnetism" (2: 263)[45]

Columbus also prefigured Humboldt stylistically: Humboldt credits the admiral with introducing that "modern tone of feeling" (2: 66), "a species of intellectual enjoyment wholly unknown to antiquity" (2: 79). In his very simplicity and directness, Columbus knew "how to represent with the simplicity of individualizing truth that which he has received by his own contemplation" (2: 81). Hence also the repeated association of Humboldt's own pursuit of cosmic unity with a specific experience and a specific locale, his own "Columbian encounter": Chimborazo, a "picture of regularity," where "climates are arranged in strata, as it were, above one another," where "the laws of diminishing temperature . . . are engraved in eternal strokes on the rocky walls of the Andes" (1: 12; 2: 200)—that is, where the unity and lawfulness of the universe, its cosmic character, reveals itself to the sensitive spirit. As Humboldt himself reminded his audience, the Humboldtian enterprise, introduced in the *Essai sur la géographie des plantes* and its accompanying "physical portrait of the tropics," was simply an elaboration of that primordial experience of nature's order in Guayaquil in 1803 "at the foot of Chimborazo," armed with all that the human intellect had to that point achieved, a representative of "modern cultivation."[46] This final, Humboldtian period of the history of cosmic views was inaugurated, like all the others, by an immediate experience of nature.

Humboldt defines this final period by the triumph of a single critical method, which in "modern times" characterizes all intellectual endeavors:

> That which has especially favored the progress of knowledge in the nineteenth century, and imparted to the age its principal character, is the general and beneficial endeavor not to limit our attention to that which has been recently acquired, but to test strictly, by measure and weight, all earlier acquisitions; to separate certain knowledge from mere conjectures founded on analogy, and thus to subject every portion of knowledge, whether it

be physical astronomy, the study of terrestrial natural forces, geology, or archaeology, to the same strict method of criticism. (2: 355)

That is, it is the *critical method* that distinguishes the age and makes it a scientific and progressive one, in history as well as in physical science, by fixing the boundary between the truly known and the less known or unknown, between certain and uncertain knowledge, between observation and speculation, between fact and myth or prejudice. The critical method sets limits and thereby makes the progress of knowledge always a matter of expansion and generalization. The critical method defines science and its progress as the accumulation of rigorously policed facts, alongside less certain observations and mere intimations (*Ahndungen*) of the interconnection of phenomena. According to Humboldt, whereas in previous ages advances in cosmic understanding came by the intervention of external events and were not expressions of the societies that made them, the modern age is thoroughly and essentially scientific in character. In the modern age all branches of cultivation are animated at once by the same liberal impulse toward critical expansion, universalization, and generalization. Again, this critical spirit is not what we consider the distinguishing mark of the scientific attitude or method; it is not even what Humboldt's first biographers considered the principle of scientific advance (mathematical or neo-Kantian metaphysical construction). This observation only emphasizes the degree to which the progress of science for Humboldt was a general historical progress, essentially empirical, geographical, and conservative.

Scientific progress is also, Humboldt insists, a never-ending

Figure 3. *Opposite page.*
"Mata Virgem Perto de Mangaratiba." Lithograph originally published in Johann Mortiz Rugendas, *Malerische Reise in Brasilien* (Paris, 1835).

Humboldt himself never visited Brazil, but he greatly admired Rugendas's renderings of lush tropical features. Rugendas captured the character that Humboldt called the "physiognomy" of a country.

task (2: 354–55). The empirical element, the elements of contingency and indeterminacy, the unresolved, the unknown and unmeasured always remain and should forever make us modest and self-critical. For every nebular cloud resolved into stars, another appears billowing beyond in the telescope's field; for every atomistic or mechanical interpretation of physical processes, a dynamic one based on forces stands ready to supplant it; for every physical law, there are phenomena beyond its limits.[47] Physical research can never attain the last step and infringe upon the territory of religion; but neither can religion or metaphysics impose itself upon or set an end to physical research. "A more animating conviction, and one more consonant with the great destiny of our race, is, that the conquests already achieved constitute only a very inconsiderable portion of those to which free humanity will attain in future ages by the progress of mental activity and general cultivation. Every acquisition won by investigation is merely a step to the attainment of higher things in the eventful course of human affairs" (2: 355).[48]

Humboldt would also have us read the progress of cosmic views as an index to the progress of civilization itself. The "History of Cosmic Views" is also "part of the history of human civilization" (2: 107/137).[49] Reflections on the history of civilization had been central to Humboldt's new science all along. When he left Europe for the New World in 1799, he was already committed, as he told Thomas Jefferson in 1804, "to the study of humankind in its various states of cultivation and barbarity." This commitment resulted in pioneering work in what we might now call archaeology, comparative philology, comparative religion, and political economy, but which for Humboldt were auxiliary parts of a single study: "universal history." Humboldt's *Vues des Cordillères et monumens des peuples indigènes de l'Amérique* (1810 [1814]), the results of his investigations into the remains of Inca and Aztec peoples and their relationship to the rest of humankind, represented a contribution to "the philosophical study of the human mind."

These investigations included studies of monuments, sculptures, ritual objects, costumes, and an extended dissertation on hieroglyphs and their evolution into alphabetic scripts, all concerned with "the analogies observed between the peoples of the two hemispheres." The result was a "picture of the uniform and progressive development of the human mind . . . an imposing spectacle presented by human genius, ranging from the tombs of Tinian and the statues of Easter Island to the Mexican temple at Mitla, and from the malformed idols surrounding this temple to the masterpieces of the chisels of Praxiteles and Lysippus!"[50]

The historical part of *Cosmos* focuses this history of humanity through a history of cosmic awareness—an awareness of the extent and unity of nature. "True cosmical views are the result of observation and ideal combination, and of a long-continued communion with the external world; nor are they a work of a single people, but the fruits yielded by reciprocal communication, and by a great, if not general, intercourse between different nations" (2: 116). Like d'Alembert's prefatory discourse to the *Encyclopédie*, *Cosmos* combines faith in the long-term material and moral progress of humanity through the inextinguishable vitality of the human mind, with an acute awareness of the historical fragility of this progress. Innumerable obstacles might stifle the progressive advance of humanity, among them poverty, political repression, and economic inequality. Whatever might erect barriers between men, isolate them, and fetter the commerce of things and ideas threatened a culture with decline. In this respect, *Cosmos* extends the argument that Humboldt had developed at length in his *Examen critique de l'histoire de la géographie*. The object of the *Examen critique* was to demonstrate that the great discoveries of the fifteenth and sixteenth centuries were no accident: they had deep roots in the Middle Ages, and from there in Greek and Roman antiquity. The mental, then moral, liberations of the Renaissance exemplified a historical principle, that even during apparently "dark" ages, the human mind advanced, preparing for ages of enlightenment: "A conservative principle maintains the living development of intelligence, in individuals or in entire peoples."[51] *Cosmos* thereby performs the functions of both *Weltbeschreibung* and *Weltgeschichte*: "to impress [us] deeply with the belief in an ancient inner necessity, which steers all the activity of intellectual and material forces in ever-recurring circles, which only periodically expand and contract" (1: 30, 32).

<hr>

CONCLUSION

If *Cosmos* seems a somewhat discomforting work, that is because for us it exists in a liminal region, between physical science, philosophy of science, and history of science; between science and "general" or "popular" culture; between natural science and fine art. Humboldt's contemporaries had just as much difficulty judging the book: many critics, Humboldt complained, evaluated *Cosmos* as though it were an encyclopedia (the English critics were especially guilty of this), and others judged the work as a poetic essay on nature, of literary interest, without real scientific interest.[52] Humboldt would find it necessary in the introduction to volume 3 (1850–51), and again in the introduction to volume 5 (posthumously published in 1862), to define his task and "speak out freely once more about the little-understood object of *Kosmos*, about that alone which it could be, and about those demands that have been made of it from the realm of the impossible."[53] As the University of Berlin physiologist Johannes Müller (1801–58) told Humboldt after reading the first volume: "You've demonstrated that [gathering together the threads of all the sciences] is the task of a higher office within physical research itself, but I confess, one which at this point in time you alone are capable of occupying."[54] But Humboldt was not seeking disciples, cosmic scientists to follow his lead: the very aim of *Cosmos* was to "portray an epoch in the intellectual development of humankind" and so to memorialize an entire epoch as "the age of science," and as the age of Humboldt.

Errata

2: 112/143: Otté translates Humboldt's "einer Verallgemeinerung der Ansichten über die Verwandtschaft des Menschengeschlechts und seine muthmaßlich von *mehreren* Punkten ausgehenden Verbreitungsstrahlen" [a generalization of views regarding the interrelations of humankind and its rays of dispersal, presumably from *several* points] to "a generalization of views regarding the affinity of races, and their conjectural extension in various directions from *one* common point of radiation" [my emphases].

This might be a translator's error, but it may also be a translator's "correction" of Humboldt. Humboldt certainly meant that empirical evidence, to which all physical science was limited, could never lead to the cognition of the origin of humankind. Everywhere the researcher looked, there were already peoples with languages and their own origin myths, and the limitations of physical science (in this case, comparative linguistics) taught us to recognize the story of Adam and Eve as just that—the origin myth of Judaeo-Christian peoples (see 1: 364/381). Two related factors might have led to Otté's "misinterpretation": (1) Religious sensibilities. At least one German paper based its accusations of atheism on Humboldt's description of the Adam and Eve story as a myth[55]; and where *Kosmos* reads "have a purely mythical character," *Cosmos* reads "are not devoid of a mythical character." (2) Humanitarian sensibilities. British abolitionists routinely relied on the positive assertion of unitary human origins, backed by the Old Testament or by comparative anatomy, to condemn slavery, whereas Humboldt's indictment of slavery, like his indictment of European serfdom, rested on an appeal to the concept of human nature. Humboldt was perhaps the best-known abolitionist of his time in Europe, successfully convincing Friedrich Wilhelm IV in 1857 to declare that any slave who reached Prussian soil was free. (See Philip S. Foner, *Alexander von Humboldt über die Sklaverei in den USA* [Berlin: Humboldt-Universität zu Berlin, 1981].)

2: 201/239: Otté translates both "Geschichte der physischen Wissenschaften" [history of the natural sciences] and "Geschichte der Weltanschauung" [history of the contemplation of the universe] as the latter, confusing the contrast between them which Humboldt is trying to make here.

Notes

1. James J. Sheehan, *German History, 1770–1866* (Oxford: Oxford University Press, 1989), pp. 604–615. On the United Diet, see Robert Berdahl, *Politics of the Prussian Nobility* (Princeton: Princeton University Press, 1988), pp. 333–47. A detailed treatment of the Prussian reform and its evolution through the 1848 Revolution is given by Reinhard Koselleck, *Preußen zwischen Reform und Revolution*, 3rd ed. (Stuttgart: Klett-Cotta, 1981).

2. See *Briefe von Alexander von Humboldt an Varnhagen von Ense aus den Jahren 1827–58* (Leipzig: F. A. Brockhaus, 1860), pp. 53–55; Alfred Dove, "Alexander von Humboldt auf dem Höhe seiner Jahre," in Karl Bruhns, ed., *Alexander von Humboldt. Eine wissenschaftliche Biographie*, 3 vols. (Leipzig: F. A. Brockhaus, 1872), vol. 2, pp. 190–209.

3. For Alexander's attempt to manage his brother's memory as a liberal, see *Briefe an Varnhagen*, 23 May 1835 p. 29; 7 November 1837 p. 50; 4 November 1849 p. 251. In 1837, Varnhagen, another relic of reform and himself a liberal journalist, published a biographical essay on Wilhelm von Humboldt, which he sent to Alexander for comment; Alexander asked Varnhagen to add that his brother had died just after finishing plans for a national representative constitution for Prussia, including election procedures. See K. A. Varnhagen von Ense, "Wilhelm von Humboldt" (1837) *Denkwürdigkeiten und vermischte Schriften* (Mannheim, 1838), 4: 276–322.

4. For "Jacobin," see *Briefe an Varnhagen*, p. 238.

5. James J. Sheehan, *German Liberalism in the Nineteenth Century* (Chicago: University of Chicago Press, 1978), pp. 14–17, 28–29, 32. Humboldt's rather Jeffersonian vision of the free society, so unlike the industrial liberalism that dominated Prussian politics under Bismarck, was also utterly consonant with contemporary German liberalism.

6. Alexander von Humboldt, *Relation historique du voyage aux régions équinoxiales du Nouveau Continent* (Paris: Smith et Gide, 1825), vol. 3, p. 150.

7. On Humboldt's political theory and practice as applied to the revolutions in Spanish America, see Lutz Raphael, "Freiheit und Wohlstand der Nationen: Alexander von Humboldts Analysen der politischen Zustände Amerikas und das politische Denken seiner Zeit," *Historische Zeitung*, 260 (1995), pp. 749–76.

8. Max Hoffmann, *August Böckh: Lebensbeschreibung und Auswahl aus seinem wissenschaftlichen Briefwechsel* (Leipzig: B. Teubner, 1901), p. 430. Versions of this comment were made to several correspondents at the time of the king's 1844 address to the members of the town and University of Königsberg, on the occasion of the university's jubilee. The king called on his subjects to follow him forward to the new Prussia. See Heinrich von Treitschke, *Deutsche Geschichte im neunzehnten Jahrhundert*, 6th ed. (Leipzig, 1914; facsimile Königstein: Athenäum, 1981), 5: 237.

9. Dove, "Humboldt," pp. 284–89, 298–99. This is a frequent theme in Humboldt's correspondence of the period.

10. Humboldt to Varnhagen, 6 September 1844, *Briefe an Varnhagen*, p. 168. See Dove, "Humboldt," pp. 306–307. For a vivid description of Friedrich Wilhelm IV's "romantic" constitutional policy, see Koselleck, *Preußen*, pp. 362–63. Humboldt once referred to Friedrich Wilhelm IV's wish "to build a miniature Middle Ages," quoted in Berdahl, *Politics of the Prussian Nobility*, p. 315). In 1842, for the baptism of the future Edward V, Humboldt accompanied Friedrich Wilhelm IV to Windsor, where one of the principal items of business was planning an Anglo-Prussian Evangelical diocese in Jerusalem. One of Humboldt's closest friends, Christian Carl Josias Bunsen, was the chief architect of the joint ecclesiastical venture, a deep admirer of the Anglican Church as a social unifier, and in the 1840s a favorite of Friedrich Wilhelm IV.

11. Humboldt openly defended Raumer, another relic of the Reform Era, in the Akademie der Wissenschaften in February 1847, when Raumer was forced to resign his seat for praising Friedrich II's religious tolerance in an address to the Akademie in Friedrich Wilhelm IV's presence. Dove, "Humboldt," 2: 310–13.

12. Hoffmann, *August Böckh*, p. 422.

13. Hoffmann, *August Böckh*, pp. 118–19 (August Böckh, *Gesammelte kleine Schriften* [Leipzig: B. Teubner, 1858], 1:328).

14. Dove, "Humboldt," 2. 401 13, Justinus Korner to Georg von Cotta, 10 [May?] 1853 and 8 December 1853, in *Briefe an Cotta. Vom Vormärz bis Bismarck, 1833–1863*, ed. Herbert Schiller (Berlin: Cotta, 1934), pp. 497, 501.

15. Dove, "Humboldt," 2: 431–32; *Briefe an Varnhagen*, 8 July 1854, p. 288. Humboldt had been appointed to the Council of State—a royal advisory body consisting of the members of the royal house, certain officials, and "especially trusted subjects," and viewed by the Crown as a substitute for a representative assembly—by Friedrich Wilhelm IV in 1840. (Hans Schneider, *Der preußische Staatsrat, 1817–1918* [Munich: C. H. Beck, 1952], appendix). On Stahl as chief ideologist of religious German conservatism, see Sheehan, *German History*, pp. 595–96; see also Berdahl, *Prussian Nobility*, chap. 10. On the Staatsrat, see Koselleck, *Preußen*, pp. 264–76.

16. References to *Cosmos* are usually made to the Otté translation; where I have altered the Otté translation to better render the original German, the German page citation follows the English, separated by a solidus.

17. Although hostile biographers hinted at Wilhelm von Humboldt's atheism, Prince Metternich thanked Humboldt especially for this citation from his brother as an acknowledgment of the truth of Christianity in the first volume. Metternich to Humboldt, 10 May 1846, in *Briefe an Varnhagen*, pp. 218–20.

18. Humboldt to Georg von Cotta, Berlin, 20 February 1848, *Briefe an Cotta*, p. 34.

19. Humboldt told Varnhagen (2 October 1845, *Briefe an Varnhagen*, p. 183) that he was adding this passage to the French translation in response to the accusations of materialism in the *Westminster Review*; either Otté borrowed the passage from the French (she certainly consulted the French translation: see 1: 7), or Humboldt communicated this revision and others directly to Otté and Bohn. (In her preface, Otté states that Humboldt and Bohn corresponded about the translation.)

20. Humboldt to Varnhagen, 3 June 1845, *Briefe an Varnhagen*, p. 172. Ludwig Feuerbach (1804–72) and Bruno Bauer (1809–82) were among the most popular and notorious "Young Hegelian" critics of traditional religion. Philipp Conrad Marheineke (1780–1846) was the most famous of Hegelian theologians, professor of theology at Berlin, and editor of Hegel's *Lectures on the Philosophy of Religion* (1842), where Marheineke daringly praised Bauer after Bauer had been forbidden by the Ministry of Religion and Education to lecture at Prussian universities. See *The Young Hegelians: An Anthology*, ed. Lawrence Stepelevich (Cambridge: Cambridge University Press, 1983). For Humboldt's awareness of orthodox hostility to the natural sciences and of the persecutions of dissenters and Young Hegelians, see *Briefe an Varnhagen*, 21 March 1842, p. 111; 6 April 1842, p. 117; 3–4 June 1845, pp. 172–73.

21. *Physiologische Briefe* (1845–47), p. xxii; cited in Frederick Gregory, *Scientific Materialism in Nineteenth-Century Germany* (Dordrecht: D. Reidel, 1977), p. 192.

22. Humboldt uses "life in civil society" (*bürgerliches Leben*) here as it was used by German Romantics and German officialdom, as the sphere of the "economy" in the larger Aristotelian sense of local, material, daily life and the confrontation of selfish interests. See Frederick Beiser, *Enlightenment, Revolution, and Romanticism. The Genesis of Modern German Political Thought* (Cambridge: Harvard University Press, 1992), pp. 232–36.

23. Humboldt's comparison of philosophical systems with short-lived meteors comes from the erstwhile nature philosopher F. W. J. Schelling, whom Humboldt here quotes with some sarcasm. See Humboldt to Boeckh, [1843], in Hoffmann, *August Böckh*, p. 425.

24. For a comprehensive treatment of natural philosophy and the powers of the mind as problems of public order in the late eighteenth century, see Simon Schaffer, "Natural Philosophy and Public Spectacle in the Eighteenth Century," *History of Science*, 21 (1983), pp. 21–43, and "Genius in Romantic Natural Philosophy," in *Romanticism and the Sciences*, ed. N. Jardine and A. Cunningham (Cambridge: Cambridge University Press, 1990), pp. 82–98.

25. Humboldt to C. G. Ehrenberg, Teplitz, 28 July 1836, in Alexander von Humboldt, *Aus meinem Leben*, ed. Kurt-R. Biermann, 2nd ed. (Leipzig: Urania Verlag, 1989), p. 208.

26. Humboldt, *Magasin encyclopédique* 4 (1796): 463; Humboldt to

David Friedländer, in *Die Jugendbriefe Alexander von Humboldts*, ed. Fritz Lange and Ilse Jahn (Berlin: Akademie Verlag, 1973), pp. 658–59. Various forms of the same declaration were repeated in letters to Willdenow, p. 664; Karsten, p. 666; Freiesleben, p. 680; and von Moll, p. 682.

27. Alexander von Humboldt, *Essai sur la géographie des plantes* (Paris: F. Schoell, 1805); *Relation historique du voyage aux régions équinoxiales du Nouveau Continent* (Paris: F. Schoell, 1814), vol. 1, introduction; *Fragmens de géologie et de climatologie asiatiques* (Paris: Gide, 1831), vol. 1, p. xi.

28. Alexander von Humboldt, "Beobachtungen über die Gesetze der Wärmeabnahme in den höhern Regionene der Atmosphäre, und über die untern Gränze der Schneelinie," *Annalen der Physik* 24 (1806), pp. 1–2.

29. Humboldt, *Géographie des plantes*, pp. 41–42.

30. Evidently under pressure from English arbiters of science like John Herschel and William Whewell over the sensitive issue of the relationship between Newton and Kepler, Humboldt was forced to refine his claim that "the final object of human investigation in *empirical science* is the discovery of laws" (1: 30/32) in the introduction to volume 3 by adding a sharp distinction between the recognition of empirical regularities (e.g., Kepler's Laws) and the discovery of dynamic laws (e.g., Newton's Laws of Gravity), between the recognition of lawfulness and of causal connection. (3: 6, 7n) The phenomenological character of Humboldt's science and the idea of *Cosmos*, however, emphasizing the progressive, always empirical-aesthetic cognition of necessity, was fundamentally opposed to any firm distinction between dynamic and empirical laws.

31. Alexander von Humboldt, "Sur les lois que l'on observe dans la distribution des formes végétales," in *Dictionnaire des sciences naturelles*, ed. F. Cuvier (Strasbourg: Levrault, 1823), 18: 431; Alexander von Humboldt, "Nouvelles recherches sur la distribution des formes végétales," *Annales de chimie et de physique*, 16 (1821): 279–80.

32. Humboldt, "Nouvelles recherches," pp. 267, 279–80.

33. See, e.g., Humboldt, *Relation historique du voyage*, vol. 1, chap. xiii; vol. 2, 383–84. In the German edition of *Essai sur la géographie des plantes*, Humboldt described investigations into the geographical relation of phenomena as giving insight into "the history of our planet" rather than into "general physics."

34. Alexander von Humboldt, "Die Entbindung des Wärmestoffs als geo-gnostisches Phänomen betrachtet," in *Versuche über die Zusammensetzung des Luftkreises und über einige andere Gegenstände der Naturlehre* (Braunschweig: Vieweg, 1799), pp. 190–91; Michael Dettelbach, "Humboldtian Science," in *Cultures of Natural History*, ed. N. Jardine, J. Secord, and E. Spary (Cambridge: Cambridge University Press, 1995), pp. 294–95.

35. Humboldt, *Relation historique du voyage*, vol. 1, introduction.

36. On the relationship between the "uranological" and the "telluric" parts of physical cosmography see *Cosmos*, 1: 44–49, and also the introduction to volume 3. Humboldt introduces the concept of a *physische Weltbeschreibung* through its terrestrial part (*physische Erdbeschreibung*) in order to rehearse the distinction between systematic natural history, on the one hand, and a dynamically concerned comparative geography, on the other, before applying this distinction to the *physische Beschreibung* of the heavens. Classifying mathematics as a rational "organ," or instrument, applied to astronomical observations like a taxonomy to zoological observations, giving it a linguistic and phenomenological rather than ontological significance, further describes the "natural historical" character of astronomy. In a physical cosmography, Humboldt insists, the uranological part must not be subordinated to the telluric, as has long been the custom in "geographies," but rather should be seen as a part of "the history of nature," indeed the exemplary part, capable of illustrating in all its simplicity how "physical description of the earth" is a causal and historical science. The uranological part independently and archetypally exhibits the lawfulness and physical historicity of nature, the relation between "description" and "history of nature"; and it makes *Cosmos* a study of this lawfulness and genetic unity itself, of the "well-orderedness of the world," rather than simply a comparative geography.

37. See Malcolm Nicolson, "Alexander von Humboldt, Humboldtian Science, and the Origins of the Study of Vegetation," *History of Science,* 25 (1987), pp. 167–94; see also Edward Bunkse, "Humboldt and an Aesthetic Tradition in Geography," *Geographical Review*, 71 (1981), pp. 127–46.

38. For a vivid introduction to the interest in recreating nature and its effects in the arts of landscape painting, landscape gardening, architecture, stage decoration, diorama, panorama, and photography, in the first decades of the nineteenth century, see Paul Johnson, *The Birth of the Modern, 1815–1830* (New York: Harper Collins, 1991), pp. 153–64. On landscape gardening and public parks in the German lands, see Paul Ortwin Rave, *Gärten der Goethezeit* (Berlin: Hausel, 1981), esp. pp. 135–52 (on Prince Pückler); on Prince Pückler's landscape designs and their social agenda, see Herman Graf von Arnim, *Ein Fürst unter den Gärtnern: Prince Pückler als Landschaftskünstler und der Muskauer Park* (Frankfurt am Main: Ullstein, 1981) and *Fürst Pückler und die Gartenbaukunst,* ed. Ruth B. Emde and Winfried Herrmann (Dortmund: Harenberg, 1995), pp. 62–65, 134–39; on the palm house which once stood in the royal park on the Pfaueninsel outside of Potsdam, see Michael Seiler, *Das Palmenhaus auf der Pfaueninsel* (Berlin: Haude and Spener, 1989). On panorama and diorama, see Heinz Buddemeier, *Panorama, Diorama, Photographie: Entstehung und Wirkung neuer*

Medien im neunzehnten Jahrhundert (Munich: W. Fink, 1970) and Helmut and Alison Gernsheim, *L. J. M. Daguerre: the history of the diorama and the daguerrotype* (New York: Dover, 1968). The recent exhibition of Frederick Edwin Church's 5-1/2-x-10 ft. landscape "The Heart of the Andes" at the Metropolitan Museum of Art, complete with the original window-like frame and opera-glasses, reconstructed one such mid-nineteenth-century public landscape exhibition, directly influenced by *Cosmos*. The painting drew 12,000 spectators during its first exhibition in New York in May, 1859. For Church's reading of Humboldt, see Kevin J. Avery, *Church's Great Picture "The Heart of the Andes"* (New York: Metropolitan Museum of Art, 1993), pp. 12–19. For another landscape painter directly inspired by Humboldt, see the lithographs by Albert Berg, *Études physiognomiques sur la végétation de l'Amérique tropicale* (London: Paul and Dominic Colnagh, 1854). In addition to calling for first-hand studies of vegetational physiognomy, Humboldt also took an active interest in the cultivation of techniques for faithfully recording natural forms. He was a member of the commission of the French Academy of Sciences (chaired by François Arago), which examined and enthusiastically supported daguerrotypy in 1839, and in the 1850s, Humboldt became an enthusiast of "auto-galvanoplasty" (electroplating) as a means of reproducing plant forms directly from nature (Hanno Beck, *Alexander von Humboldt*, 2 vols. [Wiesbaden: Franz Steiner, 1959–61], vol. 2, p. 210).

39. The link between this section and Humboldt's interpretation of the significance of his 1827–28 lectures is especially evident in his Singakademie lectures. There his reflections on "incitements to the study of nature" came at the very end of the series, explicitly as a comment on the wide interest and enthusiastic audiences that his course on *physische Erdbeschreibung* had attracted (Alexander von Humboldt, *Humboldt über das Universum*, ed. Jürgen Hamel and Klaus-Harro Tiemann [Frankfurt am Main: Insel Verlag, 1993], p. 210). Humboldt also pressed his interpretation of the scientific character of the age specifically on his fellow naturalists: he read an expanded version of these reflections to the eleventh meeting of the Association of German Scientists and Physicians at Breslau in 1833, and at the fourteenth meeting in Jena in 1836, he illustrated his lecture "on the varieties of enjoyment afforded by Nature and the scientific development of general laws" (the first section of *Cosmos*) with the Brazilian landscapes of Moritz Rugendas (*Alexander von Humboldt. Chronologische Übersicht über wichtige Daten seines Lebens,* ed. Kurt-R. Biermann et al., 2nd ed. [Berlin: Akademie Verlag, 1983]).

40. For Whewell and Comte, see Richard Yeo, *Defining Science: William Whewell, Natural Knowledge and Public Debate in Early Victorian Britain* (Cambridge: Cambridge University Press, 1993), pp. 50–51, 145–49; and in general, see Wolf Lepenies, *Das Ende der*

Naturgeschichte (Munich: Heuser, 1976); see also Dietrich von Engelhardt, *Historisches Bewußtsein in der Naturwissenschaft von der Aufklärung bis zum Positivismus* (Munich: Karl Alber, 1979).

41. Humboldt to Varnhagen, 27 October 1834, *Briefe an Varnhagen*, p. 20; Humboldt to William Whewell, 14 August 1840, in Isaac Todhunter, *William Whewell, D.D.* (London: Macmillan, 1876), p. 148. Simon Schaffer brought this letter to my attention.

42. Humboldt to Boeckh, 26 December 1845, in Hoffmann, *August Böckh*, pp. 430–31.

43. Humboldt's history of the progress of cosmic views thus practically forces the reader to consult a good historical atlas not only because of the depth of Humboldt's scholarship, but also because the very principle on which cosmic views advanced, according to Humboldt (at least until the seventeenth century) was that of geographical expansion and accumulating geographical experience.

44. The foreshadowing extends to the disbelief and rejection with which the reports of Alexander's companions were received by Hellenist and Roman scholars—Megasthenes prefiguring d'Acosta. Like most of the themes in *Cosmos*, Humboldt had described the seventeenth- and eighteenth-century disdain for the sixteenth-century narratives of the Conquista and his own vindication of these eyewitness accounts in the exposition of his American findings (Alexander von Humboldt, *Vues des Cordillères et monuments des peuples indigènes de l'Amérique*, 2 vols. (Paris: Librairie Grecque-Latine-Allemonde, 1816), 1: pp. 9–11, p. 3).

45. For Humboldt and the historiography of Columbus and the Columbian encounter, see Anthony Pagden, *European Encounters with the New World* (New Haven: Yale University Press, 1993), pp. 104–15. With specific reference to reproducing the vivid, Columbian encounter with the "combined action of forces," see Mary Louise Pratt, *Imperial Eyes: Travel Writing and Transculturation* (London: Routledge, 1992), pp. 120–27.

46. Humboldt, *Géographie des plantes*, pp. vi–vii, 43.

47. On the telescopic resolvability of nebulae and the hypothesis of a universal luminous fluid, see Humboldt to H. C. Schumacher, 8 September 1849, *Briefwechsel zwischen Alexander von Humboldt und Heinrich Christian Schumacher*, ed. Kurt-R. Biermann (Berlin: Akademie Verlag, 1980), pp. 128–30. Warnings against hypostatizing the *atomistische Bildersprache* of mechanical philosophy is another constant in Humboldt's work, from his *Essai sur la géographie des plantes* to *Cosmos* (e.g., 1: 57–58). In the introduction to volume 5, published posthumously in 1862, Humboldt acknowledged the mechanical theory of heat—that work and heat are interconvertible—but considered it a new *Naturphilosophie* and remained skeptical about the reduction of nature to a single principle, especially one clothed in a *symbolische Atomistik*.

48. The German original (2: 399) places the emphasis on "free humanity will achieve in future ages" by placing the phrase at the end

of the sentence, and neither does the German version specify "mental activity," simply "activity," which does not distinguish between the material and intellectual work to be done in freedom. The word that Otté occasionally translates as *eventful* (German *verhängnißvollen*) appears frequently in volume 2 to indicate the ramified connections of historical events ("momentous"), which Otté's translation fails to render.

49. The German words *Bildung, Gesittung, Vermenschlichung, Civilisation,* and *Cultur* (the last two are late-eighteenth-century adoptions from the French) are still difficult to render into English; Otté uses *civilization* and *cultivation* interchangeably, never *culture,* and once uses *humanization* for *Vermenschlichung.* For a contemporary and systematic articulation of these terms, see Wilhelm von Humboldt, "Über die Verschiedenheit des menschlichen Sprachbaues," in *Über die Kawi-Sprache,* 1: xxxvii–xxxviii, a work that Humboldt certainly knew because he edited it for publication after his brother's death.

50. Humboldt to Thomas Jefferson, 24 May 1804, in *Briefe aus Amerika,* ed. Ulrike Moheit (Berlin: Akademie Verlag, 1993), p. 292; see also Humboldt, *Vues des Cordillères,* p. 3.

51. Alexander von Humboldt, *Examen critique de l'histoire de la géographie du Nouveau Continent et des progrès de l'astronomie nautique aux XVe et XVIe siècles,* 5 vols. (Paris: Gide, 1835–39), 1: xvii–xviii; 3: pp. 154–58.

52. Humboldt to Georg von Cotta, Berlin, 20 February 1848, *Briefe an Cotta,* p. 4. Humboldt's parting statement in volume 2 is clearly aimed at those who, like the "English" critics, saw *Cosmos* as an encyclopedia of the sciences and evaluated it according to its treatment of each discipline—astronomy, geology, and so forth. "Much that, according to other views than mine, regarding the composition of a book of nature, may have appeared wanting, will there [in the third volume] find its place" (2: 354). Likewise, those who criticized Humboldt's partiality to French and German science failed to see that *Cosmos* was in part the portrait of a mind: "The subjective aspect ought to predominate: one should be able after my death to read from my writings with whom I've lived, who affected me" (Humboldt to J. F. Encke, quoted in Dove, "Humboldt," 2: 372). Such critics missed the point of *Cosmos*: not the actual unity of nature, but the ability of the mind of the properly cultivated individual to grasp that unity. See the introduction to volume 1 for selections from the British reviews.

53. Humboldt to Boeckh, [1849], in Hoffmann, *August Böckh,* p. 440.

54. Johannes Müller to Humboldt, 30 October 1846, in Dove, "Humboldt," 2: 389.

55. Humboldt to Karl Varnhagen, 3 June 1845, *Briefe an Varnhagen,* p. 172.

COSMOS

VOLUME 2

CONTENTS OF VOL. II.

PART I.

INCITEMENTS TO THE STUDY OF NATURE.

PART II.

SUMMARY.

I. *Poetic Delineation of Nature.*—The principal results of observation referring to a purely objective mode of treating a scientific description of nature have already been treated of in the picture of nature; we now, therefore, proceed to consider the reflection of the image conveyed by the external senses to the feelings and a poetically-framed imagination. The mode of feeling appertaining to the Greeks and Romans. On the reproach advanced against these nations having entertained a less vivid sentiment for nature. The expression of such a sentiment is more rare among them, solely in consequence of natural descriptions being used as mere accessories in the great forms of lyric and epic poetry, and all things being brought in the ancient Hellenic forms of art within the sphere of humanity, and being made subservient to it. Pæans to Spring, Homer, Hesiod. Tragic authors: fragments of a lost work of Aristotle. Bucolic poetry, Nonnus, Anthology— p. 27. Romans: Lucretius, Virgil, Ovid, Lucan, Lucilius the younger. A subsequent period, in which the poetic element appears only as an incidental adornment of thought; the Mosella, a poem of Ausonius. Roman prose writers; Cicero in his letters, Tacitus, Pliny. Description of Roman villas—p. 38. Changes in the mode of feeling and in their representation produced by the diffusion of Christianity and by an anchorite life. Minucius Felix in Octavius. Passages taken from the writings of the Fathers of the Church: Basil the Great in the wilderness on the Armenian river Iris, Gregory Nyssa, Chrysostom. Melancholy and sentimental tone of feeling—p. 38–43. Influence of the difference of races manifested in the different tone of feeling pervading the natural descriptions of the nations of Hellenic, Italian, North Germanic, Semitic, Persian, and Indian descent. The florid poetic literature of the three last-named races shows that the animated feeling for nature evinced by the North Germanic races is not alone to be ascribed to a long deprivation of all enjoyment of nature through a protracted winter. The opinions of Jacob and Wilhelm Grimm on the chivalric poetry of the Minnesingers and of the German animal epos; Celto-Irish descriptions of nature—p. 48. East and west Arian nations (Indians and Persians). The Ramayana and Mahabharata; Sakuntala and Kalidasa's Messenger of Clouds. Persian literature in the Iranian Highlands does not ascend beyond the period of the Sassanidæ—p. 54. (A fragment of Theodor Goldstücker.) Finnish epic and songs, collected by Elias Lönnrot from the lips of the Karelians—p. 56. Aramæic nations: natural poetry of the Hebrews, in which we trace the reflection of Monotheism—p. 57–60. Ancient Arabic poetry. Descriptions in Antar of the Bedouin life in the desert. Descriptions of nature in Amru'l Kais—p. 61. After the downfall of the Aramæic, Greek, and Roman power, there appears Dante Alighieri, whose poetic creations breathe from time to time the deepest sentiment of admiration for the terrestrial life of nature. Petrarch, Boiardo, and Vittoria Colonna. The *Ætna Dialogus* and the picturesque delineation of the luxuriant vegetation of the New World in the *Historiæ Venetæ* of Bembo. Christopher Columbus—p. 66. Camoens's *Lusiad*—p. 68. Spanish poetry: the *Araucana* of Don Alonso de Ercilla. Fray Luis de Leon and Calderon, with the remarks on the same of Ludwig Tieck. Shakspeare, Milton, Thomson—p. 74. French prose writers : Rousseau, Buffon, Bernardin de St. Pierre, and Chateaubriand—p. 75–77. Review of the narratives of the older travelers of the Middle Ages, John Mandeville, Hans Schiltberger, and Bernhard von Breitenbach; contrast with modern travelers. Cook's companion, George Forster—p. 80. The blame sometimes justly applied to descriptive poetry as an independent form does not refer to the attempt either to give a picture of distant zones visited by the writer, or to convey to others, by the force of applicable words, an image of the results yielded by a direct contemplation of nature. All parts of the vast sphere of creation, from the equator to the frigid zones, are endowed with the happy power of exercising a vivid impression on the human mind—p. 82.

II. *Landscape painting* in its animating influence on the study of nature. In classical antiquity, in accordance with the respective mental direction of different nations, landscape painting and the poetic delineation of a particular region were neither of them independent objects of art. The elder Philostratus. Scenography. Ludius. Evidences of landscape painting among the Indians in the brilliant period of Vikramaditya. Herculaneum and Pompeii. Painting among Christians, from Constantine the Great to the beginning of the Middle Ages; part of landscape painting in the historical pictures of the brothers Van Eyck. The seventeenth century the most brilliant epoch of landscape painting. Miniatures on manuscripts — p. 87. Development of the elements of painting. (Claude Lorraine, Ruysdael, Gaspard and Nicolas Poussin, Everdingen, Hobbima, and Cuyp.) Subsequent striving to give natural truthfulness to the representation of vegetable forms. Representation of tropical vegetation. Franz Post, the companion of Prince Maurice of Nassau. Eckhout. Requirement for a representation of the physiognomy of nature. The great and still imperfectly completed cosmical event of the independence of Spanish and Portuguese America, and the foundation of constitutional freedom in regions of the chain of Cordilleras between the tropics, where there are populous cities situated at an elevation of 14,000 feet above the level of the sea, together with the increasing civilization of India, New Holland, the Sandwich Islands, and Southern Africa, will undoubtedly impart a new impulse and a more exalted character to landscape painting, no less than to meteorology and descriptive geography. Importance and application of Barker's panoramas. The conception of the unity of nature and the feeling of the harmonious accord pervading the Cosmos will increase in force among men in proportion to the multiplication of the means for representing all natural phenomena in delineating pictures—p. 98.

III. *Cultivation of Exotic Plants.*—Impression of the physiognomy of vegetable forms, as far as plantations are capable of producing such an impression. Landscape gardening. Earliest plantation of parks in Central and Southern Asia. Trees and groves sacred to the gods—p. 102. The gardens of the nations of Eastern Asia. Chinese gardens under the victorious dynasty of Han. Poem on a garden, by the Chinese statesman See-ma-kuang, at the close of the eleventh century. Prescripts of Lieu-tscheu. Poem of the Emperor Kien-long, descriptive of nature. Influence of the connection of Buddhist monastic establishments on the distribution of beautiful characteristic vegetable forms —p. 105.

B. *History of the Physical Contemplation of the Universe.*—The histo ry of the recognition of the universe is wholly different from the history of the natural sciences, as given in our elementary works on physics, and on the morphology of plants and animals. This is the history of our conception of the unity of phenomena, and of the reciprocal connection existing among the natural forces of the universe. Mode of treating a history of the Cosmos: *a.* The independent efforts of reason to gain a knowledge of natural laws; *b.* Cosmical events which have suddenly enlarged the horizon of observation; *c.* The invention of new means of sensuous perception. Languages. Points of radiation from which civilization has been diffused. Primitive physics and the natural science of barbarous nations obscured by civilization—p. 118.

Principal Momenta of a History of a Physical Contemplation of the Universe.

I. *The basin of the Mediterranean* the starting-point of the attempts to extend the idea of the Cosmos. Subdivisions in the form of the basin. Importance of the form of the Arabian Gulf. Intersection of two geognostic systems of elevation from N.E. to S.W., and from S.S.E. to N.N.W. Importance of the latter direction of the lines of intersection considered with reference to general international intercourse. Ancient civilization of the nations dwelling round the Mediterranean. The Valley of the Nile, the ancient and modern kingdom of the Egyptians. The Phœnicians, a race who favored general intercourse, were the means of diffusing alphabetical writing (Phœnician signs), coins as medium of currency, and the original Babylonian weights and measures. The science of numbers, arithmetic. The art of navigating by night. West African colonies—p. 130.

Pelasgian Tyrrhenians and Etruscans (Rasenæ). Peculiar tendency of the Etrurian races to maintain an intimate communion with natural forces; the fulguratores and aquileges—p. 140.

Other anciently civilized races dwelling around the Mediterranean. Traces of cultivation in the East, under the Phrygians and Lycians; and in the West, under the Turduli and the Turdetani. Dawn of Hellenic power. Western Asia the great thoroughfare of nations emigrating from the East; the Ægean island world the connecting link between Greece and the far East. Beyond the 48th degree of latitude, Europe and Asia are fused together, as it were, by flat steppes. Pherecydes of Syros, and Herodotus, considered the whole of North Scythian Asia as appertaining to Sarmatian Europe. Maritime power, and Doric and Ionic habits of life transmitted to the colonial cities. Advance toward the East, to the Euxine and Colchis; first acquaintance with the western shore of the Caspian Sea, confounded, according to Hecatæus, with the encircling Eastern Ocean. Inland trade and barter carried on by the chain of Scytho-scolotic races with the Argippæans, Issedones, and the Arismaspes, rich in gold. Meteorological myth of the Hyperboreans. Opening of the port of Gadeira toward the west, which had long been closed to the Greeks. Navigation of Colæus of Samos. A glance into the boundless; an unceasing striving for the far distant; accurate knowledge of the great natural phenomenon of the periodic swelling of the sea—p. 153.

II. *Campaigns of the Macedonians under Alexander the Great, and the long-enduring Influence of the Bactrian Empire.*—With the exception of the one great event of the discovery and opening of tropical America eighteen and a half centuries later, there was no other period in which a richer field of natural views, and a more abundant mass of materials

for the foundations of cosmical knowledge, and of comparative ethnological study, were presented at once to one single portion of the human
race. The use of these materials, and the intellectual elaboration of
matter, are facilitated and rendered of more importance by the direction imparted by the Stagirite to empirical investigation, philosophical
speculation, and to the strict definitions of a language of science. The
Macedonian expedition was, in the strictest sense of the word, a scientific expedition. Callisthenes of Olynthus, the pupil of Aristotle, and
friend of Theophrastus. The knowledge of the heavens, and of the
earth and its products, was considerably increased by intercourse with
Babylon, and by the observations that had been made by the dissolved
Chaldean order of priests—p. 169.

III. *Increase of the Contemplation of the Universe under the Ptolemies.*—Grecian Egypt enjoyed the advantage of political unity, while its
geographical position, and the entrance to the Arabian Gulf, brought
the profitable traffic of the Indian Ocean within a few miles of the southeastern shores of the Mediterranean. The kingdom of the Seleucidæ
did not enjoy the advantages of a maritime trade, and was frequently
shaken by the conflicting nationality of the different satrapies. Active
traffic on rivers and caravan tracks with the elevated plateaux of the
Seres, north of the Uttara-Kuru and the Valley of the Oxus. Knowledge
of monsoons. Reopening of the canal connecting the Red Sea with the
Nile above Bubastus. History of this water route. Scientific institutions under the protection of the Lagides; the Alexandrian Museum,
and two collections of books in Bruchium and at Rhakotis. Peculiar
direction of these studies. A happy generalization of views manifests
itself, associated with an industrious accumulation of materials. Eratosthenes of Cyrene. The first attempt of the Greeks, based on imperfect data of the Bematists, to measure a degree between Syene and
Alexandria. Simultaneous advance of science in pure mathematics,
mechanics, and astronomy. Aristyllus and Timochares. Views entertained regarding the structure of the universe by Aristarchus of Samos,
and Seleucus of Babylon or of Erythræa. Hipparchus, the founder of
scientific astronomy, and the greatest independent astronomical observer
of antiquity. Euclid. Apollonius of Perga, and Archimedes—p. 179.

IV. *Influence of the Universal Dominion of the Romans and of their
Empire on the Extension of Cosmical Views.*—Considering the diversity
in the configuration of the soil, the variety of the organic products, the
distant expeditions to the Amber lands, and under Ælius Gallus to Arabia, and the peace which the Romans long enjoyed under the monarchy
of the Cæsars, they might, indeed, during four centuries, have afforded
more animated support to the pursuit of natural science; but with the
Roman national spirit perished social mobility, publicity, and the maintenance of individuality—the main supports of free institutions for the
furtherance of intellectual development. In this long period, the only
observers of nature that present themselves to our notice are Dioscorides, the Cilician, and Galen of Pergamus. Claudius Ptolemy made the
first advance in an important branch of mathematical physics, and in
the study of optics, based on experiments. Material advantages of the
extension of inland trade to the interior of Asia, and the navigation of
Myos Hormos to India. Under Vespasian and Domitian, in the time
of the dynasty of Han, a Chinese army penetrates as far as the eastern
shores of the Caspian Sea. The direction of the stream of migration in
Asia is from east to west, while in the new continent it inclines from
north to south. Asiatic migrations begin, a century and a half before

our era, with the inroads of the Hiungnu, a Turkish race, on the fair-
haired, blue-eyed, probably Indo-Germanic race of the Yueti and Usun,
near the Chinese Wall. Roman embassadors are sent, under Marcus
Aurelius, to the Chinese court by way of Tonkin. The Emperor Claudius received an embassy of the Rashias of Ceylon. The great Indian
mathematicians, Warahamihira, Brahmagupta, and probably also Aryabhatta, lived at more recent periods than those we are considering; but
the elements of knowledge, which had been earlier discovered in India
in wholly independent and separate paths, may, before the time of Diophantus, have been in part conveyed to the West by means of the extensive universal commerce carried on under the Lagides and the Cæsars. The influence of these widely-diffused commercial relations is
manifested in the colossal geographical works of Strabo and Ptolemy.
The geographical nomenclature of the latter writer has recently, by a
careful study of the Indian languages and of the history of the west Iranian Zend, been recognized as a historical memorial of these remote
commercial relations. Stupendous attempt made by Pliny to give a
description of the universe; the characteristics of his encyclopedia of
nature and art. While the long-enduring influence of the Roman dominion manifested itself in the history of the contemplation of the universe as an element of union and fusion, it was reserved for the diffusion of Christianity (when that form of faith was, from political motives,
forcibly raised to be the religion of the state of Byzantium) to aid in
awakening an idea of the unity of the human race, and by degrees to
give to that idea its proper value amid the miserable dissensions of religious parties—p. 199.

V. *Irruption of the Arabs.*—Effect of a foreign element on the process of development of European civilization. The Arabs, a Semitic
primitive race susceptible of cultivation, in part dispel the barbarism
which for two hundred years had covered Europe, which had been
shaken by national convulsions; they not only maintain ancient civilization, but extend it, and open new paths to natural investigation.
Geographical figure of the Arabian peninsula. Products of Hadramaut,
Yemen, and Oman. Mountain chains of Dschebel-Akhdar, and Asyr.
Gerrha, the ancient emporium for Indian wares, opposite to the Phœnician settlements of Aradus and Tylus. The northern portion of the
peninsula was brought into animated relations of contact with other
cultivated states, by means of the spread of Arabian races in the Syro-
Palestinian frontier mountainous districts and the lands of the Euphrates. Pre-existing indigenous civilization. Ancient participation in the
general commerce of the universe. Hostile advances to the West and
to the East. Hyksos and Ariæus, prince of the Himyarites, the allies
of Minus on the Tigris. Peculiar character of the nomadic life of the
Arabs, together with their caravan tracks and their populous cities—p.
200–208. Influence of the Nestorians, Syrians, and of the pharmaceu-
tico-medicinal school at Edessa. Taste for intercourse with nature and
her forces. The Arabs were the actual founders of the physical and
chemical sciences. The science of medicine. Scientific institutions in
the brilliant epoch of Almansur, Haroun Al-Raschid, Mamun, and Mo-
tasem. Scientific intercourse with India. Employment made of the
Tscharaka and the Susruta, and of the ancient technical arts of the
Egyptians. Botanical gardens at Cordova, under the Calif Abdurrahman the poet—p. 208–217. Efforts made at independent astronomical
observations and the improvement in instruments. Ebn Junis employs
the pendulum as a measure of time. The work of Alhazen on the re-

fraction of rays. Indian planetary tables. The disturbance in the moon's longitude recognized by Abul Wefa. Astronomical Congress of Toledo, to which Alfonso of Castille invited Rabbis and Arabs. Observatory at Meragha, of Ulugh Beig, the descendant of Timur, at Samarcand, and its influence. Measurement of a degree in the plain between Tadmor and Rakka. The Algebra of the Arabs has originated from two currents, Indian and Greek, which long flowed independently of one another. Mohammed Ben Musa, the Chowarezmier. Diophantus, first translated into Arabic at the close of the tenth century, by Abul Wefa Buzjani. By the same path which brought to the Arabs the knowledge of Indian Algebra, they likewise obtained in Persia and on the Euphrates the Indian numerals and the knowledge of the ingenious device of *Position*, or the employment of the value of position. They transmitted this custom to the revenue officers in Northern Africa, opposite to the coasts of Sicily. The probability that the Christians of the West were acquainted with Indian numerals earlier than the Arabs, and that they were acquainted, under the name of the system of the Abacus, with the employment of nine ciphers, according to their position-value. The value of position was known in the Suanpan, derived from the interior of Asia, as well as in the Tuscan Abacus. Would a permanent dominion of the Arabs, taking into account their almost exclusive predilection for the scientific (natural, descriptive, physical, and astronomical) results of Greek investigation, have been beneficial to a general and free mental cultivation, and to the creative power of art?—p. 219–228.

VI. *Period of the great Oceanic Discoveries.*—America and the Pacific. Events and extension of scientific knowledge which prepared the way for great geographical discoveries. As the acquaintance of the nations of Europe with the western portion of the globe constitutes the main object of this section, it is absolutely necessary to divide in an incontestable manner the first discovery of America in its northern and temperate zone by the Northmen, from the rediscovery of the same continent in its tropical regions. While the Califate of Bagdad flourished under the Abbassides, America was discovered and investigated to the 41½° north latitude by Leif, the son of Erik the Red. The Färoë Islands and Iceland, accidentally discovered by Naddod, must be regarded as intermediate stations, and as starting points for the expeditions to the Scandinavian portions of America. The eastern coasts of Greenland in Scoresby's Land (Svalbord), the eastern coasts of Baffin's Bay to 72° 55', and the entrance of Lancaster Sound and Barrow's Straits, were all visited—Earlier (?) Irish discoveries. The White Men's Land between Virginia and Florida. Whether, previously to Naddod and Ingolf's colonization of Iceland, this island was inhabited by Irish (Westmen from American Great Ireland), or by Irish missionaries (*Papar*, the *Clerici* of Dicuil), driven by the Northmen from the Färoë Islands? The national treasures of the most ancient records of Northern Europe, endangered by disturbances at home, were transferred to Iceland, which three and a half centuries earlier enjoyed a free social Constitution, and were there preserved to future ages. We are acquainted with the commercial relations existing between Greenland and New Scotland (the American Markland) up to 1347; but as Greenland had lost its republican Constitution as early as 1261, and, as a crown fief of Norway, had been interdicted from holding intercourse with strangers, and therefore also with Iceland, it is not surprising that Columbus, when he visited Iceland in 1477, should have obtained no tidings of the new continent situated to the west. Commercial relations existed, however, as late as 1484, between the Norwegian port of Bergen and Greenland— p. 228–238.

Widely different, in a cosmical point of view, from the isolated and barren event of the first discovery of the new continent by the Northmen, was its rediscovery in its tropical regions by Christopher Columbus, although that navigator, seeking a shorter route to Eastern Asia, had not the object of discovering a new continent, and, like Amerigo Vespucci, believed to the time of his death that he had simply reached the eastern shores of Asia. The influence exercised by the nautical discoveries of the close of the fifteenth and the beginning of the sixteenth century on the rich abundance of the ideal world, can not be thoroughly understood until we have thrown a glance on the ages which separate Columbus from the blooming period of cultivation under the Arabs. That which gave to the age of Columbus the peculiar character of an uninterrupted and successful striving for an extended knowledge of the earth, was the appearance of a small number of daring minds (Albertus Magnus, Roger Bacon, Duns Scotus, and William of Occam), who incited to independent thought and to the investigation of separate natural phenomena; the revived acquaintance with the works of Greek literature; the invention of the art of printing; the missionary embassies to the Mogul princes, and the mercantile travels to Eastern Asia and South India (Marco Polo, Mandeville, and Nicolo de' Conti); the improvement of navigation; and the use of the mariner's compass or the knowledge of the north and south pointing of the magnetic needle, which we owe to the Chinese through the Arabs—p. 238–254. Early expeditions of the Catalans to the western shores of Tropical Africa; discovery of the Azores; general atlas of Picigano, of 1367. Relations of Columbus to Toscanelli and Martin Alonso Pinzon. The more recently known chart of Juan de la Cosa. The South Pacific and its islands—p. 255–273. Discovery of the magnetic line of no variation in the Atlantic Ocean. Inflection observed in the isothermal lines a hundred nautical miles to the west of the Azores. A *physical* line of demarkation is converted into a *political* one; the line of demarkation of Pope Alexander VI., of the 4th of May, 1493. Knowledge of the distribution of heat; the line of perpetual snow is recognized as a function of geographical latitude. Movement of the waters in the Atlantic Ocean. Great beds of sea-weed—p. 273–285. Extended view into the world of space; an acquaintance with the stars of the southern sky; more a sensuous than a scientific knowledge. Improvement in the method of determining the ship's place; the political requirement for establishing the position of the papal line of demarkation increased the endeavor to discover practical methods for determining longitude. The discovery and first colonization of America, and the voyage to the East Indies round the Cape of Good Hope, coincide with the highest perfection of art, and with the attainment of intellectual freedom by means of religious reform, the forerunner of great political convulsions. The daring enterprise of the Genoese seaman is the first link in the immeasurable chain of mysterious events. Accident, and not the deceit or intrigues of Amerigo Vespucci, deprived the Continent of America of the name of Columbus. Influence of the New World on political institutions, and on the ideas and inclinations of the people of the Old Continent— p. 285–301.

VII. *Period of great Discoveries in the Regions of Space.*—The application of the telescope: a more correct view of the structure of the universe prepared the way for these discoveries. Nicholas Copernicus was engaged in making observations with the astronomer Brudzewski at Cracow when Columbus discovered America. Ideal connection between the sixteenth and seventeenth centuries, by Peurbach and Regiomontanus. Copernicus never advanced his system of the universe as an hypothesis, but as incontrovertible truth—p. 301–313. Kepler and the empirical planetary laws which he discovered—p. 313–317. Invention of the telescope; Hans Lippershey, Jacob Adriaansz (Metius), and Zacharias Jansen. The first fruits of telescopic vision : mountains of the moon ; clusters of stars and the Milky Way; the four satellites of Jupiter; the triple configuration of Saturn; the crescent form of Venus; solar spots; and the period of rotation of the sun. The discovery of the small system of Jupiter indicates a memorable epoch in the fate and sound foundation of astronomy. The discovery of Jupiter's satellites gave rise to the discovery of the velocity of light, and the recognition of this velocity led to an explanation of the aberration-ellipse of the fixed stars—the perceptive evidence of the translatory movement of the earth. To the discoveries of Galileo, Simon Marius, and Johann Fabricius followed the discovery of Saturn's satellites by Huygens and Cassini, of the zodiacal light as a revolving isolated nebulous ring by Childrey, of the variation in brilliancy of the light of the fixed stars by David Fabricius, Johann Bayer, and Holwarda. A nebula devoid of stars in Andromeda described by Simon Marius—p. 317–331. While the seventeenth century owed at its commencement its main brilliancy to the sudden extension of the knowledge of the regions of space afforded by Galileo and Kepler, and at its close to the advance made in pure mathematical science by Newton and Leibnitz, the most important of the physical problems of the processes of light, heat, and magnetism, likewise experienced a beneficial progress during this great age. Double refraction and polarization; traces of the knowledge of the interference of light in Grimaldi and Hooke. William Gilbert separates magnetism from electricity. Knowledge of the periodical advance of lines without variation. Halley's early conjecture that the polar light (the phosphorescence of the earth) is a magnetic phenomenon. Galileo's thermoscope, and its employment for a series of regular diurnal observations at stations of different elevation. Researches into the radiation of heat. Torricellian tubes, and measurements of altitude by the position of the mercury in them. Knowledge of aërial currents, and the influence of the earth's rotation on them. Law of rotation of the winds conjectured by Bacon. Happy, but short-lived, influence of the Accademia del Cimento on the establishment of mathematical natural philosophy, as based on experiment. Attempts to measure the humidity of the atmosphere; condensation hygrometer. The electric process; telluric electricity; Otto von Guericke sees, for the first time, light in induced electricity. Beginnings of pneumatic chemistry; observed increase of weight in metals from oxydation; Cardanus and Jean Rey, Hooke and Mayow. Ideas on the fundamental part of the atmosphere (*spiritus nitro-aëreus*), which enters into all metallic calxes, and is necessary to all the processes of combustion, and the respiration of animals. Influence of physical and chemical knowledge on the development of geognosy (Nicolaus Steno, Scilla, Lister); the elevation of the sea's bottom and of littoral districts. In the greatest of all geognostic phenomena—the mathematical figure of the earth—we see perceptibly reflected all the conditions of a primitive age, or, in other words, the primitive fluid state of the rotating mass and its consolidation into a terrestrial spheroid. Measurements of degrees and pendulum experiments in different latitudes. Compression. The figure of the earth was known to Newton on theoretical grounds, and the force discovered, of the operation of which the laws of Kepler are a necessary consequence. The discovery of such a force, whose existence is developed in Newton's imperishable work *Principia*, was nearly simultaneous with the opening of new paths to mathematical discovery by the invention of the infinitesimal calculus— p. 331–352.

VIII. *Retrospect, Multiplicity, and intimate Connection existing among the Scientific Efforts of Modern Times.*—Retrospect of the principal momenta in the history of cosmical contemplation connected with great events. The multiplicity of the links of connection among the different branches of science in the present day increases the difficulty of separating and limiting the individual portions—Intellectual activity henceforth produces great results almost without any external incitement, and by its own internal power manifested in every direction. The history of the physical sciences gradually fuses into that of the idea of Universal Nature—p. 352–356.

COSMOS.

PART I.

INCITEMENTS TO THE STUDY OF NATURE.

THE IMAGE REFLECTED BY THE EXTERNAL WORLD ON THE IMAGIN-
ATION.—POETIC DESCRIPTION OF NATURE.—LANDSCAPE PAINTING.—
THE CULTIVATION OF EXOTIC PLANTS, WHICH CHARACTERIZE THE
VEGETABLE PHYSIOGNOMY OF THE VARIOUS PARTS OF THE EARTH'S
SURFACE.

WE are now about to proceed from the sphere of objects to
that of sensations. The main results of observation, which,
stripped of all the extraneous charms of fancy, belong to the
purely objective domain of a scientific delineation of nature,
have been considered in the former part of this work in the
mutually connected relations, by which they constitute one
sole picture of the universe. It now, therefore, remains for
us to consider the impressions reflected by the external senses
on the feelings, and on the poetic imagination of mankind.
An inner world is here opened before us, but in seeking to
penetrate its mysterious depths, we do not aspire, in turning
over the leaves of the great book of Nature, to arrive at that
solution of its problems which is required by the philosophy
of art in tracing æsthetic actions through the psychical powers
of the mind, or through the various manifestations of intellectual activity, but rather to depict the contemplation of
natural objects as a means of exciting a pure love of nature,
and to investigate the causes which, especially in recent times,
have, by the active medium of the imagination, so powerfully
encouraged the study of nature and the predilection for distant travels.* The inducements which promote such contemplations of nature are, as I have already remarked, of
three different kinds, namely, the æsthetic treatment of natural scenery by animated delineations of animal and vegetable
forms, constituting a very recent branch of literature; landscape painting, especially where it has caught the characteristic features of the animal and vegetable world; and the

* See vol. i., p. 57.

more widely-diffused cultivation of tropical floras, and the more strongly contrasting opposition of exotic and indigenous forms. Each of these might, owing to their historical relations, be made the object of a widely-extending consideration, but it appears to me more in conformity with the spirit and aim of this work merely to unfold a few leading ideas, in order to remind the reader how differently the aspect of nature has acted on the intellect and feelings of different nations at different epochs, and how, at periods characterized by general mental cultivation, the severer forms of science and the more delicate emanations of fancy have reciprocally striven to infuse their spirit into one another. In order to depict nature in its exalted sublimity, we must not dwell exclusively on its external manifestations, but we must trace its image, reflected in the mind of man, at one time filling the dreamy land of physical myths with forms of grace and beauty, and at another developing the noble germ of artistic creations.

In limiting myself to the simple consideration of the incitements to a scientific study of nature, I would not, however, omit calling attention to the fact that impressions arising from apparently accidental circumstances often—as is repeatedly confirmed by experience—exercise so powerful an effect on the youthful mind as to determine the whole direction of a man's career through life. The child's pleasure in the form of countries, and of seas and lakes,* as delineated in maps ; the desire to behold southern stars, invisible in our hemisphere ;† the representation of palms and cedars of Lebanon as depicted in our illustrated Bibles, may all implant in the mind the first impulse to travel into distant countries. If I might be permitted to instance my own experience, and recall to mind the source from whence sprang my early and fixed desire to visit the land of the tropics, I should name George Forster's *Delineations of the South Sea Islands*, the pictures of Hodge, which represented the shores of the Ganges, and which I first saw at the house of Warren Hastings, in London, and a colossal dragon-tree in an old tower of the Botanical Garden at Berlin. These objects, which I here instance by way of illustration, belong to the three classes of induce-

---

* As the configuration of the countries of Italy, Sicily, and Greece, and of the Caspian and Red Seas. See *Relation Historique du Voy. aux Régions Equinoxiales*, t. i., p. 208.
† Dante, *Purg.*, i., 25-28.

> Goder pareva il ciel di lor fiammelle :
> O settentrional vedovo sito,
> Poi che privato se' di mirar quelle !

ments which we have already named, viz., the description of nature when springing from an animated impression of terrestrial forms ; the delineative art of landscape painting ; and, lastly, the direct objective consideration of the characteristic features of natural forms. The power exercised by these incitements is, however, limited to the sphere embraced by modern cultivation, and to those individuals whose minds have been rendered more susceptible to such impressions by a peculiar disposition, fostered by some special direction in the development of their mental activity.

DESCRIPTION OF NATURE.—THE DIFFERENCE OF FEELING EXCITED BY THE CONTEMPLATION OF NATURE AT DIFFERENT EPOCHS AND AMONG DIFFERENT RACES OF MEN.

It has often been remarked that, although the enjoyment derived from the contemplation of nature was not wholly unknown to the ancients, the feeling was, nevertheless, much more rarely, and less vividly expressed than in modern times. In his considerations on the poetry of the sentiments, Schiller thus expresses himself :* " If we bear in mind the beautiful scenery with which the Greeks were surrounded, and remember the opportunities possessed by a people living in so genial a climate, of entering into the free enjoyment of the contemplation of nature, and observe how conformable were their mode of thought, the bent of their imaginations, and the habits of their lives to the simplicity of nature, which was so faithfully reflected in their poetic works, we can not fail to remark with surprise how few traces are to be met among them of the sentimental interest with which we, in modern times, attach ourselves to the individual characteristics of natural scenery. The Greek poet is certainly, in the highest degree, correct, faithful, and circumstantial in his descriptions of nature, but his heart has no more share in his words than if he were treating of a garment, a shield, or a suit of armor. Nature seems to interest his understanding more than his moral perceptions ; he does not cling to her charms with the fervor and the plaintive passion of the poet of modern times."

However much truth and excellence there may be in these

* See Schiller's *Sämmtliche Werke*, 1826, bd. xviii., s. 231, 473, 480, 486 ; Gervinus, *Neuere Gesch. der Poet. National-Litteratur der Deutschen*, 1840, bd. i., s. 135 ; Adolph Bekker, in *Charikles*, th. i., s. 219. Compare, also, Eduard Müller, *Ueber Sophokleische Naturanschauung und die tiefe Naturempfindung der Griechen*, 1842, s. 10, 26.

remarks, they must not be extended to the whole of antiquity;
and I moreover consider that we take a very limited view of
antiquity when, in contradistinction to the present time, we
restrict the term exclusively to the Greeks and Romans. A
profound feeling of nature pervades the most ancient poetry
of the Hebrews and Indians, and exists, therefore, among nations of very different descent—Semitic and Indo-Germanic.

We can only draw conclusions regarding the feelings entertained by the ancients for nature from those expressions of the
sentiment which have come down to us in the remains of their
literature, and we must, therefore, seek them with a care, and
judge of them with a caution proportionate to the infrequency
of their occurrence in the grand forms of lyric and epic poetry.
In the periods of Hellenic antiquity—the flowery season in
the history of mankind—we certainly meet with the tenderest
expressions of deep natural emotion, blended with the most
poetic representations of human passion, as delineating some
action derived from mythical history ; but specific descriptions
of nature occur only as accessories, for, in Grecian art, all
things are centered in the sphere of human life.

The description of nature in its manifold richness of form,
as a distinct branch of poetic literature, was wholly unknown
to the Greeks. The landscape appears among them merely
as the back-ground of the picture of which human figures constitute the main subject. Passions, breaking forth into action,
riveted their attention almost exclusively. An active life,
spent chiefly in public, drew the minds of men from dwelling
with enthusiastic exclusiveness on the silent workings of nature, and led them always to consider physical phenomena as
having reference to mankind, whether in the relations of external conformation or of internal development.* It was almost exclusively under such relations that the consideration
of nature was deemed worthy of being admitted into the domain of poetry under the fantastic form of comparisons, which
often present small detached pictures replete with objective
truthfulness.

At Delphi, pæans to Spring were sung,† being intended,

---

* Schnaase, *Geschichte der bildenden Künste bei den Alten*, bd. ii.,
1843, s. 128–138.

† Plut., *de E. I. apud Delphos*, c. 9 [an attempt of Plutarch's to explain
the meaning of an inscription at the entrance of the temple of Delphi.
—*Tr.*]. Regarding a passage of Apollonius Dyscolus of Alexandria
(*Mirab. Hist.*, c. 40), see Otfr. Müller's last work, *Gesch. der Griech.
Litteratur*, bd. i., 1845, s. 31.

probably, to express the delight of man at the termination of the discomforts of winter. A natural description of winter is interwoven (perhaps by the hand of some Ionian rhapsodist) in the *Works and Days* of Hesiod.* This poem, which is composed with noble simplicity, although in accordance with the rigid didactic form, gives instructions regarding agriculture, directions for different kinds of trade and labor, and ethic precepts for a blameless course of life. It is only elevated to the dignity of a lyric poem when the poet clothes the miseries of mankind, or the exquisite mythical allegory of Epimetheus and Pandora, in an anthropomorphic garb. In the theogony of Hesiod, which is composed of many ancient and dissimilar elements, we frequently find, as, for instance, in the enumeration of the Nereides,† natural descriptions of the realm of Neptune concealed under the significant names of mythical characters. The Bœotian, and, indeed, all the ancient schools of poetry, treat only of the phenomena of the external world, under the personification of human forms.

But if, as we have already remarked, natural descriptions, whether they delineate the richness and luxuriance of tropical vegetation, or portray the habits of animals, have only become a distinct branch of literature in the most recent times, this circumstance must not be regarded as a proof of the absence of susceptibility for the beauties of nature, where the perception of beauty was so intense,‡ nor must we suppose that the animated expression of a spirit of poetic contemplation was wanting to the Greeks, who have transmitted to us such inimitable proofs of their creative faculty alike in poetry and in sculpture. All that we are led by the tendency of our modern ideas to discover as deficient in this department of ancient literature is rather of a negative than of a positive kind, being evinced less in the absence of susceptibility than in that of the urgent impulse to give expression in words to the sentiment awakened by the charms of nature. Directed less to

---

* *Hesiodi Opera et Dies*, v. 502–561. Göttling, in *Hes. Carm.*, 1831, p. xix.; Ulrici, *Gesch. der Hellenischen Dichtkunst*, th. i., 1835, s. 337. Bernhardy, *Grundriss der Griech. Litteratur*, th. ii., s. 176. According to the opinion of Gottfr. Hermann (*Opuscula*, vol. vi., p. 239), "the picturesque description given by Hesiod of winter bears all the evidence of great antiquity."

† Hes., *Theog.*, v. 233–264. The Nereid Mera (*Od.*, xi., 326; *Il.*, xviii., 48) may perhaps be indicative of the phosphoric light seen on the surface of the sea, in the same manner as the same word μαῖρα designates the sparkling dog-star Sirius.

‡ Compare Jacobs, *Leben und Kunst der Alten*, bd. i., abth. i., s. vii.,

the inanimate world of phenomena than to the realities of active life, and to the inner and spontaneous emotions of the
mind, the earliest, and, at the same time, the noblest directions of the poetic spirit were epic and lyric. In these artificial forms, descriptions of nature can only occur as incidental
accessories, and not as special creations of fancy. As the influence of antiquity gradually disappeared, and as the bright
beauty of its blossoms faded, rhetorical figures became more
and more diffused through descriptive and didactic poetry.
This form of poetry, which in its earliest philosophical, half-
sacerdotal type, was solemn, grand, and devoid of ornament
—as we see exemplified in the poem of Empedocles *On Nature*—by degrees lost its simplicity and earlier dignity as it
became more strongly marked by a rhetorical character.

I may be permitted here to mention a few particular instances, in illustration of these general observations. In conformity with the character of the Epos, we find the most attractive scenes of nature introduced in the Homeric songs
merely as secondary adjuncts. "The shepherd rejoices in
the stillness of night, in the purity of the sky, and in the
starry radiance of the vault of heaven; he hears from afar
the rush of the mountain torrent, as it pursues its foaming
course swollen with the trunks of oaks that have been borne
along by its turbid waters."[*] The sublime description of the
sylvan loneliness of Parnassus, with its somber, thickly-wooded
and rocky valleys, contrasts with the joyous pictures of the
many-fountained poplar groves in the Phæacian island of
Scheria, and especially of the land of the Cyclops, "where
meadows waving with luxuriant and succulent grass encircle
the hills of unpruned vines."[†] Pindar, in a dithyrambus in
praise of Spring, recited at Athens, sings of "the earth covered
with new-born flowers, when, in the Argive Nemæa, the first
opening shoot of the palm announces the coming of balmy
Spring." Then he sings of Ætna as "the pillar of heaven,
the fosterer of enduring snow;" but he quickly turns away

[*] *Ilias*, viii., 555–559; iv., 452–455; xi., 115–119. Compare, also,
the crowded but animated description of the animal world, which pre-
cedes the review of the army, ii., 458–475.

[†] *Od.*, xix., 431–445; vi., 290; ix., 115–199. Compare, also, "the
verdant overshadowing of the grove" near Calypso's grotto, "where
even an immortal would linger with admiration, rejoicing in the beautiful view," v. 55–73; the breaking of the surf on the shores of the
Phæacian Islands, v. 400–442; and the gardens of Alcinoüs, vii., 113–
130. On the vernal dithyrambus of Pindar, see Böckh, *Pindari Opera*,
t. ii., part ii., p. 575–579.

from these terrific forms of inanimate nature to celebrate Hiero of Syracuse, and the victorious combats of the Greeks with the mighty race of the Persians.

We must not forget that Grecian scenery presents the peculiar charm of an intimate association of land and sea, of shores adorned with vegetation, or picturesquely girt round by rocks gleaming in the light of aërial tints, and of an ocean beautiful in the play of the ever-changing brightness of its deep-toned moving waves.

Although to other nations, sea and land, in the different pursuits of life to which they give rise, appeared as two separate spheres of nature, the Greeks—not only those who inhabited the islands, but also those occupying the southern portion of the continent—enjoyed, almost every where, the aspect of the richness and sublime grandeur imparted to the scenery by the contact and mutual influence of the two elements. How can we suppose that so intellectual and highly-gifted a race should have remained insensible to the aspect of the forest-crowned cliffs on the deeply-indented shores of the Mediterranean, to the silent interchange of the influences affecting the surface of the earth, and the lower strata of the atmosphere at the recurrence of regular seasons and hours, or to the distribution of vegetable forms? How, in an age when the poetic feelings were the strongest, could this active state of the senses have failed to manifest itself in ideal contemplation? The Greek regarded the vegetable world as standing in a manifold and mythical relation to heroes and to the gods, who were supposed to avenge every injury inflicted on the trees and plants sacred to them. Imagination animated vegetable forms with life, but the types of poetry, to which the peculiar direction of mental activity among the ancient Greeks limited them, gave only a partial development to the descriptions of natural scenery. Occasionally, however, even in the writings of their tragic poets, a deep sense of the beauty of nature breaks forth in animated descriptions of scenery in the midst of the most excited passions or the deepest tones of sadness. Thus, when Œdipus is approaching the grove of the Eumenides, the chorus sings, " the noble resting-place of the illustrious Colonos, where the melodious nightingale loves to tarry and pour forth its clear but plaintive notes.". Again it sings, " the verdant gloom of the thickly-mantling ivy, the narcissus steeped in heavenly dew, the golden-beaming crocus, and the hardy and ever fresh-sprouting olive-tree."* Sophocles strives

* *Œd. Colon.*, v. 668-719. Among delineations of scenery, indica-

to extol his native Colonos by placing the lofty form of the
fated and royal wanderer by the brink of the sleepless waters
of Cephisus, surrounded by soft and bright scenery. The repose of nature heightens the impression of pain called forth by
the image of the noble form of the blind sufferer, the victim
of mysterious and fatal passion. Euripides* also delights in
picturesque descriptions of " the pastures of Messenia and La-
conia, which, under an ever-mild sky, are refreshed by a thousand fountains, and by the waters of the beautiful Pamisos."

Bucolic poetry, which originated in the plains of Sicily, and
popularly inclined to the dramatic, has been justly termed a
transitional form. Its pastoral epics describe on a small scale
human beings rather than natural scenery ; and in this form
it appears in its greatest perfection in the writings of Theoc-
ritus. A soft elegiac element is peculiar to the idyl, as if it
had emanated from " the longing for some lost idea ;" as if,
in the breast of mankind, a certain touch of melancholy was
ever mingled with the deep feelings awakened by the aspect
of nature.

True Hellenic poetry expired with the freedom of the
Greeks, and became descriptive, didactic, and instructive. Astronomy, geography, hunting, and fishing were converted, in
the time of Alexander, into objects of poetic consideration, and
often adorned with a remarkable degree of metrical skill. The
forms and habits of animals are depicted with grace, and not
unfrequently with such accuracy that the particular genera
or even species may be recognized by the classifying naturalist of the present day. All these compositions are, however,
wholly wanting in that inner life—that inspired contemplation of nature—by which the external world becomes to the
poet, almost unconsciously to himself, a subject of his imagin-

tive of a deep feeling of nature, I would here further mention the description of Cithæron in the *Bacchæ* of Euripides, v. 1045 (Leake, *North.
Greece*, vol. ii., p. 370), where the messenger ascends from the Valley
of Asopus, the reference to the sunrise in the Valley of Delphos, in the
Ion of Euripides, v. 82, and the gloomy picture in the *Hymn on Delos*,
v. 11, by Callimachus, in which the holy Delos is represented as surrounded by sea-gulls, and scourged by tempestuous waves.

* According to Strabo (lib. viii., p. 366, Casaub.), who accuses the
tragedian of giving a geographically incorrect boundary to Elis. This
beautiful passage of Euripides occurs in the *Cresphontes*. The description of the excellence of the district of Messenia is intimately connected
with the exposition of its political relations, as, for instance, the division
of the land among the Heraclidæ. The delineation of nature is, therefore, here too, as Böckh ingeniously remarks, associated with human
interests.

ation.   The preponderance of the descriptive element shows
itself in the forty-eight cantos of the *Dionysiaca* of the Egyptian Nonnus, which are remarkable for their skillfully artistical versification.   The poet dwells with pleasure on the delineation of great convulsions of nature ; he makes a fire kindled by lightning on the woody banks of the Hydaspes burn
up even the fishes in the bed of the river ; and he shows how
ascending vapors occasion the meteorological processes of the
storm and electric rain.   Although capable of writing romantic poetry, Nonnus of Panopolis is remarkably unequal in his
style, being at one time animated and exciting, and at another
tedious and verbose.

A deeper feeling for nature and a greater delicacy of sensibility is manifested in some portions of the Greek Anthology,
which has been transmitted to us in such various ways and
from such different epochs.   In the graceful translation of
Jacobs, every thing that relates to animal and vegetable forms
has been collected in one section—these passages being small
pictures, consisting, in most cases, of mere allusions to individual forms.   The plane-tree, which " nourishes amid its
branches the grape swelling with juice," and which, in the
time of Dionysius the Elder, first penetrated from Asia Minor
through the Island of Diomedes to the shores of the Sicilian
Anapus, is perhaps too often introduced ; still, on the whole,
the ancient mind shows itself more inclined, in these songs
and epigrams, to dwell on the animal than on the vegetable
world.   The vernal idyl of Meleager of Gadara, in Cœlo-Syria, is a noble, and, at the same time, a more considerable composition.*

* *Meleagri Reliquiæ*, ed. Manso, p. 5.   Compare Jacobs, *Leben und
Kunst der Alten*, bd. i., abth. i., s. xv. ; abth. ii., s. 150–190.   Zenobetti
believed himself to have been the first to discover Meleager's poem on
Spring, in the middle of the eighteenth century (*Mel. Gadareni in Ver
Idyllion*, 1759, p. 5).   See *Brunckii Anal.*, t. iii., p. 105.   There are
two fine sylvan poems of Marianos in the *Anthol. Græca*, ii., 511 and
512.   Meleager's poem contrasts well with the praise of Spring in the
eclogues of Himerius, a Sophist, who was teacher of rhetoric at Athens
under Julian.   The style, on the whole, is cold and profusely ornate ;
but in some parts, especially in the descriptive portions, this writer
sometimes approximates closely to the modern way of considering nature.   *Himerii Sophistæ Eclogæ et Declamationes*, ed. Wernsdorf, 1790.
(Oratio iii., 3–6, and xxi., 5.)   It seems extraordinary that the lovely
situation of Constantinople should not have inspired the Sophists.
(Orat. vii., 5–7 ; xvi., 3–8.)   The passages of Nonnus, referred to in the
text, occur in *Dionys.*, ed. Petri Cunæi, 1610, lib. ii., p. 70 ; vi., p. 199 ;
xxiii., p. 16 and 619 ; xxvi., p. 694.   Compare, also, Ouwaroff, *Nonnus
von Panopolis, der Dichter*, 1817, s. 3, 16, 21.

On account of the renown attached from ancient times to the spot, I would not omit to mention the description of the wooded valley of Tempe, as given by Ælian,* probably in imitation of some earlier notice by Dicæarchus. It is the most detailed description of natural scenery by any of the Greek prose writers that we possess; and, although topographical, it is also picturesque, for the shady vale is animated by the Pythian procession (*theoria*), "which breaks from the sacred laurel the atoning bough." In the later Byzantine epoch, about the close of the fourth century, we meet more frequently with descriptions of scenery interwoven in the romances of the Greek prose writers, as is especially manifested in the pastoral romance of Longus,† in which, however, the tender scenes taken from life greatly excel the expression of the sensations awakened by the aspect of nature.

It is not my object in the present work to extend these references beyond what my own special recollection of particular forms of art may enable me to add to these general considerations of the poetic conception of the external world. I should here quit the flowery circle of Grecian antiquity, if, in a work to which I have ventured to prefix the title of Cosmos, I could pass over in silence the description of nature with which the pseudo-Aristotelian book of Cosmos, or *Order of the Universe*, begins. It describes "the earth as adorned with luxuriant vegetation, copiously watered, and (as the most admirable of all) inhabited by thinking beings."‡ The rhetorical color of this rich picture of nature, so totally unlike the concise and purely scientific mode of treatment characteristic of the Stagirite, is one of the many indications by which it has been judged that this work on the *Cosmos* is not his composition. It may, in fact, be the production of Apuleius,§ or of Chrysip-

---

* *Æliani Var. Hist. et Fragm.*, lib. iii., cap. 1, p. 139, Kühn. Compare A. Buttmann, *Quæst. de Dicæarcho* (Naumb., 1832, p. 32), and *Geogr. Gr. Min.*, ed. Gail, vol. ii., p. 140–145. We observe in the tragic poet Chæremon a remarkable love of nature, and especially a predilection for flowers, which has been compared by Sir William Jones to the sentiments evinced in the Indian poets. See Welcker, *Griechische Tragödien*, abth. iii., s. 1088.

† *Longi Pastoralia* (*Daphnis et Chloe*, ed. Seiler, 1843), lib. i., 9; iii., 12, and iv., 1–3; p. 92, 125, 137. Compare Villemaine, *Sur les Romans Grecs*, in his *Mélanges de Littérature*, t. ii., p. 435–448, where Longus is compared with Bernardin de St. Pierre.

‡ Pseudo-Aristot., *de Mundo*, cap. 3, 14–20, p. 392, Bekker.

§ See Stahr, *Aristoteles bei den Römern*, 1834, s. 173–177. Osann, *Beiträge zur Griech. und Röm. Litteraturgeschichte*, bd. i., 1835, s. 165–192. Stahr (s. 172) supposes, like Heumann, that the present Greek is

pus,* or of any other author.   In the place of the passages
relating to natural scenery, which we can not venture to ascribe to Aristotle, we possess, however, a genuine fragment
which Cicero has preserved to us from a lost work of Aristotle.†   It runs thus : " If there were beings who lived in
the depths of the earth, in dwellings adorned with statues and
paintings, and every thing which is possessed in rich abundance by those whom we esteem fortunate ; and if these beings
could receive tidings of the power and might of the gods, and
could then emerge from their hidden dwellings through the
open fissures of the earth to the places which we inhabit ; if
they could suddenly behold the earth, and the sea, and the
vault of heaven ; could recognize the expanse of the cloudy
firmament, and the might of the winds of heaven, and admire
the sun in its majesty, beauty, and radiant effulgence ; and,
lastly, when night vailed the earth in darkness, they could behold the starry heavens, the changing moon, and the stars
rising and setting in the unvarying course ordained from eternity, they would surely exclaim, ' there are gods, and such
great things must be the work of their hands.' "   It has been
justly observed that this passage is alone sufficient to corroborate Cicero's opinion of " the golden flow of Aristotle's eloquence,"‡ and that his words are pervaded by something of
the inspired force of Plato's genius.   Such a testimony to the
existence of the heavenly powers, drawn from the beauty and
stupendous greatness of the works of creation, is rarely to be
met with in the works of antiquity.

   That which we miss in the works of the Greeks, I will not
say from their want of susceptibility to the beauties of nature,
but from the direction assumed by their literature, is still more
rarely to be met with among the Romans.   A nation which,
in accordance with the ancient Sicilian habits, evinced a decided predilection for agriculture and other rural pursuits,
might have justified other expectations ; but, with all their

an altered translation of the Latin text of Apuleius.   The latter says
distinctly (*de Mundo*, p. 250, Bip.) " that he has followed Aristotle and
Theophrastus in the composition of his work."
   * Osann, op. cit., s. 194–266.
   † Cicero, *de Natura Deorum*, ii., 37.   A passage in which Sextus Empiricus (*adversus Physicos*, lib. ix., 22, p. 554, Fabr.) instances a similar
expression of Aristotle, deserves the more attention from the fact that
the same writer shortly before (ix., 20) alludes to another work of Aristotle (on divination and dreams) which is also lost to us.
   ‡ "Aristoteles flumen orationis aureum fundens." Cic., *Acad. Quæst.*,
ii., cap. 38.   (Compare Stahr, *Aristotelia*, th. ii., s. 161, and *Aristoteles
bei den Römern*, s. 53.)

disposition to practical activity, the Romans, with the cold
severity and practical understanding of their national character, were less susceptible of impressions of the senses than the
Greeks, and were more devoted to every-day reality than to
the idealizing poetic contemplation of nature. These differences in the habits and feelings of the Greeks and Romans
are reflected in their literature, as is ever the case with the
intellectual expression of national character. Here, too, we
must notice the acknowledged difference that exists in the organic structure of their respective languages, notwithstanding
the affinity between the races. The language of ancient La-
tium possesses less flexibility, a more limited adaptation of
words, a stronger character of "practical tendency" than of
ideal mobility. Moreover, the predilection evinced in the Au-
gustan age for imitating Greek images must have been detrimental to the free outpouring of native feelings, and to the
free expression of the natural bent of the mind ; but still there
were some powerful minds, which, inspired by love of country, were able by creative individuality, by elevation of
thought, and by the gentle grace of their representations, to
surmount all these obstacles. The great poem of nature,
which Lucretius has so richly decked with the charms of his
poetic genius, embraces the whole Cosmos. It has much affinity with the writings of Empedocles and Parmenides, the
archaic diction of the versification heightening the earnestness of the descriptions. Poetry is here closely interwoven
with philosophy, without, however, falling into that frigidity
of style which, in contrast with Plato's richly fanciful mode
of treating nature, was so severely blamed by Menander the
Rhetorician, in the sentence he pronounced on the Hymns of
Nature.* My brother has shown with much ingenuity the
striking analogies and differences which have arisen from the
amalgamation of metaphysical abstractions with poetry in the
ancient Greek didactic poems, as in the works of Lucretius,
and in the episode *Bhagavad* of the Indian Epic *Mahabhar-*

---

* *Menandri Rhetoris Comment. de Encomiis,* ex rec. Heeren, 1785,
sect. i., cap. 5, p. 38, 39. The severe critic terms the didactic poem
On Nature a frigid composition (ψυχρότερον), in which the forces of nature are brought forward divested of their personality—Apollo as light,
Hera as the concentration of all the phenomena of the atmosphere, and
Jupiter as heat. Plutarch also ridicules the so-called poems of nature,
which have only the form of poetry (*de Aud. Poet.*, p. 27, Steph.). According to the Stagirite (*de Poet.*, c. i.), Empedocles was more a physiologist than a poet, and has nothing in common with Homer but the
rhythmical measure used by both.

*ata.** The great physical picture of the universe by the Roman poet contrasts in its cold doctrine of atoms, and in its
frequently visionary geognostic hypotheses, with his vivid and
animated delineation of the advance of mankind from the recesses of the forest to the pursuit of agriculture, to the control
of natural forces, the more elevated cultivation of mind and
languages, and through the latter to social civilization.†
When, in the midst of the active and busy life of the statesman, and in a mind excited by political passion, a keen susceptibility for the beauties of nature and an animated love of
rural solitude still subsists, its source must be derived from the
depths of a great and noble character. Cicero's writings testify to the truth of this assertion. As is generally known,
many points in his book *De Legibus,* and in that *De Oratore,*
are copied from Plato's *Phædrus;*‡ yet his delineations of
Italian nature do not, on that account, lose any of their individuality. Plato extols in general terms "the dark shade of
the thickly-leaved plane-tree; the luxuriance of plants and
herbs in all the fragrance of their bloom; and the sweet summer breezes which fan the chirping swarms of grasshoppers."
In Cicero's smaller sketches of nature we find, as has lately
been remarked by an intelligent inquirer,§ all things described
as they still exist in the actual landscape; we see the Liris
shaded by lofty poplars; and as we descend from the steep

* "It may appear singular, but yet it is not the less correct, to attempt to connect poetry, which rejoices every where in variety of form,
color, and character, with the simplest and most abstract ideas. Poetry, science, philosophy, and history are not necessarily and essentially
divided; they are united wherever man is still in unison with the particular stage of his development, or whenever, from a truly poetic mood
of mind, he can in imagination bring himself back to it." Wilhelm von
Humboldt, *Gesammelte Werke,* bd. i., s. 98–102. (Compare, also, Bernhardy, *Röm. Litteratur,* s. 215–218, and Fried. Schlegel, *Sämmtliche
Werke,* bd. i., s. 108–110.) Cicero (*ad Quint. fratrem,* ii., 11) ascribes,
if not pettishly, at any rate very severely, more tact than creative talent
(*ingenium*) to Lucretius, who has been so highly praised by Virgil,
Ovid, and Quintilian.   † Lucret., lib. v., v. 930–1455.

‡ Plato, *Phædr.,* p. 230; Cicero, *de Leg.,* i., 5, 15; ii., 2, 1–3; ii., 3,
6. (Compare Wagner, *Comment. Perp.,* in Cic., *de Leg.,* 1814, p. 6;)
Cic., *de Oratore,* i., 7, 28 (p. 15, Ellendt).

§ See s. 431–434 of the admirable work by Rudolph Abeken, rector
of the Gymnasium at Osnabrück, which appeared in 1835 under the
title of *Cicero in seinen Briefen.* The important addition relative to
the birth-place of Cicero is by H. Abeken, the learned nephew of the
author, who was formerly chaplain to the Prussian embassy at Rome,
and is now taking part in the important Egyptian expedition of Professor Lepsius. See, also, on the birth-place of Cicero, Valery, *Voy. Hist.
en Italie,* t. iii., p. 421.

mountain behind the old towers of Arpinum, we see the grove
of oaks on the margin of the Fibrenus, and the island now
called *Isola di Carnello*, which is formed by the division of
the stream, and whither Cicero retired, in order, as he said, to
"give himself up to meditation, reading, and writing."   Arpinum, situated on the Volscian Hills, was the birth-place of
the great statesman, and its noble scenery no doubt exercised
an influence on his character in boyhood.   Unconsciously to
himself, the external aspect of the surrounding scenery im-
presses itself upon the soul of man with an intensity corresponding to the greater or less degree of his natural susceptibility, and becomes closely interwoven with the deep original
tendencies and the free natural disposition of his mental
powers.

In the midst of the eventful storms of the year 708 (from
the foundation of Rome), Cicero found consolation in his villas,
alternately at Tusculum, Arpinum, Cumæa, and Antium.
"Nothing can be more delightful," he writes to Atticus,*
"than this solitude—nothing more charming than this country place, the neighboring shore, and the view of the sea.   In
the lonely island of Astura, at the mouth of the river of the
same name, on the shore of the Tyrrhenian Sea, no human
being disturbs me ; and when, early in the morning, I retire
to the leafy recesses of some thick and wild wood, I do not
leave it till the evening.   Next to my Atticus, nothing is so
dear to me as solitude, in which I hold communion with philosophy, although often interrupted by my tears.   I struggle
as much as I am able against such emotions, but as yet I am
not equal to the contest."   It has frequently been remarked,
that in these letters, and in those of the younger Pliny, passages are met with which manifest the greatest harmony with
the expressions in use among modern sentimental writers ; for
my own part, I can only find in them the echoes of the same
deep-toned sadness which in every age and in every race bursts
forth from the recesses of the heavily-oppressed bosom.

Amid the general diffusion of Roman literature, an acquaintance with the great poetic works of Virgil, Horace,
and Tibullus is so common, that it would be superfluous to
dwell on individual examples of the tender and ever wakeful
sensibility to nature, by which some of these works are animated.   In Virgil's great epic, the nature of the poem tends
to make descriptions of scenery appear merely as accessories,

---

* Cic., *Ep. ad Atticum*, xii., 9 and 15.

occupying only a very small space. There is no individual portraiture of particular localities,* but a deep and intimate comprehension of nature is depicted in soft colors. Where, for instance, has the gentle play of the waves, or the stillness of night been more happily described ? And how well do these pleasing pictures contrast with the powerful description of the bursting tempest in the first book of the Georgics, and the picture in the Æneid of the voyage and landing at the Strophades, the crashing fall of the rock, or the flames emitted from Mount Ætna.†

From Ovid we might have expected, as the fruit of his long sojourn in the plains of Tomi, in Lower Mœsia, a poetic description of the marshes, of which, however, no account has been transmitted to us from antiquity. The exile did not indeed see that kind of steppe-like plain, which in summer is densely covered with juicy plants, varying from four to six feet in height, and which in every breath of wind present the aspect of a waving sea of flowering verdure. The place of his banishment was a desolate, swampy marsh-land, and the broken spirit of the poet, which gives itself vent in unmanly lamentation, was preoccupied with the recollection of the enjoyments of social life and the political occurrences at Rome, and thus remained dead to the impressions produced by the contemplation of the Scythian desert, with which he was surrounded. As a compensation, however, this highly-gifted poet, whose descriptions of nature are so vivid, has given us, besides his too frequently-repeated representations of grottoes, springs, and "calm moon-light nights," a remarkably characteristic, and even geognostically important delineation of a volcanic eruption at Methone, between Epidaurus and Trœzene. The passage to which we allude has already been cited in another part of this work.‡ Ovid shows us, as our readers will re-

* The passages from Virgil, which are adduced by Malte-Brun (*Annales des Voyages*, t. iii., 1808, p. 235–266) as local descriptions, merely show that the poet had a knowledge of the produce of different countries, as, for instance, the saffron of Mount Tmolus; that he was acquainted with the incense of the Sabeans, and with the true names of several small rivers; and that even the mephitic vapors which rise from a cavern in the Apennines, near Amsanctus, were not unknown to him.

† Virg., *Georg.*, i., 356–392; iii., 349–380; *Æn.*, iii., 191–211; iv., 246–251; xii., 684–689.

‡ Compare Ovid, *Met.*, i., 568–576; iii., 155–164; iii., 407–412; vii., 180–188; xv., 296–306; *Trist.*, lib. i., *El.* 3, 60; lib. iii., *El.* 4, 49; *El.* 12, 15; *Ex Ponto*, lib. iii., *Ep.* 7–9, as instances of separate pictures of natural scenery. There is a pleasant description of a spring at Hymettus, beginning with the verse,

member, "how, by the force of the impregnated vapor, the earth was distended like a bladder filled with air, or like the skin of the goat."

It is especially to be regretted that Tibullus should have left no great composition descriptive of the individual character of nature. Among the poets of the Augustan age, he belongs to the few who, being happily strangers to the Alexandrian learning, and devoted to seclusion and a rural life, drew with feeling, and therefore with simplicity, from the resources of their own mind. Elegies,* of which the landscape only constitutes the back-ground, must certainly be regarded as mere pictures of social habits; but the *Lustration of the Fields*, and the Sixth Elegy of the first book, show us what was to have been expected from the friend of Horace and of Messala.

Lucan, the grandson of the rhetorician M. Annæus Seneca, certainly resembles the latter too much in the rhetorical ornation of his diction, but yet we find among his works an admirable and vividly truthful picture of the destruction of a Druidic forest† on the now treeless shores of Marseilles. The half-severed oaks support themselves for a time by leaning tottering against each other, and, stripped of their leaves, suffer the first ray of light to pierce their awful and sacred gloom. He who has long lived amid the forests of the New World must feel how vividly the poet, with a few touches, has depicted the luxuriant growth of trees, whose colossal remains lie buried in some of the turf moors of France. In the didactic poem of *Ætna* by Lucilius the younger, a friend of L. Annæus Seneca, we certainly meet with a truthful description of the phenomena attending the eruption of a volcano; but the conception has much less of individuality than the work entitled *Ætna Dialogus*,‡ by Bembo, of which we have already spoken in terms of praise.

"Est prope purpureos colles florentis Hymetti"

(Ovid, *de Arte. Am.*, iii., 687), which, as Ross has remarked, is one of the rare instances that occur of individual delineations of nature referring to a definite locality. The poet describes the fountain of Kallia, sacred to Aphrodite, so celebrated in antiquity, which breaks forth on the western side of Hymettus, otherwise so scantily supplied with water. (See Ross, Letter to Professor Vuros, in the *Griech. Medicin. Zeitschrift*, June, 1837.

* Tibullus, ed. Voss, 1811, *Eleg.*, lib. i., 6, 21–34; lib. ii., 1, 37–66.
† Lucan, *Phars.*, iii., 400–452 (vol. i., p. 374–384, Weber).
‡ The poem of Lucilius, which is very probably a part of a larger poetic work, on the natural characteristics of Sicily, was ascribed by

When, finally, at the close of the fourth century, the art of poetry, in its grander and nobler forms, faded away, as if exhausted, poetic emanations, stripped of the charms of creative fancy, turned aside to the barren realities of science and of description.   A certain oratorical polish of style could not compensate for the diminished susceptibility for nature and an idealizing inspiration.   As a production of this unfruitful age, in which the poetic element only appeared as an incidental external adornment of thought, we may instance a poem on the Moselle by Ausonius.   As a native of Aquitanian Gaul, the poet had accompanied Valentinian in his campaign against the Allemanni.   The *Mosella*, which was composed in ancient Treves,* describes in some parts, and not ungracefully, the already vine-clad hills of one of the loveliest of our rivers, but the barren topography of the country, the enumeration of the streams falling into the Moselle, and the characteristic form, color, and habits of some of the different species of fish that are found in these waters, constitute the main features of this wholly didactic composition.

In the works of the Roman prose writers, among which we have already cited some remarkable passages by Cicero, descriptions of natural scenery are as rare as in those of Greek authors.   It is only in the writings of the great historians, Julius Cæsar, Livy, and Tacitus, that we meet with some examples of the contrary, where they are compelled to de-

Wernsdorf to Cornelius Severus.   The passages especially worthy of attention are the praises of general knowledge considered as "the fruits of the mind," v. 270–280; the lava currents, v. 360–370 and 474–515; the eruptions of water at the foot of the volcano (?), v. 395; the formation of pumice, v. 425 (p. xvi.–xx., 32, 42, 46, 50, 55, ed. Jacob, 1826).

* *Decii Magni Ausonii Mosella*, v. 189–199, p. 15, 44, Böcking.   See, also, the notice of the fish of the Moselle, which is not unimportant with reference to natural history, and has been ingeniously applied by Valenciennes, v. 85–150, p. 9–12, and contrast it with Oppian (Bernhardy, *Griech. Litt.*, th. ii., s. 1049).   The *Orthinogonia* and *Theriaca* of Æmilius Macer of Verona (imitations of the works of Nicander of Colophon), which have not come to us, belonged to the same dry, didactic style of poetry which treated of the products of nature.   A natural description of the southern coast of Gaul, which is to be found in a poetical narrative of a journey by Claudius Rutilius Numatianus, a statesman under Honorius, is more attractive than the *Mosella* of Ausonius.   Rutilius, who was driven from Rome by the irruption of the Gauls, is returning to his estates in Gaul.   We unfortunately possess only a fragment of the second book of this poem, and this does not take us beyond the quarries of Carrara.   See *Rutilii Claudii Numatiani de Reditu suo* (*e Roma in Galliam Narbonensem*) *libri duo*, rec. A. W. Zumpt, 1840, p. xv., 31–219 (with a fine map by Kiepert).   Wernsdorf, *Poetæ Lat. Min.*, t. v., pt. i., p. 125.

scribe battle-fields, the crossing of rivers or difficult mountain passes in their narrations of the struggle of man against natural obstacles. In the Annals of Tacitus, I am charmed with the description of the untoward passage of Germanicus over the Amisia, and the grand geographical delineation of the mountain chains of Syria and Palestine.* Curtius has left us a fine natural picture of a woody desert to the west of Hec-atompylos, through which the Macedonian army had to pass in the marshy region of Mazanderan.† I would refer more circumstantially to this passage if our uncertainty as to the age in which this writer lived did not prevent our deciding what was due to the poet's own imagination and what was derived from historic sources.

The great encyclopedic work of the elder Pliny, which, by the richness of its contents, surpasses any other production of antiquity, will be more fully considered in the sequel, when we enter on the "history of the contemplation of the universe." The natural history of Pliny, which has exercised a powerful influence on the Middle Ages, is, as his nephew, the younger Pliny, has elegantly remarked, "manifold as nature itself." As the creation of an irresistible passion for a comprehensive, but often indiscriminate and irregular accumulation of facts, this work is unequal in style, being sometimes simple and narrative, and sometimes full of thought, animation, and rhetorical ornament, and from its very character deficient in individual delineations of nature ; although, wherever the connection existing between the active forces of the universe, the well-ordered Cosmos (*naturæ majestas*), is made

* Tac., *Ann.*, ii., 23–24; *Hist.*, v., 6. The only fragment preserved by the rhetorician Seneca (*Suasor.*, i., p. 11, Bipont) that we possess of a heroic poem, in which Ovid's friend Pedo Albinovanus describes the deeds of Germanicus, likewise describes the unfortunate passage of the Ems (Ped. Albinov., *Elegiæ*, Amst., 1703, p. 172). Seneca considers this description of the stormy waters as more picturesque than any passage to be found in the writings of the other Roman poets. He remarks, however, *Latini declamatores in Oceani descriptione non nimis viguerunt; nam aut tumide scripserunt aut curiose.*

† Curt., in *Alex. Magno.*, vi., 16. Compare Droysen, *Gesch. Alexanders des Grossen*, 1833, s. 265. In *Quæst. Natur.*, lib. iii., c. 27–30, p. 677–686, ed. Lips., 1741, of the too rhetorical Lucius Annæus Seneca, there is a remarkable description of one of the several instances of the destruction of an originally pure and subsequently sinful race, by an almost universal deluge, commencing with the words *Cum fatalis dies diluvii venerit;* and terminating thus: *peracto exitio generis humani extinctisque pariter feris in quarum homines ingenia transierant.* See, also, the description of chaotic terrestrial revolutions, in *Bhagavata-Purana*, bk. iii., c. 17 (ed. Burnouf, t. i., p. 441).

the object of contemplation, we can not mistake the indications of a true poetic inspiration.

We would gladly instance the pleasantly-situated villas on the Pincian Hill, at Tusculum and Tibur, on the promontory of Misenum, and at Puteoli and Baiæ, as proofs of a love of nature among the Romans, had not these buildings, like those of Scaurus and Mæcenas, of Lucullus and Adrian, been overstocked with edifices designed for pomp and display ; temples, theaters, and race-courses alternating with aviaries, and houses for rearing snails and dormice. The elder Scipio had surrounded his simpler country house at Liturnum with towers in the castellated style. The name of Matius, a friend of Augustus, has come down to us as that of the person who, in his love for unnatural stiffness, first caused trees to be cut in imitation of architectural and plastic patterns. The letters of the younger Pliny give us a charming description of two of his numerous villas, Laurentinum and Tusculum.* Although, in these two buildings, surrounded by cut box-trees, we meet with a greater number of objects crowded together than we, with our ideas of nature, would esteem in accordance with good taste, yet these descriptions, as well as the imitation of the Valley of Tempe in the Tiburtine villa of Adrian, show us that a love for the free enjoyment of nature was not wholly lost sight of by the Roman citizens in their love of art, and in their anxious solicitude for their personal comfort in adapting the locality of their country houses to the prevailing relations of the sun and winds. It is gratifying to be able to add that this enjoyment was less disturbed on the estates of Pliny than elsewhere by the revolting features of slavery. This wealthy man was not only one of the most learned of his age,

---

* Plin., *Epist.*, ii., 17 ; v., 6 ; ix., 17 ; Plin., *Hist. Nat.*, xii., 6 ; Hirt, *Gesch. der Baukunst bei den Alten,* bd. ii., s. 241, 291, 376. The villa Laurentina of the younger Pliny was situated near the present Torre di Paterno, in the littoral valley of Palombara, east of Ostia. See *Viaggio da Ostia a la villa di Plinio,* 1802, p. 9, and *Le Laurentin,* by Haudelcourt, 1838, p. 62. A deep feeling for nature is expressed in the few lines which Pliny wrote from Laurentinum to Minutius Fundanus : "*Mecum tantum et cum libellis loquor. Rectam sinceramque vitam! dulce otium honestumque! O mare, o littus, verum secretumque μουσεῖον! quam multa invenitis, quam multa dictatis!*" (i., 9). Hirt was persuaded that the origin in Italy, during the fifteenth and sixteenth centuries, of that stiff and systematic style of gardening long known as the French, in contradistinction to the freer mode of landscape gardening of the English, and the early taste for wearisome and regular lines, is to be ascribed to a wish of imitating that which Pliny the younger has described in his letters (*Geschichte der Baukunst bei den Alten,* th. ii., s. 366).

but he likewise entertained feelings of humane compassion for the enslaved condition of the people, a sentiment which was but seldom expressed in antiquity. On the estates of the younger Pliny no fetters were used; and the slave was permitted freely to bequeath, as a cultivator of the soil, that which he had acquired by the labor of his own hands.[*]

No description has been transmitted to us from antiquity of the eternal snow of the Alps, reddened by the evening glow or the morning dawn, of the beauty of the blue ice of the glaciers, or of the sublimity of Swiss natural scenery, although statesmen and generals, with men of letters in their retinue, continually passed through Helvetia on their road to Gaul. All these travelers think only of complaining of the wretchedness of the roads, and never appear to have paid any attention to the romantic beauty of the scenery through which they passed. It is even known that Julius Cæsar, when he was returning to his legions in Gaul, employed his time while he was passing over the Alps in preparing a grammatical work entitled *De Analogia.*[†] Silius Italicus, who died in the time of Trajan, when Switzerland was already considerably cultivated, describes the region of the Alps as a dreary and barren wilderness,[‡] at the same time that he extols with admiration the rocky ravines of Italy, and the woody shores of the Liris (Garigliano).[§] It is also worthy of notice, that the remarkable appearance of the jointed basaltic columns which are so frequently met with, associated in groups, in Central France, on the banks of the Rhine, and in Lombardy, should never have been described or even mentioned by Roman writers.

At the period when the feelings died away which had animated classical antiquity, and directed the minds of men to a visible manifestation of human activity rather than to a passive contemplation of the external world, a new spirit arose; Christianity gradually diffused itself, and, wherever it was adopted as the religion of the state, it not only exercised a beneficial influence on the condition of the lower classes by inculcating the social freedom of mankind, but also expanded

* Plin., *Epist.*, iii., 19; viii., 16.

† Suet., *in Julio Cæsare*, cap. 56. The lost poem of Cæsar (*Iter*) described the journey to Spain, when he led his army to his last military action from Rome to Cordova by land (which was accomplished in twenty-four days according to Suetonius, and in twenty-seven days according to Strabo and Appian), when the remnant of Pompey's party, which had been defeated in Africa, had rallied together in Spain.

‡ Sil. Ital., *Punica*, lib. iii., v. 477.

§ Id. ibid., lib. iv., v. 348; lib. viii., v. 399.

the views of men in their communion with nature. The eye
no longer rested on the forms of Olympic gods. The fathers
of the Church, in their rhetorically correct and often poetical-
ly imaginative language, now taught that the Creator showed
himself great in inanimate no less than in animate nature,
and in the wild strife of the elements no less than in the still
activity of organic development. At the gradual dissolution
of the Roman dominion, creative imagination, simplicity, and
purity of diction disappeared from the writings of that dreary
age, first in the Latin territories, and then in Grecian Asia
Minor. A taste for solitude, for mournful contemplation, and
for a moody absorption of mind, may be traced simultaneously
in the style and coloring of the language. Whenever a new
element seems to develop itself in the feelings of mankind, it
may almost invariably be traced to an earlier, deep-seated individual germ. Thus the softness of Mimnermus* has often
been regarded as the expression of a general sentimental direction of the mind. The ancient world is not abruptly separated from the modern, but modifications in the religious
sentiments and the tenderest social feelings of men, and changes
in the special habits of those who exercise an influence on the
ideas of the mass, must give a sudden predominance to that
which might previously have escaped attention. It was the
tendency of the Christian mind to prove from the order of the
universe and the beauty of nature the greatness and goodness
of the Creator. This tendency to glorify the Deity in his
works gave rise to a taste for natural description. The earliest and most remarkable instances of this kind are to be met
with in the writings of Minucius Felix, a rhetorician and
lawyer at Rome, who lived in the beginning of the third century, and was the cotemporary of Tertullian and Philostratus.
We follow with pleasure the delineation of his twilight ram-
bles on the shore near Ostia, which he describes as more picturesque and more conducive to health than we find it in the
present day. In the religious discourse entitled *Octavius*, we
meet with a spirited defense of the new faith against the attacks of a heathen friend.†

The present would appear to be a fitting place to introduce
some fragmentary examples of the descriptions of nature which
occur in the writings of the Greek fathers, and which are

---

* On elegiac poetry, consult Nicol. Bach, in the *Allg. Schul-Zeitung*,
1829, abth. ii., No. 134, s. 1097.

† *Minucii Felicis Octavius*, ex. rec. *Gron. Roterod.*, 1743, cap. 2, 3, p.
12, 28; cap. 16–18, p. 151–171.

probably less well known to my readers than the evidences afforded by Roman authors, of the love of nature entertained by the ancient Italians. I will begin with a letter of Basil the Great, for which I have long cherished a special predilection. Basil, who was born at Cesarea in Cappadocia, renounced the pleasures of Athens when not more than thirty years old, and, after visiting the Christian hermitages in Cœlo-Syria and Upper Egypt, retired, like the Essenes and Therapeuti before the Christian era, to a desert on the shores of the Armenian river Iris. There his second brother* Naucratius was drowned while fishing, after having led for five years the rigid life of an anchorite. He thus writes to Gregory of Nazianzum, " I believe I may at last flatter myself with having found the end of my wanderings. The hopes of being united with thee—or I should rather say my pleasant dreams, for hopes have been justly termed the waking dreams of men— have remained unfulfilled. God has suffered me to find a place, such as has often flitted before our imaginations ; for that which fancy has shown us from afar is now made manifest to me. A high mountain, clothed with thick woods, is watered to the north by fresh and ever-flowing streams. At its foot lies an extended plain, rendered fruitful by the vapors with which it is moistened. The surrounding forest, crowded with trees of different kinds, incloses me as in a strong fortress. This wilderness is bounded by two deep ravines ; on the one side, the river, rushing in foam down the mountain, forms an almost impassable barrier, while on the other all access is impeded by a broad mountain ridge. My hut is so situated on the summit of the mountain that I can overlook the whole plain, and follow throughout its course the Iris, which is more beautiful, and has a more abundant body of water, than the Strymon near Amphipolis. The river of my wilderness, which is more impetuous than any other that I know of, breaks against the jutting rock, and throws itself foaming into the abyss below : an object of admiration to the mountain wanderer, and a source of profit to the natives, from the numerous fishes that are found in its waters. Shall

---

* On the death of Naucratius, about the year 357, see *Basilii Magni, Op. omnia,* ed. Par., 1730, t. iii., p. xlv. The Jewish Essenes, two centuries before our era, led an anchorite life on the western shores of the Dead Sea, *in communion with nature.* Pliny, in speaking of them, uses the graceful expression (v. 15), *"mira gens, socia palmarum."* The Therapeuti lived originally in monastic communities, in a charming district near Lake Mœris (Neander, *Allg. Geschichte der Christl. Religion und Kirche,* bd. i., abth. i., 1842, s. 73, 103).

I describe to thee the fructifying vapors that rise from the moist earth, or the cool breezes wafted over the rippled face of the waters ? Shall I speak of the sweet song of the birds, or of the rich luxuriance of the flowering plants ? What charms me beyond all else is the calm repose of this spot. It is only visited occasionally by huntsmen ; for my wilderness nourishes herds of deer and wild goats, but not bears and wolves. What other spot could I exchange for this ? Alcmæon, when he had found the Echinades, would not wander further."* In this simple description of scenery and of forest life, feelings are expressed which are more intimately in unison with those of modern times than any thing that has been transmitted to us from Greek or Roman antiquity. From the lonely Alpine hut to which Basil withdrew, the eye wanders over the humid and leafy roof of the forest below. The place of rest which he and his friend Gregory of Nazianzum had long desired, is at length found.† The poetic and mythical allusion at the close of the letter falls on the Christian ear like an echo from another and earlier world.

Basil's Homilies on the Hexæmeron also give evidence of his love of nature. He describes the mildness of the constantly clear nights of Asia Minor, where, according to his expression, the stars, "those everlasting blossoms of heaven," elevate the soul from the visible to the invisible.‡ When, in the myth of the creation, he would praise the beauty of the sea, he describes the aspect of the boundless ocean plain, in all its varied and ever-changing conditions, "gently moved by the breath of heaven, altering its hue as it reflects the beams of light in their white, blue, or roseate hues, and caressing the

* *Basilii M. Epist.*, xiv., p. 93 ; *Ep.* ccxxiii., p. 339. On the beautiful letter to Gregory of Nazianzum, and on the poetic frame of mind of St. Basil, see Villemain, *De l'Eloquence Chrétienne dans le Quatrième Siècle*, in his *Mélanges Historiques et Littéraires*, t. iii., p. 320–325. The Iris, on whose shores the family of the great Basil had formerly possessed an estate, rises in Armenia, and, after flowing through the plains of Pontus, and mingling with the waters of the Lycus, empties itself into the Black Sea.

† Gregory of Nazianzum did not, however, suffer himself to be enticed by the description of Basil's hermitage, preferring Arianzus in the *Tiberina Regio*, although his friend had complainingly designated it as an impure βάραθρον. See *Basilii Epist.*, ii., p. 70, and *Vita Sancti Bas.*, p. xlvi. and lix. of the edition of 1730.

‡ *Basilii Homil. in Hexæm.*, vi., 1, and iv., 6 ; Bas., *Op. Omnia*, ed. Jul. Garnier, 1839, t. i., p. 54–70. Compare with this the expression of deep sadness in the beautiful poem of Gregorius of Nazianzum, bearing the title *On the Nature of Man* (*Gregor. Naz., Op. omnia*, ed. Par., 1611, t. ii., Carm. xiii., p. 85).

shores in peaceful sport." We meet with the same sentimental and plaintive expressions regarding nature in the writings of Gregory of Nyssa, the brother of Basil the Great.
" When," he exclaims, " I see every ledge of rock, every valley and plain, covered with new-born verdure, the varied
beauty of the trees, and the lilies at my feet decked by nature
with the double charms of perfume and of color ; when in the
distance I see the ocean, toward which the clouds are onward
borne, my spirit is overpowered by a sadness not wholly devoid of enjoyment. When in autumn the fruits have passed
away, the leaves have fallen, and the branches of the trees,
dried and shriveled, are robbed of their leafy adornments, we
are instinctively led, amid the everlasting and regular change
in nature, to feel the harmony of the wondrous powers pervading all things. He who contemplates them with the eye
of the soul, feels the littleness of man amid the greatness of
the universe."*

While the Greek Christians were led by their adoration
of the Deity through the contemplation of his works to a poetic delineation of nature, they were at the same time, during
the earlier ages of their new belief, and owing to the peculiar
bent of their minds, full of contempt for all works of human
art. Thus Chrysostom abounds in passages like the following : " If the aspect of the colonnades of sumptuous buildings
would lead thy spirit astray, look upward to the vault of
heaven, and around thee on the open fields, in which herds
graze by the water's side ; who does not despise all the creations of art, when, in the stillness of his spirit, he watches
with admiration the rising of the sun, as it pours its golden
light over the face of the earth ; when, resting on the thick
grass beside the murmuring spring, or beneath the somber
shade of a thick and leafy tree, the eye rests on the far-reced-
ing and hazy distance ?"† Antioch was at that time sur-

* The quotation given in the text from Gregory of Nyssa is composed
of several fragments literally translated. They occur in *S. Gregorii
Nysseni, Op.*, ed. Par., 1615, t. i., p. 49, C ; p. 589, D ; p. 210, C ; p.
780, C ; t. ii., p. 860, B ; p. 619, B ; p. 619, D ; p. 324, D. " Be gentle
toward the emotions of sadness," says Thalassius, in one of the apho-
risms which were so much admired by his cotemporaries (*Biblioth. Patrum*, ed. Par., 1624, t. ii., p. 1180, C).

† See *Joannis Chrysostomi Op. omnia*, Par., 1838 (8vo, t. ix., p. 687,
A ; t. ii., p. 821, A, and 851, E ; t. i., p. 79). Compare, also, *Joannis
Philoponi in cap.* 1, *Geneseos de Creatione Mundi libri septem*, Viennæ
Aust., 1630, p. 192, 236, and 272, as also *Georgii Pisidæ Mundi Opifici-
um*, ed. 1596, v. 367–375, 560, 933, and 1248. The works of Basil and
of Gregory of Nazianzum soon arrested my attention, after I began to

rounded by hermitages, in one of which lived Chrysostom. It seemed as if Eloquence had recovered her element, freedom, from the fount of nature in the mountain regions of Syria and Asia Minor, which were then covered with forests.

But in those subsequent ages—so inimical to intellectual culture—when Christianity was diffused among the Germanic and Celtic nations, who had previously been devoted to the worship of nature, and had honored under rough symbols its preserving and destroying powers, intimate intercourse with nature, and a study of its phenomena were gradually considered suspicious incentives to witchcraft. This communion with nature was regarded in the same light as Tertullian, Clement of Alexandria, and almost all the older fathers of the Church, had considered the pursuit of the plastic arts. In the twelfth and thirteenth centuries, the Councils of Tours (1163) and of Paris (1209) interdicted to monks the sinful reading of works on physics.* Albertus Magnus and Roger Bacon were the first who boldly rent asunder these fetters of the intellect, and thus, as it were, absolved Nature, and restored her to her ancient rights.

We have hitherto depicted the contrasts manifested according to the different periods of time in the closely allied literature of the Greeks and Romans. But differences in the mode of thought are not limited to those which must be ascribed to the age alone, that is to say, to passing events which are constantly modified by changes in the form of government, social manners, and religious belief; for the most striking differences are those generated by varieties of races and of intellectual development. How different are the manifestations of an animated love for nature and a poetic coloring of natural descriptions among the nations of Hellenic, Northern Germanic, Semitic, Persian, or Indian descent! The opinion has been re-

collect descriptions of nature; but I am indebted to my friend and colleague H. Hase, Member of the Institute, and Conservator of the King's Library at Paris, for all the admirable translations of Chrysostom and Thallasius that I have already given.

* On the *Concilium Turonense*, under Pope Alexander III., see Ziegelbauer, *Hist. Rei Litter. ordinis S. Benedicti*, t. ii., p. 248, ed. 1754; and on the Council at Paris in 1209, and the Bull of Gregory IX., from the year 1231, see Jourdain, *Recherches Crit. sur les Traductions d'Aristote*, 1819, p. 204-206. The perusal of the physical works of Aristotle was forbidden under penalty of severe penance. In the *Concilium Lateranense* of 1139, *Sacror. Concil. nova Collectio*, ed. Ven., 1776, t. xxi., p. 528, the practice of medicine was interdicted to monks. See, on this subject, the learned and agreeable work of the young Wolfgang von Göthe, *Der Mensch und die Elementarische Natur*, 1844, s. 10.

peatedly expressed, that the love of nature evinced by northern
nations is to be referred to an innate longing for the pleasant
fields of Italy and Greece, and for the wonderful luxuriance
of tropical vegetation, when contrasted with their own prolonged deprivation of the enjoyment of nature during the
dreary season of winter.   We do not deny that this longing
for the land of palms diminishes as we approach Southern
France or the Spanish peninsula, but the now generally adopted and ethnologically correct term of *Indo-Germanic* nations
should remind us that too general an influence ought not to
be ascribed to northern winters.   The luxuriant poetic literature of the Indians teaches us that within and near the tropics, south of the chain of the Himalaya, ever-verdant and ev-
er-blooming forests have at all times powerfully excited the
imaginations of the East Arian nations, and that they have
always been more inclined toward poetic delineations of nature
than the true Germanic races who have spread themselves
over the inhospitable north as far as Iceland.   The happier
climates of Southern Asia are not, however, exempt from a
certain deprivation, or, at least, an interruption of the enjoyment of nature ; for the seasons are abruptly divided from
each other by an alternation of fructifying rain and arid destructive drought.   In the West Arian plateaux of Persia,
the barren wilderness penetrates in many parts in the form
of bays into the surrounding highly fruitful lands.   A margin
of forest land often constitutes the boundary of these far-extending seas of steppe in Central and Western Asia.   In this
manner the relations of the soil present the inhabitants of these
torrid regions with the same contrast of barrenness and vegetable abundance in a horizontal plane as is manifested in a
vertical direction by the snow-covered mountain chains of India and of Afghanistan.   Great contrasts in seasons, vegetation, and elevation are always found to be exciting elements
of poetic fancy, where an animated love for the contemplation
of nature is closely interwoven with the mental culture and
the religious aspirations of a people.

Pleasure in the contemplation of nature, which is consonant
with the characteristic bent of mind of the Germanic nations,
is in the highest degree apparent in the earliest poems of the
Middle Ages, as may be proved by many examples from the
chivalric poetry of the Minnesingers, in the period of the Ho-
henstauffen dynasty.   However numerous may be the historical points of contact connecting it with the romanesque songs
of the Provencals, we can not overlook the genuine Germanic

spirit every where breathing through it. A deep and all-pervading enjoyment of nature breathes through the manners and social arrangements of the Germanic races, and through the very spirit of freedom by which they are characterized.* Although moving and often born in courtly circles, the wandering Minnesingers never relinquished the habit of communing with nature. It was thus that their productions were often marked by a fresh, idyllic, and even elegiac tone of feeling. In order to form a just appreciation of the result of such a disposition of mind, I avail myself of the labors of my valued friends Jacob and Wilhelm Grimm, who have so profoundly investigated the literature of our German middle ages. "Our national poets during that age," writes the latter of the two brother inquirers, "have never devoted themselves to a description of nature, having no object but that of conveying to the imagination a glowing picture of the scene. A love of nature was assuredly not wanting to the ancient German Minnesingers, although they have left us no other expression of the feeling than what was evolved in lyric poems from their connection with historical events, or from the sentiments appertaining to the subject of which they treated. If we begin with the oldest and most remarkable monuments of the popular Epos, we shall find that neither the *Niebelungen* nor *Gudrun*† contain any description of natural scenery, even where the occasion seems specially to prompt its introduction. In the otherwise circumstantial description of the hunt, during which Siegfried was murdered, the flowering heath and the cool spring under the linden are only casually touched upon. In *Gudrun*, which evinces to a certain extent a more delicate finish, the feeling for nature is somewhat more apparent. When the king's daughter and her attendants, reduced to a condition of slavery, are carrying the garments of their cruel masters to the sea-shore, the time is indicated, when the winter is just melting away, and the song of rival birds has already begun. Snow and rain are falling, and the hair of the

* Fried. Schlegel, *Ueber nordische Dichtkunst*, in his *Sämmtliche Werke*, bd. x., s. 71 and 90. I may further cite, from the very early times of Charlemagne, the poetic description of the *Thiergarten* at Aix, inclosing both woods and meadows, and which occurs in the life of the great emperor, by Angilbertus, abbot of St. Riques. (See Pertz, *Monum.*, vol. i., p. 393–403.)

† See the comparison of the two epics, the poem of the Niebelungen (describing the vengeance of Chriemhild, the wife of Siegfried), and that of Gudrun, the daughter of King Hetel, in Gervinus, *Geschichte der Deutschen Litt.*, bd. i., s. 354–331.

maidens is disheveled by the rough winds of March. As Gudrun, hoping for the arrival of her liberators, is leaving her couch, and the sea begins to shine in the light of the rising morning star, she distinguishes the dark helmets and shields of her friends. This description is conveyed in but few words, but it calls before the mind a visible picture, and heightens the feeling of suspense preceding the occurrence of an important historical event. Homer, in a similar manner, depicts the island of the Cyclops and the well-ordered gardens of Alcinoüs, in order to produce a visible picture of the luxuriant profusion of the wilderness in which the giant monsters dwell, and of the splendid abode of a powerful king. Neither of the poets purposes to give an individual delineation of nature."

"The rugged simplicity of the popular epic contrasts strongly with the richly-varied narratives of the chivalric poets of the thirteenth century, who all exhibited a certain degree of artistical skill, although Hartmann von Aue, Wolfram von Eschenbach, and Gotfried von Strasburg* were so much distinguished above the rest in the beginning of the century, that they may be called great and classical. It would be easy to collect examples of a profound love of nature from their comprehensive works, as it occasionally breaks forth in similitudes; but the idea of giving an independent delineation of nature does not appear to have occurred to them. They never arrested the plot of the story to pause and contemplate the tranquil life of nature. How different are the more modern poetic compositions! Bernardin de St. Pierre makes use of events merely as frames for his pictures. The lyric poets of the thirteenth century, when they sang of *Minne* or love, which they did not, however, invariably choose as their theme, often speak of the genial month of May, of the song of the nightingale, or of the drops of dew glittering on the flowers of the heath, but these expressions are always used solely with reference to the feelings which they are intended to reflect. In like manner, when emotions of sadness are to be delineated, allusion is made to the sear and yellow leaf, the songless birds, and the seed buried beneath the snow. These thoughts recur incessantly, although not without gracefulness and diversity of expression. The tender Walther von der Vogelweide and the meditative Wolfram von Eschenbach, of whose poems we unfortunately possess but a few lyrical songs, may be adduced as brilliant examples of the cultivators of this species of writing."

* On the romantic description of the grotto of the lovers, in the *Tristan* of Gotfried of Strasburg, see Gervinus, op. cit., bd. i., s. 450.

" The question, whether contact with Southern Italy, or the intercourse opened by means of the crusades with Asia Minor, Syria, and Palestine, may not have enriched Germanic poetry with new images of natural scenery, must be answered generally in the negative, for we do not find that an acquaintance with the East gave any different direction to the productions of the Minnesingers. The Crusaders had little connection with the Saracens, and differences ever reigned among the various nations who were fighting for one common cause. One of the most ancient of the lyric poets was Friedrich von Hausen, who perished in the army of Barbarossa. His songs contain many allusions to the Crusades, but they simply express religious views, or the pain of being separated from the beloved of his heart. Neither he, nor any of those who took part in the crusades, as Reinmar the elder, Rubin, Neidhardt, and Ulrich von Lichtenstein, ever take occasion to speak of the country surrounding them. Reinmar came to Syria as a pilgrim, and, as it would appear, in the retinue of Duke Leopold VI. of Austria. He laments that he can not shake off the thoughts of home, which draw his mind away from God. The date-tree is occasionally mentioned when reference is made to the palm-branches which the pilgrims should bear on their shoulders. I do not remember an instance in which the noble scenery of Italy seems to have excited the imaginative fancy of the Minnesingers who crossed the Alps. Walther von der Vogelweide, who had made distant travels, had, however, not journeyed further into Italy than to the Po ; but Freidank* had been in Rome, and yet he merely remarks that grass grows on the palaces of those who once held sway there."

The German *Animal Epos*, which must not be confounded with the " animal fables" of the East, has arisen from a habit of social familiarity with animals, and not from any special purpose of giving a representation of them. This kind of epos, of which Jacob Grimm has treated in so masterly a

---

* *Vridankes Bescheidenheit*, by Wilhelm Grimm, 1834, s. l. and cxxviii. I have taken all that refers to the German national Epos and the Minnesingers from a letter of Wilhelm Grimm to myself, dated October, 1845. In a very old Anglo-Saxon poem on the names of the Runes, first made known by Hickes, we find the following characteristic description of the birch-tree: " Beorc is beautiful in its branches: it rustles sweetly in its leafy summit, moved to and fro by the breath of heaven." The greeting of the day is simple and noble : " The day is the messenger of the Lord, dear to man, the glorious light of God, a joy and trusting comfort to rich and poor, beneficent to all!" See, also, Wilhelm Grimm, *Ueber Deutsche Runen*, 1821, s. 94, 225, and 234.

manner in the introduction to his edition of Reinhart Fuchs, manifests a genuine delight in nature. The animals, not chained to the ground, passionately excited, and supposed to be gifted with voice, form a striking contrast with the still life of the silent plants, and constitute the ever-animated principle of the landscape. "Ancient poetry delights in considering natural life with human eyes, and thus lends to animals, and sometimes even to plants, the senses and emotions of human beings, giving at the same time a fantastic and child-like interpretation of all that had been observed in their forms and habits. Herbs and flowers that may have been gathered and used by gods and heroes are henceforward named after them. It seems, on reading the German *Animal Epos*, as if the fragrance of some ancient forest were wafted from its pages."*

We might formerly have been disposed to number among the memorials of the Germanic poetry of natural scenery the remains of the Celto-Irish poems, which for half a century flitted like vapory forms from nation to nation under the name of Ossian; but the charm has vanished since the literary fraud of the talented Macpherson has been discovered by his publication of the fictitious Gaelic original text, which was a mere retranslation of the English work. There are undoubtedly ancient Irish Fingal songs, designated as Finnian, which do not date prior to the age of Christianity, and, probably, not even from so remote a period as the eighth century; but these popular songs contain little of that sentimental delineation of nature which imparted so powerful a charm to the productions of Macpherson.†

We have already observed that, although sentimental and romantic excitement of feeling may be considered as in a high degree characteristic of the Indo-Germanic races of Northern Europe, it can not be alone referred to climate, or, in other words, to a longing, increased by protracted deprivation. We have already remarked how the literature of the Indians and Persians, which has been developed under the genial glow of southern climes, presents the most charming descriptions, not

* Jacob Grimm, in *Reinhart Fuchs*, 1834, s. ccxciv. (Compare, also, Christian Lassen, in his *Indische Alterthumskunde*, bd. i., 1843, s. 296.)

† (*Die Unächtheit der Lieder Ossian's und des Macpherson'schen Ossian's insbesondere, von Talvj*, 1840.) The first publication of Ossian by Macpherson was in 1760. The Finnian songs are, indeed, heard in the Scottish Highlands as well as in Ireland, but they have been carried, according to O'Reilly and Drummond, from the latter country to Scotland.

only of organic, but of inanimate nature ; of the transition from drought to tropical rain ; of the appearance of the first cloud on the deep azure of the pure sky, when the long-desired Etesian winds are first heard to rustle amid the feathery foliage of the lofty palms.

The present would appear a fitting place to enter somewhat further into the domain of Indian delineations of nature. " If we suppose," writes Lassen, in his admirable work on Indian antiquity,* "that a part of the Arian race emigrated to India from their native region in the northwestern portion of the continent, they would have found themselves surrounded by a wholly unknown and marvelously luxuriant vegetation. The mildness of the climate, the fruitfulness of the soil, and its rich and spontaneous products, must have imparted a brighter coloring to the new life opened before them. Owing to the originally noble characteristics of the Arian race, and the possession of superior mental endowments, in which lay the germ of all the nobleness and greatness to which the Indians have attained, the aspect of external nature gave rise in the minds of these nations to a deep meditation on the forces of nature, which has proved the means of inducing that contemplative tendency which we find so intimately interwoven in the most ancient poetry of the Indians. The all-powerful impression thus produced on the minds of the people is most clearly manifested in the fundamental dogma of their belief—the recognition of the divine in nature. The freedom from care, and the ease of supporting existence in such a climate, were also conducive to the same contemplative tendency. Who could devote themselves with less hinderance to a profound meditation of earthly life, of the condition of man after death, and of the divine essence, than the anchorites, dwelling amid forests,† the Brahmins of India, whose ancient schools consti-

---

* Lassen, *Ind. Alterthumskunde*, bd. i., s. 412–415.

† Respecting the Indian forest-hermits, Vanaprestiæ (Sylvicolæ) and Sramâni (a name which has been altered into Sarmani and Germani), see Lassen, " *de nominibus quibus veteribus appellantur Indorum philosophi*," in the *Rhein. Museum für Philologie*, 1833, s. 178–180. Wilhelm Grimm recognizes something of Indian coloring in the description of the magic forest by a priest named Lambrecht, in the *Song of Alexander*, composed more than 1200 years ago, in immediate imitation of a French original. The hero comes to a wonderful wood, where maidens, adorned with supernatural charms, spring from large flowers. He remains so long with them that both flowers and maidens fade away. (Compare Gervinus, bd. i., s. 282, and Massmann's *Denkmäler*, bd. i., s. 16.) These are the same as the maidens of Edrisi's Eastern magic island of Vacvac, called in the Latin version of the *Masudi Chothbeddin*,

tute one of the most remarkable phenomena of Indian life,
and must have exercised a special influence on the mental
development of the whole race ?"

In referring here, as I did in my public lectures, under the
guidance of my brother and other learned Sanscrit scholars,
to individual instances of that animated and frequently-ex-
pressed feeling for nature which breathes through the descriptive portions of Indian poetry, I would begin with the *Vedas*,
the most ancient and most valuable memorials of the civilization of the East Arian nations.   The main subject of these
writings is the veneration and praise of nature.   The hymns
of the *Rig-Veda* contain the most charming descriptions of
the " roseate hue of early dawn," and of the aspect of the
" golden-handed sun."   The great heroic poems of Ramayana
and Mahabharata are of more recent date than the Vedas,
but more ancient than the Puranas ; the adoration of nature
being associated with the narrative in accordance with the
character of epic creations.   In the Vedas, the locality of the
scenes which had been glorified by holy beings was seldom
indicated, but in the heroic poems the descriptions of nature
are mostly individual, and refer to definite localities, from
whence they derive that animation and life which is ever imparted when the writer draws his materials from the impressions he has himself experienced.   There is a rich tone of
coloring throughout the description of the journey of Rama
from Ayodhya to the residence of Dschanaka, in his life in
the primitive forest, and in the picture of the anchorite life of
the Panduides.

The name of Kalidasa was early and widely known among
the Western nations.   This great poet flourished in the highly-
cultivated court of Vikramaditya, and was consequently the
cotemporary of Virgil and Horace.   The English and German
translations of the Sacontala have added to the admiration
which has been so freely yielded to this poet,* whose tender-

---

puellas Vasvakienses (Humboldt, *Examen Crit. de la Géographie*, t. i.,
p. 53).

* Kalidasa lived at the court of Vikramaditya about fifty-six years
before our era.   It is highly probable that the age of the two great
heroic poems, *Ramayana* and *Mahabharata*, is much more ancient than
that of the appearance of Buddha, that is to say, prior to the middle of
the sixth century before Christ.   (Burnouf, *Bhagavata-Purana*, t. i.,
p. cxi. and cxviii.; Lassen, *Ind. Alterthumskunde*, bd. i., s. 356 and 492.)
George Forster, by the translation of *Sakuntala*, *i. e.*, by his elegant
German translation of the English version of Sir William Jones (1791),
contributed very considerably to the enthusiasm for Indian poetry

ness of feeling and richness of creative fancy entitle him to a high place in the ranks of the poets of all nations. The charm of his descriptions of nature is strikingly exemplified in the beautiful drama of *Vikrama and Urvasi*, where the king wanders through the thickets of the forest in search of the nymph Urvasi ; in the poem of *The Seasons ;* and in that of *The Messenger of Clouds* (*Meghaduta*). This last poem describes with admirable truth to nature the joy with which, after long drought, the first appearance of a rising cloud is hailed as the harbinger of the approaching season of rain. The expression, " truth to nature," of which I have just made use, can alone justify me in referring, in connection with the Indian poem of *The Messenger of the Clouds*, to a picture of the beginning of the rainy season, which I sketched* in South America, at a period when Kalidasa's *Meghaduta* was not known to me even through the translation of Chézy. The mysterious meteorological processes which take place in the atmosphere in the formation of vapors, in the form of the clouds, and in the luminous electric phenomena, are the same between the tropics in both continents ; and the idealizing art, whose province it is to exalt reality into a picture, will lose none of its charm from the fact that the analyzing spirit of observation of a later age may have succeeded in confirming the truthfulness of an ancient and simply graphic delineation.

We now turn from the East Arians or Brahminical Indians, and the marked bent of their minds toward the contemplation of the picturesque beauties of nature,† to the West which then first showed itself in Germany. I take pleasure in recalling some admirable lines of Göthe's, which appeared in 1792 :

> " Willst du die Blüthe des frühen, die Früchte des späteren Jahres,
> Willst du was reizt und entzückt, willst du, was sättigt und nährt,
> Willst du den Himmel, die Erde mit einem Namen begreifen ;
> Nenn' ich Sakontala, Dich, und so ist alles gesagt."

The most recent German translation of this Indian drama is that by Otto Böhtlingk (Bonn, 1842), from the important original text discovered by Brockhaus.

* Humboldt (*Ueber Steppen und Wüsten*), in the *Ansichten der Natur*, 2te Ausgabe, 1826, bd. i., s. 33–37.

† In order to render more complete the small portion of the text which belongs to Indian literature, and to enable me (as I did before with relation to Greek and Roman literature) to indicate the different works referred to, I will here introduce some notices on the more general consideration of the love of nature evinced by Indian writers, and kindly communicated to me in manuscript by Herr Theodor Goldstücker, a distinguished and philosophical scholar thoroughly versed in Indian poetry :

" Among all the influences affecting the intellectual development of

Arians or Persians, who had separated in different parts of
the Northern Zend, and who were originally disposed to com-

the Indian nation, the first and most important appears to me to have
been that which was exercised by the rich aspect of the country. A
deep sentiment for nature has at all times been a fundamental characteristic of the Indian mind. Three successive epochs may be pointed
out in which this feeling has manifested itself. Each of these has its
determined character deeply implanted in the mode of life and tendencies of the people. A few examples may therefore suffice to indicate
the activity of the Indian imagination, which has been evinced for
nearly three thousand years. The first epoch of the expression of a
vivid feeling for nature is manifested in the Vedas; and here we would
refer in the *Rig-Veda* to the sublime and simple descriptions of the dawn
of day (*Rig-Veda-Sanhitá*, ed. Rosen, 1838, Hymn xlvi., p. 88; Hymn
xlviii., p. 92; Hymn xcii., p. 184; Hymn cxiii., p. 233 : see, also, Höfer, *Ind. Gedichte*, 1841, Lese i., s. 3) and of 'the golden-handed sun'
(*Rig-Veda-Sanhitá*, Hymn xxii., p. 31 ; Hymn xxxv., p. 65). The adoration of nature which was connected here, as in other nations, with
an early stage of the religious belief, has in the Vedas a peculiar significance, and is always brought into the most intimate connection with
the external and internal life of man. The second epoch is very different. In it a popular mythology was formed, and its object was to mold
the sagas contained in the Vedas into a shape more easily comprehended by an age far removed in character from that which had gone by,
and to associate them with historical events which were elevated to
the domain of mythology. The two great heroic poems, the *Ramayana* and the *Mahabharata*, belong to this second epoch. The last-named
poem had also the additional object of rendering the Brahmins the
most influential of the four ancient Indian castes. The *Ramayana* is
therefore the more beautiful poem of the two : it is richer in natural
feeling, and has kept within the domain of poetry, not having been
obliged to take up elements alien and almost hostile to it. In both
poems, nature does not, as in the Vedas, constitute the whole picture,
but only a part of it. Two points essentially distinguish the conception
of nature at the period of the heroic poems from that which the Vedas
exhibit, without reference to the difference which separates the language of adoration from that of narrative. One of these points is the localization of the descriptions, as, for instance, according to Wilhelm von
Schlegel, in the first book of the *Ramayana* or *Balakanda*, and in the
second book, or *Ayodhyakanda.* See, also, on the differences between
these two great epics, Lassen, *Ind. Alterthumskunde*, bd. i., s. 482.
The next point, closely connected with the first, refers to the subject
which has enriched the natural description. Mythical narration, especially when of a historical character, necessarily gave rise to greater
distinctness and localization in the description of nature. All the writers of great epics, whether it be Valmiki, who sings the deeds of Rama,
or the authors of the *Mahabharata*, who collected the national traditions under the collective title of Vyasa, show themselves overpowered,
as it were, by emotions connected with their descriptions of external
nature. Rama's journey from Ayodhya to Dschanaka's capital, his life
in the forest, his expedition to Lanka (Ceylon), where the savage Ra-
vana, the robber of his bride, Sita, dwells, and the hermit life of the
Panduides, furnish the poet with the opportunity of following the original bent of the Indian mind, and of blending with the narration of he-

bine a spiritualized adoration of nature with the dualistic belief in Ahrimanes and Ormuzd.   What we usually term Per-

roic deeds the rich pictures of a luxuriant nature.   (*Ramayana*, ed.
Schlegel, lib. i., cap. 26, v. 13–15; lib. ii., cap. 56, v. 6–11: compare
Nalus, ed. Bopp, 1832, *Ges.*, xii., v. 1–10.)   Another point in which
the second epoch differs from that of the Vedas in regard to the feeling
for external nature is in the greater richness of the subject treated of,
which is not, like the first, limited to the phenomena of the heavenly
powers, but comprehends the whole of nature—the heavens and the
earth, with the world of plants and of animals, in all its luxuriance and
variety, and in its influence on the mind of men.   In the third epoch
of the poetic literature of India, if we except the *Puranas*, which have
the particular object of developing the religious principle in the minds
of the different sects, external nature exercises undivided sway, but the
descriptive portion of the poems is based on scientific and local observation.   By way of specifying some of the great poems belonging to this
epoch, we will mention the *Bhatti-kávya* (or Bhatti's poem), which,
like the *Ramayana*, has for its subject the exploits and adventures of
Rama, and in which there occur successively several admirable descriptions of a forest life during a term of banishment, of the sea and of its
beautiful shores, and of the breaking of the day in Ceylon (Lanka).
(*Bhatti-kávya*, ed. Calc., Part i., canto vii., p. 432; canto x., p. 715;
canto xi., p. 814.   Compare, also, *Fünf Gesänge des Bhatti-kávya*, 1837,
s. 1–18, by Professor Schütz of Bielefeld; the agreeable description of
the different periods of the day in Magha's *Sisupalabdha*, and the *Nais-
chada-tscharita* of Sri Harscha, where, however, in the story of Nalus
and Damayanti, the expression of the feeling for external nature passes
into a vague exaggeration.   This extravagance contrasts with the noble
simplicity of the *Ramayana*, as, for instance, where Visvamitra is described as leading his pupil to the shores of the Sona.   (*Sisupaladha*,
ed. Calc., p. 298 and 372.   Compare Schütz, op. cit., s. 23–28; *Nais-
chada-tscharita*, ed. Calc., Part i., v. 77–129; and *Ramayana*, ed. Schlegel, lib. i., cap. 35, v. 15–18.)   Kalidasa, the celebrated author of *Sakuntala*, has a masterly manner of representing the influence which the
aspect of nature exercises on the minds and feelings of lovers.   The
forest scene which he has portrayed in the drama of *Vikrama and Urvasi* may rank among the finest poetic creations of any period.   (*Vi-
kramorvasi*, ed. Calc., 1830, p. 71; see the translation in Wilson's *Select Specimens of the Theater of the Hindus*, Calc., 1827, vol. ii., p. 63.)
Particular reference should be made in the poem of *The Seasons* to the
passages referring to the rainy season and to spring.   (*Ritusanhára*, ed.
Bohlen, 1840, p. 11–18 and 37–45, and s. 80–88, 107–114 of Bohlen's
translation.)   In the *Messenger of Clouds*, likewise the work of Kalidasa, the influence of external nature on the feelings of men is also the
leading subject of the composition.   This poem (the *Meghaduta*, or
Messenger of Clouds, which has been edited by Gildemeister and Wilson, and translated both by Wilson and by Chézy) describes the grief of
an exile on the mountain Ramagiri.   In his longing for the presence of
his beloved, from whom he is separated, he entreats a passing cloud
to convey to her tidings of his sorrows, and describes to the cloud the
path which it must pursue, depicting the landscape as it would be reflected in a mind agitated with deep emotion.   Among the treasures
which the Indian poetry of the third period owes to the influence of
nature on the national mind, the highest praise must be awarded to the

sian literature does not go further back than the time of the Sassanides ; the most ancient monuments of their poetry have perished. It was not until the country had been subjugated by the Arabs, and had lost its original characteristics, that it again acquired a national literature among the Samanides, Gaznevides, and Seldschukes. The flourishing period of their poetry, extending from Firdusi to Hafiz and Dschami, scarcely lasted more than four or five hundred years, and hardly reaches to the time of the voyage of Vasco de Gama. We must not forget, in seeking to trace the love of nature evinced by the Indians and Persians, that these nations, if we judge according to the amount of cultivation by which they are respectively characterized, appear to be separated alike by time and space. Persian literature belongs to the Middle Ages, while the great literature of India appertains in the strictest sense to antiquity.

In the Iranian elevated plateaux nature has not the same luxuriance of arborescent vegetation, or the remarkable diversity of form and color, by which the soil of Hindostan is embellished. The chain of the Vindhya, which long continued to be the boundary line of the East Arian nations, falls within the tropical region, while the whole of Persia is situated beyond the tropics, and a portion of its poetry belongs even to the northern districts of Balkh and Fergana.

The four paradises celebrated by the Persian poets* were the pleasant valley of Soghd near Samarcand, Maschanrud near Hamadan, Scha'abi Bowan near Kal'eh Sofid in Fars, and Ghute, the plain of Damascus. Both Iran and Turan are wanting in woodland scenery, and also, therefore, in the hermit life of the forest, which exercised so powerful an influence on the imagination of the Indian poets. Gardens refreshed by cool springs, and filled with roses and fruit-trees, can form no substitute for the wild and grand natural scenery of Hindostan. It is no wonder, then, that the descriptive poetry of Persia was less fresh and animated, and that it was

---

Gitagovinda of Dschayadeva. (Rückert, in the *Zeitschrift für die Kunde des Morgenlandes*, bd. i., 1837, s. 129–173; *Gitagovinda Jayadevæ poetæ indici drama lyricum*, ed. Chr. Lassen, 1836.) We possess a masterly rhythmical translation of this poem by Rückert, which is one of the most pleasing, and, at the same time, one of the most difficult in the whole literature of the Indians. The spirit of the original is rendered with admirable fidelity, while a vivid conception of nature animates every part of this great composition."

* Journal of the *Royal Geogr. Soc. of London*, vol. x., 1841, p. 2, 3; Rückert, *Makamen Hariri's*, s. 261.

often heavy and overcharged with artificial adornment. If, in accordance with the opinion of the Persians themselves, we award the highest praise to that which we may designate by the terms spirit and wit, we must limit our admiration to the productiveness of the Persian poets, and to the infinite diversity of forms imparted to the materials which they employ ; depth and earnestness of feeling are wholly absent from their writings.*

Descriptions of natural scenery do but rarely interrupt the narrative in the historical or national Epos of Firdusi. It seems to me that there is much beauty and local truthfulness in the description of the mildness of the climate and the force of the vegetation, extolled in the praise of the coast-land of Mazanderan, which is put into the mouth of a wandering bard. The king, Kei Kawus, is represented as being excited by this praise to enter upon an expedition to the Caspian Sea, and even to attempt a new conquest.† The poems on Spring by Enweri, Dschelaleddin Rumi (who is esteemed the greatest mystic poet of the East), Adhad, and the half-Indian Feisi, generally breathe a tone of freshness and life, although a petty striving to play on words not unfrequently jars unpleasantly on the senses.‡ As Joseph von Hammer has remarked, in his great work on the history of Persian poetry, Sadi, in his *Bostan and Gulistan* (Fruit and Rose Gardens), may be regarded as indicating an age of ethical teaching, while Hafiz, whose joyous views of life have caused him to be compared to Horace, may be considered by his love-songs as the type of a high development of lyrical art ; but that, in both, bombastic affectation too frequently mars the descriptions of nature.§ The darling subject of Persian poetry, the " loves of

* Göthe, in his *Commentar zum west-östlichen Divan*, bd. vi., 1828, s. 73, 78, and 111.

† See *Le Livre des Rois*, publié par Jules Mohl, t. i., 1838, p. 487.

‡ See Jos. von Hammer, *Gesch. der schönen Redekünste Persiens*, 1818, s. 96, concerning Ewhadeddin Enweri, who lived in the twelfth century, and in whose poem on the *Schedschai* a remarkable allusion has been discovered to the mutual attraction of the heavenly bodies; s. 183, concerning Dschelaleddin Rumi, the mystic; s. 259, concerning Dschelaleddin Ahdad; and s. 403, concerning Feisi, who stood forth at the court of Akbar as a defender of the religion of Brahma, and in whose *Ghazuls* there breathes an Indian tenderness of feeling.

§ " Night comes on when the ink-bottle of heaven is overturned," is the inelegant expression of Chodschah Abdulla Wassaf, a poet who has, however, the merit of having been the first to describe the great astronomical observatory of Meragha, with its lofty gnomon. Hilali, of Asterabad, makes the disk of the moon glow with heat, and regards

the nightingale and the rose," recurs with wearying frequency, and a genuine love of nature is lost in the East amid the artificial conventionalities of the language of flowers.

On passing northward from the Iranian plateaux through Turan (Tûirja* in the Zend) to the Uralian Mountains, which separate Europe and Asia, we arrive at the primitive seat of the Finnish race ; for the Ural is as much a land of the ancient Fins as the Altai is of the ancient Turks. Among the Finnish tribes who have settled far to the west in the lowlands of Europe, Elias Lönnrot has collected from the lips of the Karelians, and the country people of Olonetz, a large number of Finnish songs, in which "there breathes," according to the expression of Jacob Grimm, "an animated love of nature rarely to be met with in any poetry but that of India."†
An ancient Epos, containing nearly three thousand verses, treats of a fight between the Fins and Laps, and the fate of a demi-god named Vaino. It gives an interesting account of Finnish country life, especially in that portion of the work where Ilmarine, the wife of the smith, sends her flocks into the woods, and offers up prayers for their safety. Few races exhibit greater or more remarkable differences in mental cultivation, and in the direction of their feelings, according as they have been determined by the degeneration of servitude, warlike ferocity, or a continual striving for political freedom, than the Fins, who have been so variously subdivided, although retaining kindred languages. In evidence of this, we need only refer to the now peaceful population among whom the Epos above referred to was found ; to the Huns, once celebrated for conquests that disturbed the then existing order of things, and who have long been confounded with the Monguls ; and, lastly, to a great and noble people, the Magyars.

After having considered the extent to which intensity in the love of nature and animation in the mode of its expression may be ascribed to differences of race, to the peculiar influence of the configuration of the soil, the form of government, and the character of religious belief, it now remains for us to throw a glance over those nations of Asia who offer the

the evening dew as "the sweat of the moon." (Jos. von Hammer, s. 247 and 371.)

* Tûirja or Turan are names whose etymology is still unknown. Burnouf (*Yacna*, t. i., p. 427–430) has acutely called attention to the Bactrian satrapy of Turiua or Turiva, mentioned in Strabo (lib. xi., p. 517, Cas.). Du Theil and Groskurd would, however, substitute the reading of Tapyria. See the work of the latter, th. ii., s. 410.

† *Ueber ein Finnisches Epos*, Jacob Grimm, 1845, s. 5.

strongest contrast to the Arian or Indo-Germanic races, or, in other words, to the Indians and Persians.

The Semitic or Aramæic nations afford evidence of a profound sentiment of love for nature in the most ancient and venerable monuments of their poetic feeling and creative fancy. This sentiment is nobly and vividly manifested in their pastoral effusions, in their hymns and choral songs, in all the splendor of lyric poetry in the Psalms of David, and in the schools of the seers and prophets, whose exalted inspiration, almost wholly removed from the past, turns its prophetic aspirations to the future.

The Hebraic poetry, besides all its innate exalted sublimity, presents the nations of the West with the special attraction of being interwoven with numerous reminiscences connected with the local seat of the religion professed by the followers of the three most widely-diffused forms of belief, Judaism, Christianity, and Mohammedanism. Thus missions, favored by the spirit of commerce, and the thirst for conquest evinced by maritime nations, have combined to bear the geographical names and natural descriptions of the East as they are preserved to us in the books of the Old Testament, far into the forests of the New World, and to the remote islands of the Pacific.

It is a characteristic of the poetry of the Hebrews, that, as a reflex of monotheism, it always embraces the universe in its unity, comprising both terrestrial life and the luminous realms of space. It dwells but rarely on the individuality of phenomena, preferring the contemplation of great masses. The Hebrew poet does not depict nature as a self-dependent object, glorious in its individual beauty, but always as in relation and subjection to a higher spiritual power. Nature is to him a work of creation and order, the living expression of the omnipresence of the Divinity in the visible world. Hence the lyrical poetry of the Hebrews, from the very nature of its subject, is grand and solemn, and when it treats of the earthly condition of mankind, is full of sad and pensive longing. It is worthy of remark, that Hebrew poetry, notwithstanding its grandeur, and the lofty tone of exaltation to which it is often elevated by the charm of music, scarcely ever loses the restraint of measure, as does the poetry of India. Devoted to the pure contemplation of the Divinity, it remains clear and simple in the midst of the most figurative forms of expression, delighting in comparisons which recur with almost rhythmical regularity.

As descriptions of nature, the writings of the Old Testa-

ment are a faithful reflection of the character of the country
in which they were composed, of the alternations of barrenness and fruitfulness, and of the Alpine forests by which the
land of Palestine was characterized.   They describe in their
regular succession the relations of the climate, the manners
of this people of herdsmen, and their hereditary aversion to
agricultural pursuits.   The epic or historical narratives are
marked by a graceful simplicity, almost more unadorned than
those of Herodotus, and most true to nature ; a point on which
the unanimous testimony of modern travelers may be received
as conclusive, owing to the inconsiderable changes effected in
the course of ages in the manners and habits of a nomadic
people.   Their lyrical poetry is more adorned, and develops a
rich and animated conception of the life of nature.   It might
almost be said that one single psalm (the 104th) represents
the image of the whole Cosmos :   " Who coverest thyself with
light as with a garment : who stretchest out the heavens like
a curtain : who layeth the beams of his chambers in the waters : who maketh the clouds his chariot : who walketh upon
the wings of the wind : who laid the foundations of the earth,
that it should not be removed forever.   He sendeth the springs
into the valleys, which run among the hills.   They give drink
to every beast of the field : the wild asses quench their thirst.
By them shall the fowls of the heaven have their habitation,
which sing among the branches.   He causeth the grass to
grow for the cattle, and herb for the service of man : that he
may bring forth food out of the earth ; and wine that maketh
glad the heart of man, and oil to make his face shine, and
bread which strengtheneth man's heart.   The trees of the
Lord are full of sap ; the cedars of Lebanon which he hath
planted ; where the birds make their nests : as for the stork,
the fir-trees are her house."   " The great and wide sea" is
then described, " wherein are things creeping innumerable,
both small and great beasts.   There go the ships : there is
that leviathan, whom thou hast made to play therein."   The
description of the heavenly bodies renders this picture of nature complete : " He appointed the moon for seasons : the sun
knoweth his going down.   Thou makest darkness, and it is
night ; wherein all the beasts of the forest do creep forth.
The young lions roar after their prey, and seek their meat
from God.   The sun ariseth, they gather themselves together,
and lay them down in their dens.   Man goeth forth unto his
work and to his labor unto the evening."
    We are astonished to find, in a lyrical poem of such a lim-

ited compass, the whole universe—the heavens and the earth
—sketched with a few bold touches. The calm and toilsome
labor of man, from the rising of the sun to the setting of the
same, when his daily work is done, is here contrasted with
the moving life of the elements of nature. This contrast and
generalization in the conception of the mutual action of natural phenomena, and this retrospection of an omnipresent invisible power, which can renew the earth or crumble it to dust,
constitute a solemn and exalted rather than a glowing and
gentle form of poetic creation.

Similar views of the Cosmos occur repeatedly in the Psalms*
(Psalm lxv., 7–14, and lxxiv., 15–17), and most fully, perhaps, in the 37th chapter of the ancient, if not ante-Mosaic
Book of Job. The meteorological processes which take place
in the atmosphere, the formation and solution of vapor, according to the changing direction of the wind, the play of its
colors, the generation of hail and of the rolling thunder, are
described with individualizing accuracy ; and many questions
are propounded which we in the present state of our physical
knowledge may indeed be able to express under more scientific definitions, but scarcely to answer satisfactorily. The
Book of Job is generally regarded as the most perfect specimen of the poetry of the Hebrews. It is alike picturesque in
the delineation of individual phenomena, and artistically skillful in the didactic arrangement of the whole work. In all
the modern languages into which the Book of Job has been
translated, its images, drawn from the natural scenery of the
East, leave a deep impression on the mind. "The Lord
walketh on the heights of the waters, on the ridges of the
waves towering high beneath the force of the wind." "The
morning red has colored the margins of the earth, and variously formed the covering of clouds, as the hand of man molds
the yielding clay." The habits of animals are described, as,
for instance, those of the wild ass, the horse, the buffalo, the
rhinoceros, and the crocodile, the eagle and the ostrich. We
see "the pure ether spread, during the scorching heat of the
south wind, as a melted mirror over the parched desert."†

* Noble echoes of the ancient Hebraic poetry are found in the eleventh century, in the hymns of the Spanish Synagogue poet, Salomo ben
Jehudah Gabirol, which contain a poetic paraphrase of the pseudo-Aristotelian book, *De Mundo*. See *Die Religiöse Poesie der Juden in
Spanien*, by Michael Sachs, 1845, s. 7, 217, and 229. The sketches
drawn from nature, and found in the writings of Mose ben Jakob ben
Esra (s. 69, 77, and 285), are full of vigor and grandeur.

† I have taken the passages in the Book of Job from the translation

Where nature has but sparingly bestowed her gifts, the senses
of man are sharpened, and he marks every change in the moving clouds of the atmosphere around him, tracing in the solitude of the dreary desert, as on the face of the deep and moving sea, every phenomenon through its varied changes, back
to the signs by which its coming was proclaimed. The climate of Palestine, especially in the arid and rocky portions of
the country, is peculiarly adapted to give rise to such observations.

The poetic literature of the Hebrews is not deficient in variety of form; for while the Hebrew poetry breathes a tone
of warlike enthusiasm from Joshua to Samuel, the little book
of the gleaner Ruth presents us with a charming and exqui-
sitely simple picture of nature. Göthe,* at the period of his
enthusiasm for the East, spoke of it " as the loveliest specimen of epic and idyl poetry which we possess."

Even in more recent times, we observe in the earliest literature of the Arabs a faint reflection of that grand, contemplative consideration of nature which was an original characteristic of the Semitic races. I would here refer to the picturesque delineation of Bedouin desert life, which the gram-
marian Asmai has associated with the great name of Antar,
and has interwoven with other pre-Mohammedan sagas of
heroic deeds into one great work. The principal character in
this romantic novel is the Antar (of the race of Abs, and son
of the princely leader Scheddad and of a black slave), whose
verses have been preserved among the prize poems (*Moallakât*) hung up in the Kaaba. The learned English translator,
Terrick Hamilton, has remarked the Biblical tone which
breathes through the style of Antar.† Asmai makes the son

---

and exposition of Umbreit (1824), s. xxix.–xlii., and 290–314. (Compare, generally, Gesenius, *Geschichte der Hebr. Sprache und Schrift*, s.
33; and *Jobi Antiquissimi Carminis Hebr. Natura atque Virtutes*, ed.
Ilgen, p. 28.) The longest and most characteristic description of an animal which we meet with in Job is that of the crocodile (xl., 25—xli.,
26), and yet it contains one of the evidences of the writer being himself a native of Palestine. (Umbreit, s. xli. and 308.) As the river-
horse of the Nile and the crocodile were formerly found throughout the
whole Delta of the Nile, it is not surprising that the knowledge of such
strangely-formed animals should have spread into the contiguous region
of Palestine.

  * Göthe, in his *Commentar zum west-östlichen Divan*, s. 8.

  † *Antar*, a Bedouin romance, translated from the Arabic by Terrick
Hamilton, vol. i., p. xxvi.; Hammer, in the *Wiener Jahrbüchern der
Litteratur*, bd. vi., 1819, s. 229; Rosenmüller, in the *Charakteren der
vornehmsten Dichter aller Nationen*, bd. v. (1798), s. 251.

of the desert go to Constantinople, and thus a picturesque contrast of Greek culture and nomadic ruggedness is introduced. The small space occupied in the earliest Arabic poems by natural delineations of the country will excite but little surprise when we remember, as has been remarked by my friend Freytag of Bonn, who is so celebrated for his knowledge of this branch of literature, that the principal subjects of these poems are narrations of deeds of arms, and praise of hospitality and fidelity, and that scarcely any of the bards were natives of Arabia Felix. A wearying uniformity of grassy plains and sandy deserts could not excite a love of nature, except under peculiar and rare conditions of mind.

Where the soil is not adorned by woods and forests, the phenomena of the atmosphere, as winds, storms, and the long-wished-for rain, occupy the mind more strongly, as we have already remarked. For the sake of referring to a natural image of this kind in the Arabian poets, I would especially notice Antar's *Moallakât*, which describes the meadows rendered fruitful by rain, and visited by swarms of buzzing insects ;* the fine description of storms in Amru'l Kais, and in the seventh book of the celebrated *Hamasa ;*† and, lastly, the picture in the *Nahegha Dhohyani* of the rising of the Euphrates, when its waves bear in their course masses of reeds and trunks of trees.‡ The eighth book of *Hamasa*, inscribed "Travel and Sleepiness," naturally attracted my special attention ; I soon found, however, that "sleepiness"§ was limited to the first fragment of the book, and that the choice of the subject was the more excusable, as the composition is referred to a night journey on a camel.

* *Antara cum schol. Sunsenii*, ed. Menil., 1816, v. 15.
† *Amrulkeisi Moallakât*, ed. E. G. Hengstenberg, 1823 ; *Hamasa*, ed. Freytag, Part i., 1828, lib. vii., p. 785. Compare, also, the pleasing work entitled *Amrilkais, the Poet and King*, translated by Fr. Rückert, 1843, p. 29 and 62, where southern showers of rain are twice described with exceeding truth to nature. The royal poet visited the court of the Emperor Justinian, several years before the birth of Mohammed, to seek aid against his enemies. See *Le Divan d'Amro 'lkais*, accompagné d'une traduction par le Baron MacQuckin de Slane, 1837, p. 111.
‡ *Nabeghah Dhobyani*, in Silvestre de Sacy's *Chrestom. Arabe*, 1806, t. iii., p. 47. On the early Arabian literature generally, see Weil's *Die Poet. Litteratur der Araber vor Mohammed*, 1837, s. 15 and 90, as well as Freytag's *Darstellung der Arabischen Verskunst*, 1830, s. 372–392. We may soon expect an excellent and complete version of the Arabian poetry, descriptive of nature, in the writings of Hamasa, from our great poet, Friedrich Rückert.
§ *Hamasæ Carmina*, ed. Freytag, Part i., 1828, p. 788. "Here finishes," it is said in p. 796, "the chapter on travel and sleepiness."

I have endeavored, in this section, to manifest, in a fragmentary manner, the different influence exercised by the external world, or the aspect of animate and inanimate nature
at different periods of time, on the thoughts and mode of feeling of different races. I have extracted from the history of
literature the characteristic expressions of the love of nature.
My object, therefore, as throughout the whole of this work,
has been, to give general rather than complete views, by the
selection of examples illustrative of the peculiar characteristics
of different epochs and different races of men. I have noticed
the changes manifested in the literature of the Greeks and
Romans, to the gradual decay of those feelings which gave
an imperishable luster to classical antiquity in the West, and
I have traced in the writings of the early fathers of the Christian Church the beautiful expression of a love of nature, developed in the calm seclusion of an anchorite life. In considering the Indo-Germanic races (using the term in its strictest
definition), we have passed from the German poetry of the
Middle Ages to that of the highly-civilized ancient East Arians (Indians), and of the less favored West Arians, or inhabitants of ancient Iran. After a rapid glance at the Celtic
Gaelic songs and the recently-discovered Finnish Epos, I have
delineated the rich life of nature that breathes forth from the
exalted compositions of the Hebrews and Arabs—races of Semitic or Aramæic origin ; and thus we have traced the images reflected by the external world on the imagination of
nations dwelling in the north and southeast of Europe, in
Western Asia, in the Persian plateaux, and in the Indian
tropical regions. I have been induced to pursue this course
from the idea that, in order to comprehend nature in all its
vast sublimity, it would be necessary to present it under a
two-fold aspect, first objectively, as an actual phenomenon,
and next subjectively, as it is reflected in the feelings of mankind.

When the glory of the Aramæic, Greek, and Roman dominion, or, I might almost say, when the ancient world had
passed away, we find in the great and inspired founder of a
new era, Dante Alighieri, occasional manifestations of the
deepest sensibility to the charms of the terrestrial life of nature, whenever he abstracts himself from the passionate and
subjective control of that despondent mysticism which constituted the general circle of his ideas. The period in which
he lived followed immediately that of the decline of the Sua-
bian•Minnesingers, of whom I have already spoken. At the

close of the first canto of his *Purgatorio*,* Dante depicts with
inimitable grace the morning fragrance, and the trembling
light on the mirror of the gently-moved and distant sea (*il
tremolar della marina*) ; and in the fifth canto, the bursting
of the clouds, and the swelling of the rivers, when, after the
battle of Campaldino, the body of Buonconte da Montefeltro
was lost in the Arno.†   The entrance into the thick grove of
the terrestrial paradise is drawn from the poet's remembrance
of the pine forest near Ravenna, "*la pineta in sul lito di
chiassi*,"‡ where the matin song of the birds resounds through
the leafy boughs.   The local fidelity of this picture of nature
contrasts in the celestial paradise with the "stream of light
flashing innumerable sparks,§ which fall into the flowers on
the shore, and then, as if inebriated with their sweet fragrance, plunge back into the stream, while others rise around
them."   It would almost seem as if this fiction had its origin
in the poet's recollection of that peculiar and rare phosphores-
cent condition of the ocean, when luminous points appear to
rise from the breaking waves, and, spreading themselves over
the surface of the waters, convert the liquid plain into a moving sea of sparkling stars.   The remarkable conciseness of the
style of the *Divina Commedia* adds to the depth and earnestness of the impression which it produces.

In lingering on Italian ground, although avoiding the frigid
pastoral romances, I would here refer, after Dante, to the
plaintive sonnet in which Petrarch describes the impression

---

* Dante, *Purgatorio*, canto i., v. 115 :

> "L' alba vinceva l' ora mattutina
> Che fuggia 'nnanzi, sì che di lontano
> Conobbi il tremolar della marina" ....

† *Purg.*, canto v., v. 109–127 :

> "Ben sai come nell' aer si raccoglie
> Quell' umido vapor, che in acqua riede,
> Tosto che sale, dove 'l freddo il coglie" ....

‡ *Purg.*, canto xxviii., v. 1–24.
§ *Parad.*, canto xxx., v. 61–69 :

> "E vidi lume in forma di riviera
> Fulvido di fulgori intra due rive
> Dipinte di mirabil primavera.
>
> Di tal fiumana uscian faville vive
> E d' ogni parte si mettean ne' fiori,
> Quasi rubin, che oro circonscrive.
>
> Poi come inebriate dagli odori,
> Riprofondavan se nel miro gurge
> E s' una entrava, un altra n' uscia fuori."

I do not make any extracts from the Canzones of the *Vita Nuova*, because the similitudes and images which they contain do not belong to
the purely natural range of terrestrial phenomena.

made on his mind by the charming Valley of Vaucluse, after death had robbed him of Laura ; the smaller poems of Boiardo, the friend of Hercules d'Este ; and, more recently, the stanzas of Vittoria Colonna.*

When classical literature acquired a more generally-diffused vigor by the intercourse suddenly opened with the politically degenerated Greeks, we meet with the earliest evidence of this better spirit in the works of Cardinal Bembo, the friend and counselor of Raphael, and the patron of art ; for in the *Ætna Dialogus*, written in the youth of the author, there is a charming and vivid sketch of the geographical distribution of the plants growing on the declivities of the mountain, from the rich corn-fields of Sicily to the snow-covered margin of the crater. The finished work of his maturer age, the *Historiæ Venetæ*, characterizes still more picturesquely the climate and vegetation of the New Continent.

Every thing concurred at this period to fill the imaginations of men with grand images of the suddenly-extended boundaries of the known world, and of the enlargement of human powers, which had been of simultaneous occurrence. As, in antiquity, the Macedonian expeditions to Paropanisus and the wooded alluvial valleys of Western Asia awakened impressions derived from the aspect of a richly-adorned exotic nature, whose images were vividly reflected in the works of

---

* I would here refer to Boiardo's sonnet, beginning,

Ombrosa selva, che il mio duolo ascolti,

and the fine stanzas of Vittoria Colonna, which begin,

Quando miro la terra ornata e bella,
Di mille vaghi ed odorati fiori . . . .

A fine and very characteristic description of the country seat of Fracastoro, on the hill of Incassi (Mons Caphius), near Verona, is given by this writer (who was equally distinguished in medicine, mathematics, and poetry), in his *Naugerius de Poetica Dialogus.* Hieron. Fracastorii, Op. 1591, Part i., p. 321–326. See, also, in a didactic poem by the same writer, lib. ii., v. 208–219 (Op., p. 636), the pleasing passage on the culture of the *Citrus* in Italy. I miss with astonishment any expression of feeling connected with the aspect of nature in the letters of Petrarch, either when, in 1345 (three years, therefore, before the death of Laura), he attempted the ascent of Mont Ventour from Vaucluse, in the eager hope of beholding from thence a part of his native land ; when he ascended the banks of the Rhine to Cologne; or when he visited the Gulf of Baiæ. He lived more in the world of his classical remembrances of Cicero and the Roman poets, or in the emotions of his ascetic melancholy, than in the actual scenes by which he was surrounded. (See *Petrarchæ Epist. de Rebus Familiaribus*, lib. iv., 1, v. 3 and 4; p. 119, 156, and 161, ed. Lugdun., 1601). There is, however, an exceedingly picturesque description of a great tempest which he observed near Naples in 1343 (lib. v., 5, p. 165).

highly-gifted writers, even for centuries afterward, so, in like manner, did the discovery of America act in exercising a second and stronger influence on the western nations than that of the crusades. The tropical world, with all the luxuriance of its vegetation on the plains, with all the gradations of its varied organisms on the declivities of the Cordilleras, and with all the reminiscences of northern climates associated with the inhabited plateaux of Mexico, New Granada, and Quito, was now first revealed to the eyes of Europeans. Fancy, without whose aid no truly great work can succeed in the hands of man, lent a peculiar charm to the delineations of nature sketched by Columbus and Vespucci. The first of these discoverers is distinguished for his deep and earnest sentiment of religion, as we find exemplified in his description of the mild sky of Paria, and of the mass of water of the Orinoco, which he believed to flow from the eastern paradise, while the second is remarkable for the intimate acquaintance he evinces with the poets of ancient and modern times, as shown in his description of the Brazilian coast. The religious sentiment thus early evinced by Columbus became converted, with increasing years, and under the influence of the persecutions which he had to encounter, into a feeling of melancholy and morbid enthusiasm.

In the heroic ages of the Portuguese and Castilian races, it was not thirst for gold alone, as has been asserted from ignorance of the national character at that period, but rather a general spirit of daring, that led to the prosecution of distant voyages. The names of Hayti, Cubagua, and Darien acted on the imaginations of men in the beginning of the sixteenth century in the same manner as those of Tinian and Otaheite have done in more recent times, since Anson and Cook. If the narrations of far-distant lands then drew the youth of the Spanish peninsula, Flanders, Lombardy, and Southern Germany, to rally around the victorious standard of an imperial leader on the ridges of the Andes, or the burning plains of Uraba and Coro, the milder influence of a more modern civilization, when all portions of the earth's surface were more generally accessible, gave other motives and directions to the restless longing for distant travels. A passionate love of the study of nature, which originated chiefly in the north, glowed in the breast of all ; intellectual expansion of views became associated with enlargement of knowledge ; while the poetic and sentimental tone of feeling, peculiar to the epoch of which we speak, has, since the close of the last century, been identified with literary compositions, whose forms were unknown to former ages.

On casting a retrospective glance on the great discoveries which prepared the way for this modern tone of feeling, our attention is especially attracted by the descriptions of nature which we owe to the pen of Columbus. It is only recently that we have been in possession of his own ship's journal, his letters to the Chancellor Sanchez, to the Donna Juana de la Torre, governess of the Infant Don Juan, and to Queen Isabella. I have already attempted, in my critical investigation of the history of the geography of the fifteenth and sixteenth centuries,* to show with what depth of feeling for nature the great discoverer was endowed, and how he described the earth and the new heaven opened to his eyes (*viage nuevo al nuevo cielo i mundo que fasta entonces estaba en occulto*) with a beauty and simplicity of expression which can only be adequately appreciated by those who are conversant with the ancient vigor of the language at the period in which he wrote.

The physiognomy and forms of the vegetation, the impenetrable thickets of the forests, " in which one can scarcely distinguish the stems to which the several blossoms and leaves belong," the wild luxuriance of the flowering soil along the humid shores, and the rose-colored flamingoes, which, fishing' at early morn at the mouth of the rivers, impart animation to the scenery, all, in turn, arrested the attention of the old mariner as he sailed along the shores of Cuba, between the small Lucayan islands and the Jardinillos, which I too have visited. Each newly-discovered land seems to him more beautiful than the one last described, and he deplores his inability to find words in which to express the sweet impressions awakened in his mind. Wholly unacquainted with botany (although, through the influence of Arabian and Jewish physicians, some superficial knowledge of plants had been diffused in Spain), he was led, by a simple love of nature, to individualize all the unknown forms he beheld. Thus, in Cuba alone, he distinguishes seven or eight different species of palms, more beautiful and taller than the date-tree (*variedades de palmas superiores a las nuestras en su belleza y altura*). He informs his learned friend Anghiera that he has seen pines and palms (*palmeta et pineta*) wonderfully associated together in one and the same plain ; and he even so acutely observed the vegetation around him, that he was the first to notice that

* Humboldt, *Examen Critique de l'Histoire de la Géographie du nouveau Continent*, t. iii., p 227–248.

there were pines on the mountains of Cibao whose fruits are not fir-cones, but berries like the olives of the *Axarafe de Sevilla;* and further, as I have already remarked, Columbus* already separated the genus Podocarpus from the family of Abietineæ.

"The beauty of the new land," says the discoverer, "far surpasses the *Campiña de Cordova.* The trees are bright, with an ever-verdant foliage, and are always laden with fruit. The plants on the ground are high and flowering. The air is warm as that of April in Castile, and the nightingale sings more melodiously than words can describe. At night the song of other smaller birds resounds sweetly, and I have also heard our grasshoppers and frogs. Once I came to a deeply-inclosed harbor, and saw a high mountain that had never been seen by any mortal eye, and from whence gentle waters (*lindas aguas*) flowed down. The mountain was covered with firs and variously-formed trees adorned with beautiful blossoms. On sailing up the stream, which empties itself into the bay, I was astonished at the cool shade, the clear, crystal-like water, and the number of the singing birds. I felt as if I could never leave so charming a spot, as if a thousand tongues would fail to describe all these things, and as if my hand were spell-bound and refused to write (*para hacer relacion a los Reyes de las cosas que vian no bastaran mil lenguas a referillo, ni la mano para lo escribir, que le parecia questaba encantado*)."|

We here learn, from the journal of a wholly unlettered seaman, the power which the beauty of nature, in its individual forms, may exercise on a susceptible mind. Feelings ennoble language; for the style of the Admiral, especially when, at the age of sixty-seven, on his fourth voyage, he relates his wonderful dream‡ on the shore of Veragua, if not more eloquent, is at any rate more interesting than the allegorical, pastoral romances of Boccacio, and the two poems of *Arcadia* by Sannazaro and Sydney, than Garcilasso's *Salicio y Nemoroso,* or than the *Diana* of Jorge de Montemayor. The

* See vol. i., p. 282.

† Journal of Columbus on his first voyage (Oct. 29, 1492; Nov. 25–29; Dec. 7–16; Dec. 21). See, also, his letter to Doña Maria de Guzman, ama del Principe D. Juan, Dec., 1500, in Navarrete, *Coleccion de los Viages que hiciéron por mar los Españoles,* t. i., p. 43, 65, 72, 82, 92, 100, and 266.

‡ Navarrete, op. cit., p. 303–304, *Carta del Almirante a los Reyes escrita en Jamaica a 7 de Julio,* 1503; Humboldt, *Examen Crit.*, t. iii., p. 231–236.

elegiac idyllic element unfortunately predominated too long in the literature of the Spaniards and Italians. It required all the freshness of delineation which characterized the adventures of Cervantes's Knight of La Mancha to atone for the Galatea of the same author. Pastoral romance, however it may be ennobled by the beauty of language and tenderness of sentiment manifested in the works of the above-named great writers, must, from its very nature, remain cold and wearisome, like the allegorical and artificial productions of the Middle Ages. Individuality of observation can alone lead to a truthful representation of nature ; thus it is supposed that the finest descriptive stanzas in the *Gerusalemma Liberata** may be traced to impressions derived from the poet's recollection of the beautiful scenery of Sorrento by which he was surrounded.

The power of stamping descriptions of nature with the impress of faithful individuality, which springs from actual observation, is most richly displayed in the great national epic of Portuguese literature. It seems as if a perfumed Eastern air breathed throughout this poem, which was written under a tropical sky in the rocky grotto near Macao, and in the Moluccas. Although I would not venture to assume that my opinion could serve as a confirmation of the bold expression of Friedrich Schlegel, that " the *Lusiad* of Camoens far surpasses Ariosto in richness of color and luxuriance of fancy,"† I may be permitted to add, as an observer of nature, that in the descriptive portions of the work, the enthusiasm of the poet, the ornaments of diction, and the sweet tones of melancholy never impede the accurate representation of physical phenomena, but rather, as is always the case where art draws from a pure source, heighten the animated impression of the greatness and truth of the delineations. Camoens abounds in inimitable descriptions of the never-ceasing connection between the air and sea—between the varying form of the cloudy canopy, its meteorological processes, and the different conditions

---

* Tasso, canto xvi., stanze 9–16.

† See Friedrich Schlegel's *Sämmtl. Werke*, bd. ii., s. 96 ; and on the disturbing mythological dualism, and the mixture of antique fable with Christian contemplations, see bd. x., s. 54. Camoens has tried, in stanzas 82–84, which have not met with sufficient admiration, to justify this mythological dualism. Tethys avows, in a naïve manner, but in verses inspired by the noblest conception of poetry, " that she herself, Saturn, Jupiter, and all the host of gods, are vain fables, created by the blind delusion of mortals, and serving only to lend a charm to song— *A Sancta Providencia que em Jupiter aqui se representa.*"

of the surface of the ocean. He describes this surface when, curled by gentle breezes, the short waves flash beneath the play of the reflected beams of light, and again when the ships of Coelho and Paul de Gama contend in a fearful storm against the wildly-agitated elements.* Camoens is, in the strictest sense of the word, a great sea painter. He had served as a soldier, and fought in the Empire of Morocco, at the foot of Atlas, in the Red Sea, and on the Persian Gulf; twice he had doubled the Cape, and, inspired by a deep love of nature, he passed sixteen years in observing the phenomena of the ocean on the Indian and Chinese shores. He describes the electric fires of St. Elmo (the Castor and Pollux of the ancient Greek mariners), "the living light,† sacred to the seaman." He depicts the threatening water-spout in its gradual development, "how the cloud woven from fine vapor revolves in a circle, and, letting down a slender tube, thirstily, as it were, sucks up the water, and how, when the black cloud is filled, the foot of the cone recedes, and, flying upward to the sky, gives back in its flight, as fresh water, that which it had drawn from the waves with a surging noise."‡ "Let the book-learned," says the poet, and his taunting words might almost be applied to the present age, "try to explain the hidden wonders of this world, since, trusting to reason and science alone, they are so ready to pronounce as false what is heard from the lips of the sailor, whose only guide is experience."

The talent of the enthusiastic poet for describing nature is not limited to separate phenomena, but is very conspicuous in the passages in which he comprehends large masses at one glance. The third book sketches, in a few strokes, the form

* *Os Lusiadas de Camões*, canto i., est. 19; canto vi., est. 71–82. See, also, the comparison in the description of a tempest raging in a forest, canto i., est. 35.

† The fire of St. Elmo, "*o lume vivo que a maritima gente tem por santo, em tempo de tormenta*" (canto v., est. 18). One flame, the Helena of the Greek mariners, brings misfortune (Plin., ii., 37) ; two flames, Castor and Pollux, appearing with a rustling noise, "like fluttering birds," are good omens (Stob., *Eclog. Phys.*, i., p. 514; Seneca, *Nat. Quæst.*, i., 1). On the eminently graphical character of Camoens's descriptions of nature, see the great Paris edition of 1818, in the *Vida de Camões*, by Dom Joze Maria de Souza, p. cii.

‡ The water-spout in canto v., est. 19–22, may be compared with the equally poetic and faithful description of *Lucretius*, vi., 423–442. On the fresh water, which, toward the close of the phenomenon, appears to fall from the upper part of the column of water, see Ogden *On Water Spouts* (from observations made in 1820, during a voyage from Havana to Norfolk), in Silliman's *American Journal of Science*, vol. xxix., 1836, p. 254–260.

of Europe,* from the coldest north to " the Lusitanian realm,
and the strait where Hercules achieved his last labor." Allusion is constantly made to the manners and civilization of
the nations who inhabit this diversified portion of the earth.
From the Prussians, Muscovites, and the races " *que o Rhe-
no frio lava,*" he hastens to the glorious plains of Hellas,
" *que creastes os peitos eloquentes, e os juizos de alta phanta-
sia.*" In the tenth book he takes a more extended view.
Tethys leads Gama to a high mountain, to reveal to him the
secrets of the mechanism of the earth (*machina do mundo*),
and to disclose the course of the planets (according to the
Ptolemaic hypothesis).† It is a vision in the style of Dante,
and as the earth forms the center of the moving universe, all
the knowledge then acquired concerning the countries already
discovered, and their produce, is included in the description
of the globe.‡ Europe is no longer, as in the third book, the
sole object of attention, but all portions of the earth are in
turns passed in review ; even " the land of the Holy Cross"
(Brazil) is named, and the coasts discovered by Magellan, "by
birth but not by loyalty a son of Lusitania."

If I have specially extolled Camoens as a sea painter, it
was in order to indicate that the aspect of a terrestrial life
appears to have attracted his attention less powerfully. Sismondi has justly remarked that the whole poem bears no
trace of graphical description of tropical vegetation, and its
peculiar physiognomy. Spices and other aromatic substances,

---

* Canto iii., est. 7–21. In my references I have always followed the
text of Camoens according to the editio princeps of 1572, which has
been given afresh in the excellent and splendid editions of Dom Joze
Maria de Souza-Botelho (Paris, 1818). In the German quotations I
have generally used the translation of Donner (1833). The principal
aim of the *Lusiad* of Camoens is to do honor to his nation. It would be
a monument well worthy of his fame, and of the nation whom he extols,
if a hall were constructed in Lisbon, after the noble examples of the
halls of Schiller and Göthe in the Grand Ducal Palace of Weimar, and
if the twelve grand compositions of my talented and deceased friend
Gérard, which adorn the Souza edition, were executed in large dimensions, in fresco, on well-lighted walls. The dream of the King Dom
Manoel, in which the rivers Indus and Ganges appear to him ; the
Giant Adamastor hovering over the Cape of Good Hope ("*Eu sóu
aquelle occulto e grande Cabo, a quem chamais vós outros Tormentorio*") ;
the murder of Ignes de Castro, and the lovely Ilha de Venus, would
all produce the most admirable effect.

† Canto x., est. 79–90. Camoens, like Vespucci, speaks of the part
of the heavens nearest to the southern pole as poor in stars (canto v.,
est. 14). He is also acquainted with the ice of the southern seas (canto
v., est. 27). ‡ Canto x., est. 91–141.

together with useful products of commerce, are alone noticed. The episode of the magic island* certainly presents the most charming pictures of natural scenery, but the vegetation, as befits an *Ilha de Venus*, is composed of "myrtles, citrons, fragrant lemon-trees, and pomegranates," all belonging to the climate of Southern Europe. We find a greater sense of enjoyment from the littoral woods, and more attention devoted to the forms of the vegetable kingdom, in the writings of the greatest navigator of his day, Columbus; but then, it must be admitted, while the latter notes down in his journal the vivid impressions of each day as they arose, the poem of Camoens was written to do honor to the great achievements of the Portuguese. The poet, accustomed to harmonious sounds, could not either have felt much disposed to borrow from the language of the natives strange names of plants, or to have interwoven them in the description of landscapes, which were designed as back-grounds for the main subjects of which he treated.

By the side of the image of the knightly Camoens has often been placed the equally romantic one of a Spanish warrior, who served under the banners of the great Emperor in Peru and Chili, and sang in those distant climes the deeds in which he had himself taken so honorable a share. But in the whole epic poem of the *Araucana*, by Don Alonso de Ercilla, the aspect of volcanoes covered with eternal snow, of torrid sylvan valleys, and of arms of the sea extending far into the land, has not been productive of any descriptions which may be regarded as graphical. The exaggerated praise which Cervantes takes occasion to expend on Ercilla in the ingenious satirical review of Don Quixote's books, is probably merely the result of the rivalry subsisting between the Spanish and Italian schools of poetry, but it would almost appear to have deceived Voltaire and many modern critics. The Araucana is certainly penetrated by a noble feeling of nationality. The description of the manners of a wild race, who perish in struggling for the liberty of their country, is not devoid of animation, but Ercilla's style is not smooth or easy, while it is overloaded with proper names, and is devoid of all trace of poetic enthusiasm.†

* Canto ix., est. 51–63. (Consult Ludwig Kriegk, *Schriften zur allgemeinen Erdkunde*, 1840, s. 338.) The whole Ilha de Venus is an allegorical fable, as is clearly shown in est. 89; but the beginning of the relation of Dom Manoel's dream describes an Indian mountain and forest district (canto iv., est. 70).

† A predilection for the old literature of Spain, and for the enchanting region in which the *Araucana* of Alonso de Ercilla y Zuñiga was

This enthusiastic poetic inspiration is to be traced, however, in many strophes of the *Romancero Caballeresco ;** in the religious melancholy pervading the writings of Fray Luis de Leon, as, for instance, in his description of the charming night, when he celebrates the eternal lights (*resplandores eternales*) of the starry heavens ;† and in the compositions of Calderon.

composed, has led me to read through the whole of this poem (which, unfortunately, comprises 42,000 verses) on two occasions, once in Peru, and again recently in Paris, when, by the kindness of a learned traveler, M. Ternaux Compans, I received, for the purpose of comparing it with Ercilla, a very scarce book, printed in 1596 at Lima, and containing the nineteen cantos of the *Arauco domado* (*compuesto por el Licenciado Pedro de Oña natural de los Infantes de Engol en Chile*). Of the epic poem of Ercilla, which Voltaire regarded as an *Iliad*, and Sismondi as a newspaper in rhyme, the first fifteen cantos were composed between 1555 and 1563, and were published in 1569 ; the later cantos were first printed in 1590, only six years before the wretched poem of Pedro de Oña, which bears the same title as one of the master-works of Lope de Vega, in which the Cacique Caupolican is also the principal personage. Ercilla is unaffected and true-hearted, especially in those parts of his composition which he wrote in the field, mostly on the bark of trees and the skins of animals, for want of paper. The description of his poverty, and of the ingratitude which he, like others, experienced at the court of King Philip, is extremely touching, particularly at the close of the 37th canto :

> " Climas pasè, mudè constelaciones,
> Golfos innavegables navegando,
> Estendiendo Señor, vuestra corona
> Hasta casi la austral frigida zona."

" The flower of my life is past ; led by a late-earned experience, I will renounce earthly things, weep, and no longer sing." The natural descriptions of the garden of the sorcerer, of the tempest raised by Eponamon, and the delineation of the ocean (Part i., p. 80, 135, and 173 ; Part ii., p. 130 and 161, in the edition of 1733), are wholly devoid of life and animation. Geographical registers of words are accumulated in such a manner that, in canto xxvii., twenty-seven proper names follow each other in a single stanza of eight lines. Part ii. of the *Araucana* is not by Ercilla, but is a continuation, in twenty cantos, by Diego de Santistevan Osorio, appended to the thirty-seven cantos of Ercilla.

* See, in *Romancero de Romances Caballerescos é Historicos ordenado,* por D. Augustin Duran, Part i., p. 189, and Part ii., p. 237, the fine strophes commencing *Yba declinando el dia—Su curso y ligeras horas,* and those on the flight of King Rodrigo, beginning

> "*Cuando las pintadas aves*
> *Mudas están, y la tierra*
> *A tenta escucha los rios.*"

† Fray Luis de Leon, *Obras Proprias y Traducciones, dedicadas a Don Pedro Portocareró,* 1681, p. 120 : Noche serena. A deep feeling for nature also manifests itself occasionally in the ancient mystic poetry of the Spaniards (as, for instance, in Fray Luis de Granada, Santa Teresa de Jesus, and Malon de Chaide) ; but the natural pictures are generally only the external investment under which the ideal religious conception is symbolized.

"At the period when Spanish comedy had attained its fullest development," says my friend Ludwig Tieck, one of the profoundest critics of dramatic literature, "we often find, in the romanesque and lyrical meter of Calderon and his cotemporaries, dazzlingly beautiful descriptions of the sea, of mountains, gardens, and sylvan valleys, but these are always so interwoven with allegorical allusions, and adorned with so much artificial brilliancy, that we feel we are reading harmoniously rhythmical descriptions, recurring continually with only slight variations, rather than as if we could breathe the free air of nature, or feel the reality of the mountain breath and the valley's shade." In the play of *Life is a Dream* (*la vida es sueño*), Calderon makes the Prince Sigismund lament the misery of his captivity in a number of gracefully-drawn contrasts with the freedom of all organic nature. He depicts birds "which flit with rapid wings across the wide expanse of heaven;" fishes, "which but just emerged from the mud and sand, seek the wide ocean, whose boundlessness seems scarcely sufficient for their bold course. Even the stream which winds its tortuous way among flowers finds a free passage across the meadow; and I," cries Sigismund, in despair, "I, who have more life than these, and a freer spirit, must content myself with less freedom!" In the same manner Don Fernando speaks to the King of Fez, in *The Steadfast Prince*, although the style is often disfigured by antitheses, witty comparisons, and artificially-turned phrases from the school of Gongora.* I have referred to these individual examples because they show, in dramatic poetry, which treats chiefly of events, passions, and characters, that descriptions become merely the reflections, as it were, of the disposition and tone of feeling of the principal personages. Shakspeare, who, in the hurry of his animated action, has hardly ever time or opportunity for entering deliberately into the descriptions of natural scenery, yet paints them by accidental reference, and in allusion to the feelings of the principal characters, in such a manner that we seem to see them and live in them. Thus, in the *Midsummer Night's Dream*, we live in the wood; and in the closing scenes of the *Merchant of Venice*, we see the moonshine which brightens the warm summer's night, without there being actually any direct description of either. "A true description of nature occurs, howev-

* Calderon, in *The Steadfast Prince*, on the approach of the fleet, Act i., scene 1; and on the sovereignty of the wild beasts in the forests, Act iii., scene 2.

er, in *King Lear*, where the seemingly mad Edgar represents
to his blind father, Gloucester, while on the plain, that they
are ascending Dover Cliff. The description of the view, on
looking into the depths below, actually excites a feeling of
giddiness."*

If, in Shakspeare, the inward animation of the feelings and
the grand simplicity of the language gave such a wonderful
degree of life-like truth and individuality to the expression of
nature, in Milton's exalted poem of *Paradise Lost* the descriptions are, from the very nature of the subject, more magnificent than graphic. The whole richness of the poet's fancy
and diction is lavished on the descriptions of the luxuriant
beauty of Paradise, but, as in Thomson's charming didactic
poem of *The Seasons*, vegetation could only be sketched in
general and more indefinite outlines. According to the judgment of critics deeply versed in Indian poetry, Kalidasa's
poem on a similar subject, the *Ritusanhara*, which was written more than fifteen hundred years earlier, individualizes,
with greater vividness, the powerful vegetation of tropical regions, but it wants the charm which, in Thomson's work,
springs from the more varied division of the year in northern
latitudes, as the transition of the autumn rich in fruits to the
winter, and of the winter to the reanimating season of Spring;
and from the images which may thus be drawn of the labors
or pleasurable pursuits of men in each part of the year.

If we proceed to a period nearer our own time, we observe
that, since the latter half of the eighteenth century, delineative prose especially has developed itself with peculiar vigor.
Although the general mass of knowledge has been so excessively enlarged from the universally-extended study of nature,
it does not appear that, in those susceptible of a higher degree of poetic inspiration, intellectual contemplation has sunk
under the weight of accumulated knowledge, but rather that,
as a result of poetic spontaneity, it has gained in comprehen-
siveness and elevation; and, learning how to penetrate deeper into the structure of the earth's crust, has explored in the
mountain masses of our planet the stratified sepulchers of extinct organisms, and traced the geographical distribution of
animals and plants, and the mutual connection of races.
Thus, among those who were the first, by an exciting appeal
to the imaginative faculties, powerfully to animate the senti-

---

* I have taken the passages distinguished in the text by marks of
quotation, and relating to Calderon and Shakspeare, from unpublished
letters addressed to myself by Ludwig Tieck.

ment of enjoyment derived from communion with nature, and consequently, also, to give impetus to its inseparable accompaniment, the love of distant travels, we may mention in France Jean Jacques Rousseau, Buffon, and Bernardin de St. Pierre, and, exceptionally to include a still living author, I would name my old friend Auguste de Chateaubriand;* in Great Britain, the intellectual Playfair; and in Germany, Cook's companion on his second voyage of circumnavigation, the eloquent George Forster, who was endowed with so peculiarly happy a faculty of generalization in the study of nature.

It would be foreign to the present work were I to undertake to inquire into the characteristics of these writers, and investigate the causes which at one time lend a charm and grace to the descriptions of natural scenery contained in their universally-diffused works, and at another disturb the impressions which they were designed to call forth; but as a traveler, who has derived the greater portion of his knowledge from immediate observation, I may perhaps be permitted to introduce a few scattered remarks on a recent, and, on the whole, but little cultivated branch of literature. Buffon—great and earnest as he was—simultaneously embracing a knowledge of the planetary structures, of organization, and of the laws of light and magnetic forces, and far more profoundly versed in physical investigations than his cotemporaries supposed, shows more artificial elaboration of style and more rhetorical pomp than individualizing truthfulness when he passes from the description of the habits of animals to the delineation of natural scenery, inclining the mind to the reception of exalted impressions rather than seizing upon the imagination by presenting a visible picture of actual nature, or conveying to the senses the echo, as it were, of reality. Even throughout the most justly celebrated of his works in this department of literature, we instinctively feel that he could never have left Central Europe, and that he is deficient in personal observation of the tropical world, which he believes he is correctly describing. But that which we most especially miss in the writings of the great naturalist is a harmonious mode of connecting the representation of nature with the expression of awakened feelings; he is, in fact, deficient in almost all that flows from the mysterious analogy existing between the mental emotions of the mind and the phenomena of the perceptive world.

* [This distinguished writer died July 4th of the present year (1848).]—*Tr.*

A greater depth of feeling and a fresher spirit of animation
pervade the works of Jean Jacques Rousseau, Bernardin de
St. Pierre, and Chateaubriand. If I here allude to the per-
suasive eloquence of the first of these writers, as manifested
in the picturesque scenes of Clarens and La Meillerie on Lake
Leman, it is because, in the principal works of this zealous
but ill-instructed plant-collector—which were written twenty
years before Buffon's fanciful *Epoques de la Nature**—poetic
inspiration shows itself principally in the innermost peculiarities of the language, breaking forth as fluently in his prose as
in the immortal poems of Klopstock, Schiller, Goethe, and
Byron. Even where there is no purpose of bringing forward
subjects immediately connected with the natural sciences, our
pleasure in these studies, when referring to the limited portions of the earth best known to us, may be increased by the
charm of a poetic mode of representation.

In recurring to prose writers, we dwell with pleasure on
the small work entitled *Paul et Virginie*, to which Bernardin
de St. Pierre owes the fairer portion of his literary reputation.
The work to which I allude, which can scarcely be rivaled
by any production comprised in the literature of other countries, is the simple picture of an island in the midst of a tropical sea, in which, sometimes favored by the serenity of the
sky, and sometimes threatened by the violent conflict of the
elements, two charming creatures stand picturesquely forth
from the wild sylvan luxuriance surrounding them as with a
variegated flowery tapestry. Here, and in the *Chaumière Indienne*, and even in his *Études de la Nature*, which are un-

* The succession in which the works referred to were published is
as follows: Jean Jacques Rousseau, 1759, *Nouvelle Héloise;* Buffon,
Epoques de la Nature, 1778, but his *Histoire Naturelle*, 1749–1767 ; Bernardin de St. Pierre, *Études de la Nature*, 1784, *Paul et Virginie*, 1788,
Chaumière Indienne, 1791; George Forster, *Reise nach der Südsee*,
1777, *Kleine Schriften*, 1794. More than half a century before the
publication of the *Nouvelle Héloise*, Madame de Sévigné, in her charming letters, had already shown a vivid sense of the beauty of nature,
such as was rarely expressed in the age of Louis XIV. See the fine
natural descriptions in the letters of April 20, May 31, August 15, September 16, and November 6, 1671, and October 23 and December 28,
1689 (Aubenas, *Hist. de Madame de Sévigné*, 1842, p. 201 and 427).
My reason for referring in the text to the old German poet, Paul Flemming, who, from 1633 to 1639, accompanied Adam Olearius on his
journey to Muscovy and to Persia, is that, according to the convincing
authority of my friend, Varnhagen von Ense (*Biographische Denkw.*,
bd. iv., s. 4, 75, and 129), "the character of Flemming's compositions
is marked with a fresh and healthful vigor, while his images of nature
are tender and full of life."

fortunately disfigured by wild theories and erroneous physical opinions, the aspect of the sea, the grouping of the clouds, the rustling of the air amid the crowded bamboos, the waving of the leafy crown of the slender palms, are all sketched with inimitable truth. Bernardin de St. Pierre's master-work, *Paul et Virginie*, accompanied me to the climes whence it took its origin. For many years it was the constant companion of myself and my valued friend and fellow-traveler Bonpland, and often (the reader must forgive this appeal to personal feelings), in the calm brilliancy of a southern sky, or when, in the rainy season, the thunder re-echoed, and the lightning gleamed through the forests that skirt the shores of the Orinoco, we felt ourselves penetrated by the marvelous truth with which tropical nature is described, with all its peculiarity of character, in this little work. A like power of grasping individualities, without destroying the general impression of the whole, and without depriving the subject of a free innate animation of poetical fancy, characterizes, even in a higher degree, the intellectual and sensitive mind of the author of *Atala, René, Les Martyres*, and *Les Voyages à l'Orient*. In the works of his creative fancy, all contrasts of scenery in the remotest portions of the earth are brought before the reader with the most remarkable distinctness. The earnest grandeur of historical associations could alone impart a character of such depth and repose to the impressions produced by a rapid journey.

In the literature of Germany, as in that of Italy and Spain, the love of nature manifested itself too long under the artificial form of idyl-pastoral romances and didactic poems. Such was the course too frequently pursued by the Persian traveler Paul Flemming, by Brockes, the sensitive Ewald von Kleist, Hagedorn, Salomon Gessner, and by Haller, one of the greatest naturalists of any age, whose local descriptions possess, it must, however, be owned, a more clearly-defined outline and more objective truth of coloring. The elegiac-idyllic element was conspicuous at that period in the morbid tone pervading landscape poetry, and even in Voss, that noble and profound student of classical antiquity, the poverty of the subject could not be concealed by a higher and more elegant finish of style. It was only when the study of the earth's surface acquired profoundness and diversity of character, and the natural sciences were no longer limited to a tabular enumeration of marvelous productions, but were elevated to a higher and more comprehensive view of comparative geography, that this finished development of language could be employed for the purpose of giving animated pictures of distant regions.

The earlier travelers of the Middle Ages, as, for instance, John Mandeville (1353), Hans Schiltberger of Munich (1425), and Bernhard von Breytenback (1486), delight us even in the present day by their charming simplicity, their freedom of style, and the self-confidence with which they step before a public, who, from their utter ignorance, listen with the greater curiosity and readiness of belief, because they have not as yet learned to feel ashamed of appearing ignorant, amused, or astonished. The interest attached to the narratives of travels was then almost wholly dramatic, and the necessary and easily introduced admixture of the marvelous gave them almost an epic coloring. The manners of foreign nations are not so much described as they are rendered incidentally discernible by the contact of the travelers with the natives. The vegetation is unnamed and unheeded, with the exception of an occasional allusion to some pleasantly-flavored or strangely-formed fruit, or to the extraordinary dimensions of particular kinds of stems or leaves of plants. Among animals, they describe, with the greatest predilection, first, those which exhibit most resemblance to the human form, and, next, those which are the wildest and most formidable. The cotemporaries of these travelers believed in all the dangers which few of them had shared, and the slowness of navigation and the want of means of communication caused the Indies, as all the tropical regions were then called, to appear at an immeasurable distance. Columbus* was not yet justified in writing to Queen Isabella, " the world is small, much smaller than people suppose."

The almost forgotten travels of the Middle Ages to which we have alluded, possessed, however, with all the poverty of their materials, many advantages in point of composition over the majority of our modern voyages. They had that character of unity which every work of art requires ; every thing was associated with one action, and made subservient to the narration of the journey itself. The interest was derived from the simple, vivid, and generally implicitly-believed relation of dangers overcome. Christian travelers, in their ignorance of what had already been done by Arabs, Spanish Jews, and Buddhist missionaries, boasted of being the first to see and

---

* Letter of the Admiral from Jamaica, July 7, 1503 : " *El mundo es poco; digo que el mundo no es tan grande como dice el vulgo*" (Navarrete, *Coleccion de Viages Esp.*, t. i., p. 300).

describe every thing.   In the midst of the obscurity in which the East and the interior of Asia were shrouded, distance seemed only to magnify the grand proportions of individual forms.   This unity of composition is almost wholly wanting in most of our recent voyages, especially where their object is the acquirement of scientific knowledge.   The narrative in the latter case is secondary to observations, and is almost wholly lost sight of.   It is only the relation of toilsome and frequently uninstructive mountain ascents, and, above all, of bold maritime expeditions, of actual voyages of discovery in unexplored regions, or of a sojourn in the dreadful waste of the icy polar zone, that can afford any dramatic interest, or admit of any great degree of individuality of delineation ; for here the desolation of the scene, and the helplessness and isolation of the seamen, individualize the picture and excite the imagination so much the more powerfully.

If, from what has already been said, it be undeniably true that in modern books of travel the action is thrown in the back-ground, being in most cases only a means of linking together successive observations of nature and of manners, yet this partial disadvantage is fully compensated for by the increased value of the facts observed, the greater expansion of natural views, and the laudable endeavor to employ the peculiar characteristics of different languages in rendering natural descriptions clear and distinct.   We are indebted to modern cultivation for a constantly-advancing enlargement of our field of view, an increasing accumulation of ideas and feelings, and the powerful influence of their mutual reaction.   Without leaving the land of our birth, we not only learn to know how the earth's surface is fashioned in the remotest zones, and by what animal and vegetable forms it is occupied, but we may even hope to have delineations presented to us which shall vividly reflect, in some degree at least, the impressions conveyed by the aspect of external nature to the inhabitants of those distant regions.   To satisfy this demand, to comply with a requirement that may be termed a species of intellectual enjoyment wholly unknown to antiquity, is an object for which modern times are striving, and it is an object which will be crowned with success, since it is the common work of all civilized nations, and because the greater perfection of the means of communication by sea and land renders the whole earth more accessible, and facilitates the comparison of the most widely-separated parts.

I have here attempted to indicate the direction in which

the power possessed by the observer of representing what he
has seen, the animating influence of the descriptive element,
and the multiplication and enlargement of views opened to us
on the vast theater of natural forces, may all serve as means
of encouraging the scientific study of nature, and enlarging its
domain.   The writer who, in our German literature, according to my opinion, has most vigorously and successfully opened
this path, is my celebrated teacher and friend, George Forster.
Through him began a new era of scientific voyages, the aim
of which was to arrive at a knowledge of the comparative
history and geography of different countries. Gifted with delicate æsthetic feelings, and retaining a vivid impression of the
pictures with which Tahiti and the other then happy islands
of the Pacific had filled his imagination, as in recent times
that of Charles Darwin,* George Forster was the first to depict in pleasing colors the changing stages of vegetation, the
relations of climate and of articles of food in their influence
on the civilization of mankind, according to differences of original descent and habitation.   All that can give truth, individuality, and distinctiveness to the delineation of exotic nature is united in his works.   We trace, not only in his admirable description of Cook's second voyage of discovery, but
still more in his smaller writings, the germ of that richer fruit
which has since been matured.† But alas! even to his noble,
sensitive, and ever-hopeful spirit, life yielded no happiness.

If the appellation of descriptive and landscape poetry has
sometimes been applied, as a term of disparagement, to those
descriptions of natural objects and scenes which in recent
times have so greatly embellished the literature of Germany,
France, England, and America, its application, in this sense,
must be referred only to the abuse of the supposed enlargement of the domain of art.   Rhythmical descriptions of natural objects, as presented to us by Delille, at the close of a
long and honorably-spent career, can not be considered as
poems of nature, using the term in its strictest definition, notwithstanding the expenditure of refined rules of diction and
versification.   They are wanting in poetic inspiration, and
consequently strangers to the domain of poetry, and are cold
and dry, as all must be that shines by mere external polish.

* See *Journal and Remarks*, by Charles Darwin, 1832–1836, in the
Narrative of the Voyages of the Adventure and Beagle, vol. iii., p. 479–
490, where there occurs an extremely beautiful description of Tahiti.
† On the merit of George Forster as a man and a writer, see Gervinus,
Gesch. der Poet. National-Litteratur der Deutschen, th. v., s. 390–392.

But when the so-called descriptive poetry is justly blamed as an independent form of art, such disapprobation does not certainly apply to an earnest endeavor to convey to the minds of others, by the force of well-applied words, a distinct image of the results yielded by the richer mass of modern knowledge. Ought any means to be left unemployed by which an animated picture of a distant zone, untraversed by ourselves, may be presented to the mind with all the vividness of truth, enabling us even to enjoy some portion of the pleasure derived from the immediate contact with nature ? The Arabs express themselves no less truly than metaphorically when they say that the best description is that by which the ear is converted into an eye.* It is one of the evils of the present day that an unhappy tendency to vapid poetic prose and to sentimental effusions has infected simultaneously, in different countries, even the style of many justly celebrated travelers and writers on natural history. Extravagances of this nature are so much the more to be regretted, where the style degenerates into rhetorical bombast or morbid sentimentality, either from want of literary cultivation, or more particularly from the absence of all genuine emotion.

Descriptions of nature, I would again observe, may be defined with sufficient sharpness and scientific accuracy, without on that account being deprived of the vivifying breath of imagination. The poetic element must emanate from the intuitive perception of the connection between the sensuous and the intellectual, and of the universality and reciprocal limitation and unity of all the vital forces of nature. The more elevated the subject, the more carefully should all external adornments of diction be avoided. The true effect of a picture of nature depends on its composition ; every attempt at an artificial appeal from the author must therefore necessarily exert a disturbing influence. He who, familiar with the great works of antiquity, and secure in the possession of the riches of his native language, knows how to represent with the simplicity of individualizing truth that which he has received from his own contemplation, will not fail in producing the impression he seeks to convey ; for, in describing the boundlessness of nature, and not the limited circuit of his own mind, he is enabled to leave to others unfettered freedom of feeling.

It is not, however, the vivid description of the richly-adorned lands of the equinoctial zone, in which intensity of light and of humid heat accelerates and heightens the development of

* Freytag's *Darstellung der Arabischen Verskunst*, 1830, s. 402.

all organic germs, that has alone imparted the powerful attraction which in the present day is attached to the study of all branches of natural science. This secret charm, excited by a deep insight into organic life, is not limited to the tropical world. Every portion of the earth offers to our view the wonders of progressive formation and development, according to ever-recurring or slightly-deviating types. Universal is the awful rule of those natural powers which, amid the clouds that darken the canopy of heaven with storms, as well as in the delicate tissues of organic substances, resolve the ancient strife of the elements into accordant harmony. All portions of the vast circuit of creation—from the equator to the coldest zones —wherever the breath of spring unfolds a blossom, the mind may rejoice in the inspiring power of nature. Our German land is especially justified in cherishing such a belief, for where is the southern nation who would not envy us the great master of poesy, whose works are all pervaded by a profound veneration for nature, which is alike discernible in *The Sorrows of Werther*, in the *Recollections of Italy*, in the *Metamorphoses of Plants*, and in so many of his poems? Who has more eloquently excited his cotemporaries to "solve the holy problem of the universe," and to renew the bond which in the dawn of mankind united together philosophy, physics, and poetry? Who has drawn others with a more powerful attraction to that land, the home of his intellect, where, as he sings,

> Ein sanfter Wind vom blauem Himmel weht,
> Die Myrte still, und hoch der Lorbeer steht!

LANDSCAPE PAINTING IN ITS INFLUENCE ON THE STUDY OF NATURE. —GRAPHICAL REPRESENTATION OF THE PHYSIOGNOMY OF PLANTS. —THE CHARACTER AND ASPECT OF VEGETATION IN DIFFERENT ZONES.

LANDSCAPE painting, and fresh and vivid descriptions of nature, alike conduce to heighten the charm emanating from a study of the external world, which is shown us in all its diversity of form by both, while both are alike capable, in a greater or lesser degree, according to the success of the attempt, to combine the visible and invisible in our contemplation of nature. The effort to connect these several elements forms the last and noblest aim of delineative art, but the present pages, from the scientific object to which they are devoted, must be restricted to a different point of view. Landscape painting can not, therefore, be noticed in any further relation than that of its representation of the physiognomy and character of different portions of the earth, and as it increases the desire for the prosecution of distant travels, and thus incites men in an equally instructive and charming manner to a free communion with nature.

In that portion of antiquity which we specially designate as classical, landscape painting, as well as poetic delineations of places, could not, from the direction of the Greek and Roman mind, be regarded as an independent branch of art. Both were considered merely as accessories ; landscape painting being for a long time used only as the back-ground of historical compositions, or as an accidental decoration for painted walls. In a similar manner, the epic poet delineated the locality of some historical occurrence by a picturesque description of the landscape, or of the back-ground, I would say, if I may be permitted here again to use the term, in front of which the acting personages move. The history of art teaches us how gradually the accessory parts have been converted into the main subject of description, and how landscape painting has been separated from historical painting, and gradually established as a distinct form ; and, lastly, how human figures were employed as mere secondary parts to some mountain or forest scene, or in some sea or garden view. The separation of these two species—historical and landscape painting—has been thus effected by gradual stages, which have tended to favor the advance of art through all the various phases of its development. It has been justly remarked, that painting generally remained subordinate to sculpture among the ancients, and that the feeling for the picturesque beauty of scenery which the artist endeavors to reproduce from his canvas was unknown to antiquity, and is exclusively of modern origin.

Graphic indications of the peculiar characteristics of a locality must, however, have been discernible in the most ancient paintings of the Greeks, as instances of which we may mention (if the testimony of Herodotus be correct)* that Mandrocles of Samos caused a large painting of the passage of the army over the Bosporus to be executed for the Persian king,† and that Polygnotus painted the fall of Troy in the Lesche at

* Herod., iv., 88.
† A portion of the works of Polygnotus and Mikon (the painting of the battle of Marathon in the Pokile at Athens) was, according to the testimony of Himerius, still to be seen at the end of the fourth century (of our era), consequently when they had been executed 850 years. (Letronne, *Lettres sur la Peinture Historique Murale*, 1835, p. 202 and 453.)

Delphi. Among the paintings described by the elder Philostratus, mention is made of a landscape in which smoke was seen to rise from the summit of a volcano, and lava streams to flow into the neighboring sea. In this very complicated composition of a view of seven islands, the most recent commentators* think they can recognize the actual representation of the volcanic district of the Æolian or Lipari Islands north of Sicily. The perspective scenic decorations, which were made to heighten the effect of the representation of the master-works of Æschylus and Sophocles, gradually enlarged this branch of art† by increasing the demand for an illusive imitation of inanimate objects, as buildings, woods, and rocks.

In consequence of the greater perfection to which scenography had attained, landscape painting passed among the Greeks and their imitators, the Romans, from the stage to their halls, adorned with columns, where the long ranges of wall were covered at first with more circumscribed views,‡ but shortly afterward with extensive pictures of cities, seashores, and wide tracts of pasture land, on which flocks were grazing.§ Although the Roman painter Ludius, who lived in the Augustan age, can not be said to have invented these graceful decorations, he yet made them generally popular,|| animating them by the addition of small figures.¶ Almost at the same period, and probably even half a century earlier, we find landscape painting mentioned as a much-practiced art among the Indians during the brilliant epoch of Vikramaditya.

* *Philostratorum Imagines*, ed. Jacobs et Welcker, 1825, p. 79 and 485. Both the learned editors defend, against former suspicions, the authenticity of the description of the paintings contained in the ancient Neapolitan Pinacothek (Jacobs, p. xvii. and xlvi.; Welcker, p. lv. and xlvi.). Otfried Müller conjectures that Philostratus's picture of the islands (ii., 17), as well as that of the marshy district of the Bosporus (i., 9), and of the fishermen (i., 12 and 13), bore much resemblance, in their mode of representation, to the mosaic of Palestrina. Plato speaks, in the introductory part of *Critias* (p. 107), of landscape painting as the art of pictorially representing mountains, rivers, and forests.

† Particularly through Agatharcus, or, at least, according to the rules he established. Aristot., *Poet.*, iv., 16 ; Vitruv., lib. v., cap. 7 ; lib. vii. in Præf. (ed. Alois Maxinius, 1836, t. i., p. 292; t. ii., p. 56). Compare, also, Letronne's work, op. cit., p. 271–280.

‡ On *Objects of Rhopographia*, see Welcker *ad Philostr. Imag.*, p. 397.
§ Vitruv., lib. vii., cap. 5 (t. ii., p. 91).
|| Hirt., *Gesch. der bildenden Künste bei den Alten*, 1833, s. 332 ; Letronne, p. 262 and 468.
¶ Ludius qui primus (?) instituit amœnissimam parietum picturam (Plin., xxxv., 10). The *topiaria opera* of Pliny, and the *varietates topiorum* of Vitruvius, were small decorative landscape paintings. The passage quoted in the text of Kalidasa occurs in the *Sakuntala*, act vi,

In the charming drama of *Sakuntala*, the image of his beloved is shown to King Dushmanta, who is not satisfied with that alone, as he desires that " the artist should depict the places which were most dear to his beloved—the Malini River, with a sand-bank on which the red flamingoes are standing ; a chain of hills skirting on the Himalaya, and gazelles resting on these hills." These requirements are not easy to comply with, and they at least indicate a belief in the practicability of executing such an intricate composition.

In Rome, landscape painting was developed into a separate branch of art from the time of the Cæsars ; but, if we may judge from the many specimens preserved to us in the excavations of Herculaneum, Pompeii, and Stabiæ, these pictures of nature were frequently nothing more than bird's-eye views of the country, similar to maps, and more like a delineation of sea-port towns, villas, and artificially-arranged gardens, than the representation of free nature. That which may have been regarded as the habitably comfortable element in a landscape seems to have alone attracted the Greeks and Romans, and not that which we term the wild and romantic. Their imitations might be so far accurate as frequent disregard of perspective and a taste for artificial and conventional arrangement permitted, and their arabesque-like compositions, to which the critical Vitruvius was averse, often exhibited a rhythmically-recurring and well-conceived representation of animal and vegetable forms ; but yet, to borrow an expression of Otfried Müller,* " the vague and mysterious reflection of the mind, which seems to appeal to us from the landscape, appeared to the ancients, from the peculiar bent of their feelings, as incapable of artistic development, and their delineations were sketched with more of sportiveness than earnestness and sentiment."

We have thus indicated the analogy which existed in the process of development of the two means—descriptive diction

---

* Otfried Müller, *Archäologie der Kunst*, 1830, s. 609. Having already spoken in the text of the paintings found in Pompeii and Herculaneum as being compositions but little allied to the freedom of nature, I must here notice some exceptions, which may be considered as landscapes in the strict modern sense of the word. See *Pitture d'Ercolano*, vol. ii., tab. 45 ; vol. iii., tab. 53 ; and, as back-grounds in charming historical compositions, vol. iv., tab. 61, 62, and 63. I do not refer to the remarkable representation in the *Monumenti dell' Instituto di Corrispondenza Archeologica*, vol. iii., tab. 9, since its genuine antiquity has already been called in question by Raoal Rochette, an archæologist of much acuteness of observation.

and graphical representations—by which the attempt to render the impressions produced by the aspect of nature appreciable to the sensuous faculties has gradually attained a certain degree of independence.

The specimens of ancient landscape painting in the manner
of Ludius, which have been recovered from the excavations at
Pompeii (lately renewed with so happy a result), belong most
probably to a single and very short period, viz., that intervening between Nero and Titus,* for the city had been entirely
destroyed by an earthquake only sixteen years before the celebrated eruption of Vesuvius.

The character of the subsequent style of painting practiced
by the early Christians remained nearly allied to that of the
true Greek and Roman schools of art from the time of Constantine the Great to the beginning of the Middle Ages. A
rich mine of old memorials is opened to us in the miniatures
which adorn splendid and well-preserved manuscripts, and in
the rarer mosaics of the same period.† Rumohr makes mention of a Psalter in the Barberina Library at Rome, where,
in a miniature, David is represented "playing the harp, and
surrounded by a pleasant grove, from the branches of which
nymphs look forth to listen. This personification testifies to
the antique nature of the whole picture." Since the middle
of the sixth century, when Italy was impoverished and politically disturbed, the Byzantine art in the Eastern empire still
preserved the lingering echoes and types of a better epoch.
Such memorials as these form the transition to the creations

* In refutation of the supposition of Du Theil (*Voyage en Italie*, par
l'Abbé Barthélemy, p. 284) that Pompeii still existed in splendor under Adrian, and was not completely destroyed till toward the close of
the fifth century, see Adolph von Hoff, *Geschichte der Veränderungen
der Erdoberfläche*, th. ii., 1824, s. 195–199.

† See Waagen, *Kunstwerke und Künstler in England und Paris*, th.
iii., 1839, s. 195–201; and particularly s. 217–224, where he describes
the celebrated Psalter of the tenth century (in the Paris Library), which
proves how long the "antique mode of composition" maintained itself
in Constantinople. I was indebted to the kind and valuable communications of this profound connoisseur of art (Professor Waagen, director
of the Gallery of Paintings of my native city), at the time of my public
lectures in 1828, for interesting notices on the history of art after the
period of the Roman empire. What I afterward wrote on the gradual
development of landscape painting, I communicated in Dresden, in
the winter of 1835, to Baron von Rumohr, the distinguished and too
early deceased author of the *Italienische Forschungen*. I received
from this excellent man a great number of historical illustrations, which
he even permitted me to publish if the form of my work should render
it expedient.

of the later Middle Ages, when the love for illuminated manuscripts had spread from Greece, in the East, through southern and western lands into the Frankish monarchy, among
the Anglo-Saxons and the inhabitants of the Netherlands.
It is, therefore, a fact of no slight importance for the history
of modern art, that "the celebrated brothers Hubert and Johann van Eyck belonged essentially to a school of miniature
painters, which, since the last half of the fourteenth century,
attained to a high degree of perfection in Flanders."*

The historical paintings of the brothers Van Eyck present
us with the first instances of carefully-executed landscapes.
Neither of them ever visited Italy, but the younger brother,
Johann, enjoyed the opportunity of seeing the vegetation of
Southern Europe when, in the year 1428, he accompanied the
embassy which Philip the Good, duke of Burgundy, sent to
Lisbon when he sued for the hand of the daughter of King
John I. of Portugal. In the Museum of Berlin are preserved
the wings of the famous picture which the above-named celebrated painters—the actual founders of the great Flemish
school—executed for the Cathedral at Ghent. On these wings,
which represent holy hermits and pilgrims, Johann van Eyck
has embellished the landscape with orange and date trees and
cypresses, which, from their extreme truth to nature, impart
a solemn and imposing character to the other dark masses in
the picture. One feels, on looking at this painting, that the
artist must himself have received the impression of a vegetation fanned by gentle breezes.

In considering the master-works of the brothers Van Eyck,
we have not advanced beyond the first half of the fifteenth
century, when the more highly-perfected style of oil painting,
which was only just beginning to replace painting in tempera,
had already attained to a high degree of technical perfection.
The taste for a vivid representation of natural forms was
awakened, and, if we would trace the gradual extension and
elevation of this feeling for nature, we must bear in mind that
Antonio di Messina, a pupil of the brothers Van Eyck, transplanted the predilection for landscape painting to Venice, and
that the pictures of the Van Eyck school exercised a similar
action in Florence on Domenico Ghirlandaio and other masters.† The artists at this epoch directed their efforts to a care-

* Waagen, op. cit., th. i., 1837, s. 59; th. iii., 1839, s. 352–359. [See
Lanzi's *History of Painting*, Bohn's Standard Library, 1847, vol. i., p.
81–87.]—*Tr.*
† "Pinturicchio painted rich and well-composed landscapes as inde-

ful but almost timid imitation of nature, and the master-works of Titian afford the earliest evidence of freedom and grandeur in the representation of natural scenes ; but in this respect, also, Giorgione seems to have served as a model for that great painter. I had the opportunity for many years of admiring in the gallery of the Louvre at Paris that picture of Titian which represents the death of Peter Martyr, overpowered in a forest by an Albigense, in the presence of another Dominican monk.* The form of the forest-trees, and their foliage, the mountainous and blue distance, the tone of coloring, and the lights glowing through the whole, leave a solemn impression of the earnestness, grandeur, and depth of feelings which pervade this simple landscape composition. So vivid was Titian's admiration of nature, that not only in the pictures of beautiful women, as in the back-ground of his exquisitely-formed Venus in the Dresden Gallery, but also in those of a graver nature, as, for instance, in his picture of the poet Pietro Aretino, he painted the surrounding landscape and sky in harmony with the individual character of the subject. Annibal Caracci and Domenichino, in the Bolognese school, adhered faithfully to this elevation of style. If, however, the great epoch of historical painting belong to the sixteenth century,

pendent decorations, in the Belvidere of the Vatican. He appears to have exercised an influence on Raphael, in whose paintings there are many landscape peculiarities which can not be traced to Perugino. In Pinturicchio and his friends we also already meet with those singular, pointed forms of mountains which, in your lectures, you were disposed to derive from the Tyrolese dolomitic cones which Leopold von Buch has rendered so celebrated, and which may have produced an impression on travelers and artists from the constant intercourse existing between Italy and Germany. I am more inclined to believe that these conical forms in the earliest Italian landscapes are either very old conventional modes of representing mountain forms in antique bass-reliefs and mosaic works, or that they must be regarded as unskillfully foreshortened views of Soracte and similarly isolated mountains in the Campagna di Roma." (From a letter addressed to me by Carl Friedrich von Rumohr, in October, 1832.) In order to indicate more precisely the conical and pointed mountains in question, I would refer to the fanciful landscape which forms the back-ground in Leonardo da Vinci's universally admired picture of Mona Lisa (the consort of Francesco del Giocondo). Among the artists of the Flemish school who have more particularly developed landscape painting as a separate branch of art, we must name Patenier's successor, Henry de Bles, named *Civetta* from his animal monogram, and subsequently the brothers Matthew and Paul Bril, who excited a strong taste in favor of this particular branch of art during their sojourn in Rome. In Germany, Albrecht Altdorfer, Durer's pupil, practiced landscape painting even somewhat earlier and with greater success than Patenier.

　* Painted for the Church of San Giovanni e Paolo at Venice.

that of landscape painting appertains undoubtedly to the seventeenth. As the riches of nature became more known and more carefully observed, the feeling of art was likewise able to extend itself over a greater diversity of objects, while, at the same time, the means of technical representation had simultaneously been brought to a higher degree of perfection. The relations between the inner tone of feelings and the delineation of external nature became more intimate, and, by the links thus established between the two, the gentle and mild expression of the beautiful in nature was elevated, and, as a consequence of this elevation, belief in the power of the external world over the emotions of the mind was simultaneously awakened. When this excitement, in conformity with the noble aim of all art, converts the actual into an ideal object of fancy ; when it arouses within our minds a feeling of harmonious repose, the enjoyment is not unaccompanied by emotion, for the heart is touched whenever we look into the depths of nature or of humanity.* In the same century we find thronged together Claude Lorraine, the idyllic painter of light and aërial distance ; Ruysdael, with his dark woodland scenes and lowering skies ; Gaspard and Nicolas Poussin, with their nobly-delineated forms of trees ; and Everdingen, Hobbima, and Cuyp, so true to life in their delineations.†

In this happy period of the development of art, a noble effort was manifested to introduce all the vegetable forms yielded by the North of Europe, Southern Italy, and the Spanish Peninsula. The landscape was embellished with oranges and laurels, with pines and date-trees ; the latter (which, with the exception of the small Chamærops, originally a native of European sea-shores, was the only member of the noble family of palms known from personal observation) was generally represented as having a snake-like and scaly trunk,‡ and long

* Wilhelm von Humboldt, *Gesammelte Werke*, bd. iv., s. 37. See also, on the different gradations of the life of nature, and on the tone of mind awakened by the landscape around, Carus, in his interesting work, *Briefen über die Landschaftmalerei*, 1831, s. 45.

† The great century of painting comprehended the works of Johann Breughel, 1569-1625 ; Rubens, 1577-1640 ; Domenichino, 1581-1641 ; Philippe de Champaigne, 1602-1674 ; Nicolas Poussin, 1594-1655 ; Gaspar Poussin (Dughet), 1613-1675 ; Claude Lorraine, 1600-1682 ; Albert Cuyp, 1606-1672 ; Jan Both, 1610-1650 ; Salvator Rosa, 1615-1673 ; Everdingen, 1621-1675 ; Nicolaus Berghem, 1624-1683 ; Swanevelt, 1620-1690 ; Ruysdael, 1635-1681 ; Minderhoot Hobbima, Jan Wynants, Adriaan van de Velde, 1639-1672 ; Carl Dujardin, 1644-1687.

‡ Some strangely-fanciful representations of date palms, which have a knob in the middle of the leafy crown, are to be seen in an old pic-

served as the representative of tropical vegetation, as, in like
manner, Pinus pinea is even still very generally supposed to
furnish an exclusive characteristic of the vegetable forms of
Italy. The contour of high mountain chains was but little
studied, and snow-covered peaks, which projected beyond the
green Alpine meadows, were, at that period, still regarded by
naturalists and landscape painters as inaccessible. The physiognomy of rocky masses seems scarcely to have excited any
attempt at accurate representation, excepting where a water-
fall broke in foam over the mountain side. We may here remark another instance of the diversity of comprehension manifested by a free and artistic spirit in its intimate communion
with nature. Rubens, who, in his great hunting pieces, had
depicted the fierce movements of wild animals with inimitable animation, succeeded, as the delineator of historical events,
in representing, with equal truth and vividness, the form of
the landscape in the waste and rocky elevated plain surrounding the Escurial.*

The delineation of natural objects included in the branch of
art at present under consideration could not have gained in
diversity and exactness until the geographical field of view
became extended, the means of traveling in foreign countries
facilitated, and the appreciation of the beauty and configuration of vegetable forms, and their arrangement in groups of
natural families, excited. The discoveries of Columbus, Vasco
de Gama, and Alvarez Cabral, in Central America, Southern
Asia, and the Brazils ; the extensive trade in spices and drugs
carried on by the Spaniards, Portuguese, Italians, and Flem-
ings, and the establishment of botanical gardens at Pisa, Padua, and Bologna, between 1544 and 1568, although not yet
furnished with hot-houses properly so called, certainly made
artists acquainted with many remarkable forms of exotic products, including even some that belong to a tropical vegetation. Single fruits, flowers, and branches were painted with
much natural truth and grace by Johann Breughel, whose
reputation had been already established before the close of the
sixteenth century ; but it is not until the middle of the seventeenth century that we meet with landscapes which reproduce
the individual character of the torrid zone, as impressed upon
the artist's mind by actual observation. The merit of the
earliest attempt at such a mode of representation belongs probably, as I find from Waagen, to the Flemish painter Franz

ture of Cima da Conegliano, of the school of Bellino (Dresden Gallery,
1835, No. 40).   * Dresden Gallery, No. 917.

Post, of Haarlem, who accompanied Prince Maurice of Nassau to Brazil, where that prince, who took great interest in all subjects connected with the tropical world, was Dutch stadtholder, in the conquered Portuguese possessions, from 1637 to 1644. Post continued for many years to make studies from nature at Cape St. Augustine, in the Bay of All Saints, on the shores of the River St. Francisco, and at the lower course of the Amazon.* These studies he himself partly executed

* Franz Post, or Poost, was born at Haarlem in 1620, and died there in 1680. His brother also accompanied Count Maurice of Nassau as an architect. Of the paintings, some representing the banks of the Amazon are to be seen in the picture gallery at Schleisheim, while others are at Berlin, Hanover, and Prague. The line engravings in Barläus, *Reise des Prinzen Moritz von Nassau*, and in the royal collection of copper-plate prints at Berlin, evince a fine conception of nature in depicting the form of the coast, the nature of the ground, and the vegetation. They represent Musaceæ, Cacti, palms, different species of Ficus, with the well-known board-like excrescences at the foot of the stem, Rhizophoræ, and arborescent grasses. The picturesque Brazilian voyage is made to terminate (plate iv.), singularly enough, with a German forest of pines which surround the castle of Dillenburg. The remark in the text, on the influence which the establishment of botanic gardens in Upper Italy, toward the middle of the sixteenth century, may have exercised on the knowledge of the physiognomy of tropical forms of vegetation, leads me here to draw attention to the well-founded fact that, in the thirteenth century, Albertus Magnus, who was equally energetic in promoting the Aristotelian philosophy and the pursuit of the science of nature, probably had a hot-house in the convent of the Dominicans at Cologne. This celebrated man, who was suspected of sorcery on account of his speaking machine, entertained the King of the Romans, William of Holland, on his passage through Cologne on the 6th of January, 1259, in a large space in the convent garden, where he preserved fruit-trees and plants in flower throughout the winter by maintaining a pleasant degree of heat. The account of this banquet, exaggerated into something marvelous, occurs in the *Chronica Joannis de Beka*, written in the middle of the fourteenth century (Beka et Heda de Episcopis Ultrajectinis, recogn. ab. Arn. Buchelio, 1643, p. 79 ; Jourdain, *Recherches Critiques sur l'Age des Traductions d'Aristote*, 1819, p. 331 ; Buhle, *Gesch. der Philosophie*, th. v., s. 296). Although the ancients, as we find from the excavations at Pompeii, made use of panes of glass in buildings, yet nothing has been found to indicate the use of glass or hot houses in ancient horticulture. The mode of conducting heat by the caldaria into baths might have led to the construction of such forcing or hot houses, but the shortness of the Greek and Italian winters must have caused the want of artificial heat to be less felt in horticulture. The Adonis gardens (κῆποι Αδώνιδος), so indicative of the meaning of the festival of Adonis, consisted, according to Böckh, of plants in small pots, which were, no doubt, intended to represent the garden where Aphrodite met Adonis, who was the symbol of the quickly-fading bloom of youth, of luxuriant growth, and of rapid decay. The festivals of Adonis were, therefore, seasons of solemn lamentations for women, and belonged to the festivals in which the au-

as paintings, and partly etched with much spirit. To this period belong the remarkably large oil pictures preserved in Denmark, in a gallery of the beautiful palace of Frederiksborg, which were painted by Eckhout, who, in 1641, was also on the Brazilian coast with Prince Maurice of Nassau. In these compositions, palms, papaws, bananas, and heliconias are most characteristically delineated, as are also brightly-plumaged birds, and small quadrupeds, and the form and appearance of the natives.

These examples of a delineation of the physiognomy of natural scenery were not followed by many artists of merit before Cook's second voyage of circumnavigation. What Hodges did for the western islands of the Pacific, and my distinguished countryman, Ferdinand Bauer, for New Holland and Van Diemen's Land, has been since done, in more recent times, on a far grander scale, and in a masterly manner, by Moritz Rugendas, Count Clarac, Ferdinand Bellermann, and Edward Hildebrandt ; and for the tropical vegetation of America, and for many other parts of the earth, by Heinrich von Kittlitz, the companion of the Russian Admiral Lütke, on his voyage of circumnavigation.*

cients lamented the decay of nature. As I have spoken in the text of hot-house plants in contrast with those which grow naturally, I would add that the ancients frequently used the term "Adonis gardens" proverbially, to indicate something which had shot up rapidly, without promise of perfect maturity or duration. These plants, which were lettuce, fennel, barley, and wheat, and not variegated flowers, were forced, by extreme care, into rapid growth in summer (and not in the winter), and were often made to grow to maturity in a period of only eight days. Creuzer, in his *Symbolik und Mythologie*, 1841, th. ii., s. 427, 430, 479, und 481, supposes "that strong natural and artificial heat, in the room in which they were placed, was used to hasten the growth of plants in the Adonis gardens."   The garden of the Dominican convent at Cologne reminds us of the Greenland or Icelandic convent of St. Thomas, where the garden was kept free from snow by being warmed by natural thermal springs, as is related by the brothers Zeni, in the account of their travels (1388–1404), which, from the geographical localities indicated, must be considered as very problematical. (Compare Zurla, *Viaggiatori Veneziani*, t. ii., p. 63–69 ; and Humboldt, *Examen Critique de l'Hist. de la Géographie*, t. ii., p. 127.) The introduction in our botanic gardens of regular hot-houses seems to be of more recent date than is generally supposed. Ripe pine-apples were first obtained at the end of the seventeenth century (Beckmann's *History of Inventions*, Bohn's Standard Library, 1846, vol. i., p. 103–106); and Linnæus even asserts, in the *Musa Cliffortiana florens Hartecampi*, that the first banana which flowered in Europe was in 1731, at Vienna, in the garden of Prince Eugene.

    * These views of tropical vegetation, which designate the "physiognomy of plants," constitute, in the Royal Museum at Berlin (in the de-

He who, with a keen appreciation of the beauties of nature manifested in mountains, rivers, and forest glades, has himself traveled over the torrid zone, and seen the luxuriance and diversity of vegetation, not only on the cultivated sea-coasts, but on the declivities of the snow-crowned Andes, the Himalaya, or the Nilgherry Mountains of Mysore, or in the primitive forests, amid the net-work of rivers lying between the Orinoco and the Amazon, can alone feel what an inexhaustible treasure remains still unopened by the landscape painter between the tropics in both continents, or in the island-world of Sumatra, Borneo, and the Philippines; and how all the spirited and admirable efforts already made in this portion of art fall far short of the magnitude of those riches of nature of which it may yet become possessed. Are we not justified in hoping that landscape painting will flourish with a new and hitherto unknown brilliancy when artists of merit shall more frequently pass the narrow limits of the Mediterranean, and when they shall be enabled, far in the interior of continents, in the humid mountain valleys of the tropical world, to seize, with the genuine freshness of a pure and youthful spirit, on the true image of the varied forms of nature?

These noble regions have hitherto been visited mostly by travelers whose want of artistical education, and whose differently directed scientific pursuits afforded few opportunities of their perfecting themselves in landscape painting. Only very few among them have been susceptible of seizing on the total impression of the tropical zone, in addition to the botanical interest excited by the individual forms of flowers and leaves. It has frequently happened that the artists appointed to accompany expeditions fitted out at the national expense have been chosen without due consideration, and almost by accident, and have been thus found less prepared than such appointments required; and the end of the voyage may thus have drawn near before even the most talented among them, by a prolonged sojourn among grand scenes of nature, and by frequent attempts to imitate what they saw, had more than

partment of miniatures, drawings, and engravings), a treasure of art which, owing to its peculiarity and picturesque variety, is incomparably superior to any other collection. The title of the papers edited by Von Kittlitz is *Vegetations-Ansichten der Küstenländer und Inseln des stillen Oceans, aufgenommen 1827–1829, auf der Entdeckungs-reise der kais. Russ. Corvette Senjäwin* (Siegen, 1844). There is also great fidelity to nature in the drawings of Carl Bodmer, which are engraved in a masterly manner, and which greatly embellish the large work of the travels of Prince Maximilian of Wied in the interior of North America.

begun to acquire a certain technical mastery of their art. Voyages of circumnavigation are, besides, but seldom of a character to allow of artists visiting any extensive tracts of forest land, the upper courses of large rivers, or the summits of inland chains of mountains.

Colored sketches, taken directly from nature, are the only means by which the artist, on his return, may reproduce the character of distant regions in more elaborately finished pictures; and this object will be the more fully attained where the painter has, at the same time, drawn or painted directly from nature a large number of separate studies of the foliage of trees; of leafy, flowering, or fruit-bearing stems; of prostrate trunks, overgrown with Pothos and Orchideæ; of rocks and of portions of the shore, and the soil of the forest. The possession of such correctly-drawn and well-proportioned sketches will enable the artist to dispense with all the deceptive aid of hothouse forms and so-called botanical delineations.

A great event in the history of the world, such as the emancipation of Spanish and Portuguese America from the dominion of European rule, or the increase of cultivation in India, New Holland, the Sandwich Islands, and the southern colonies of Africa, will incontestably impart to meteorology and the descriptive natural sciences, as well as to landscape painting, a new impetus and a high tone of feeling, which probably could not have been attained independently of these local relations. In South America, populous cities lie at an elevation of nearly 14,000 feet above the level of the sea. From these heights the eye ranges over all the climatic gradations of vegetable forms. What may we not, therefore, expect from a picturesque study of nature, if, after the settlement of social discord and the establishment of free institutions, a feeling of art shall at length be awakened in those elevated regions?

All that is expressed by the passions, and all that relates to the beauty of the human form, has attained its highest perfection in the temperate northern zone under the skies of Greece and Italy. The artist, drawing from the depths of nature no less than from the contemplation of beings of his own species, derives the types of historical painting alike from free creation and from truthful imitation. Landscape painting, though not simply an imitative art, has a more material origin and a more earthly limitation. It requires for its development a large number of various and direct impressions, which, when received from external contemplation, must be fertilized by the powers of the mind, in order to be given back to the senses of others as a free work of art. The grander style of heroic landscape painting is the combined result of a profound appreciation of nature and of this inward process of the mind.

Every where, in every separate portion of the earth, nature is indeed only a reflex of the whole. The forms of organisms recur again and again in different combinations. Even the icy north is cheered for months together by the presence of herbs and large Alpine blossoms covering the earth, and by the aspect of a mild azure sky. Hitherto landscape painting among us has pursued her graceful labors familiar only with the simpler forms of our native floras, but not, on that account, without depth of feeling and richness of creative fancy. Dwelling only on the native and indigenous forms of our vegetation, this branch of art, notwithstanding that it has been circumscribed by such narrow limits, has yet afforded sufficient scope for highly-gifted painters, such as the Caracci, Gaspard Poussin, Claude Lorraine, and Ruysdael, to produce the loveliest and most varied creations of art, by their magical power of managing the grouping of trees and the effects of light and shade. That progress which may still be expected in the different departments of art, and to which I have already drawn attention, in order to indicate the ancient bond which unites natural science with poetry and artistic feeling, can not impair the fame of the master works above referred to, for, as we have observed, a distinction must be made in landscape painting, as in every other branch of art, between the elements generated by the more limited field of contemplation and direct observation, and those which spring from the boundless depth of feeling and from the force of idealizing mental power. The grand conceptions which landscape painting, as a more or less inspired branch of the poetry of nature, owes to the creative power of the mind, are, like man himself, and the imaginative faculties with which he is endowed, independent of place. These remarks especially refer to the gradations in the forms of trees from Ruysdael and Everdingen, through the works of Claude Lorraine, to Poussin and Annibal Caracci. In the great masters of art there is no indication of local limitation. But an extension of the visible horizon, and an acquaintance with the nobler and grander forms of nature, and with the luxurious fullness of life in the tropical world, afford the advantage of not simply enriching the material ground-work of landscape painting, but also of inducing more vivid impressions in the minds of less highly-gifted

painters, and thus heightening their powers of artistic creation.

I would here be permitted to refer to some remarks which
I published nearly half a century ago, in a treatise which has
been but little read, entitled *Ideen zu einer Physiognomik
der Gewächse,** and which stands in the most intimate connection with the subject under consideration.   He who comprehends nature at a single glance, and knows how to abstract
his mind from local phenomena, will easily perceive how organic force and the abundance of vital development increase
with the increase of warmth from the poles to the equator.
This charming luxuriance of nature increases, in a lesser degree, from the north of Europe to the lovely shores of the
Mediterranean than from the Iberian Peninsula, Southern
Italy, and Greece, toward the tropics.   The naked earth is
covered with an unequally woven, flowery mantle, thicker
where the sun rises high in a sky of deep azure, or is only
vailed by light and feathery clouds, and thinner toward the
gloomy north, where the returning frost too soon blights the
opening bud or destroys the ripening fruit.   While, in the cold
zones, the bark of the trees is covered with dry moss or with
lichens, the region of palms and of feathery arborescent ferns
shows the trunks of Anacardia and of the gigantic species of
Ficus embellished by Cymbidia and the fragrant Vanilla.
The fresh green of the Dracontium, and the deeply-serrated
leaves of the Pothos, contrast with the variegated blossoms of
the Orchideæ, while climbing Bauhiniæ, Passifloræ, and yel-
low-blossomed Banisteriæ, entwining the stems of forest trees,
spread far and high in air, and delicate flowers are unfolded
from the roots of the Theobromæ, and from the thick and
rough bark of the Crescentiæ and the Gustaviæ.   In the
midst of this abundance of flowers and leaves, and this luxu-
riantly wild entanglement of climbing plants, it is often difficult for the naturalist to discover to which stem different
flowers and leaves belong ; nay, one single tree adorned with
Paulliniæ, Bignoniæ, and Dendrobia, presents a mass of vegetable forms which, if disentangled, would cover a considerable
space of ground.

Each portion of the earth has, however, its peculiar and

* Humboldt, *Ansichten der Natur*, 2te Ausgabe, 1826, bd. i., s. 7, 16,
21, 36, and 42.   Compare, also, two very instructive memoirs, Friedrich von Martius, *Physiognomie des Pflanzenreiches in Brasilien*, 1824,
and M. von Olfers, *allgemeine Uebersicht von Brasilien*, in Feldner's
Reisen, 1828, th. i., s. 18-23.

characteristic beauty : to the tropics belong diversity and grandeur in the forms of plants ; to the north, the aspect of tracts of meadow-land, and the periodic and long-desired revival of nature at the earliest breath of the gentle breezes of spring. As in the Musaceæ (Pisang) we have the greatest expansion, so in the Casuarinæ and in the needle-tree we have the greatest contraction of the leaf vessels. Firs, Thujæ, and Cypresses constitute a northern flora which is very uncommon in the plains of the tropics. Their ever-verdant green enlivens the dreary winter landscape, and proclaims to the inhabitants of the north that, even when snow and ice have covered the ground, the inner life of vegetation, like Promethean fire, is never extinguished on our planet.

Every zone of vegetation has, besides its own attractions, a peculiar character, which calls forth in us special impressions. Referring here only to our own native plants, I would ask, who does not feel himself variously affected beneath the somber shade of the beech, on hills crowned with scattered pines, or in the midst of grassy plains, where the wind rustles among the trembling leaves of the birch ? As in different organic beings we recognize a distinct physiognomy, and as descriptive botany and zoology are, in the strict definition of the words, merely analytic classifications of animal and vegetable forms, so there is also a certain physiognomy of nature exclusively peculiar to each portion of the earth. The idea which the artist wishes to indicate by the expressions " Swiss nature" or " Italian skies," is based on a vague sense of some local characteristic. The azure of the sky, the form of the clouds, the vapory mist resting in the distance, the luxuriant development of plants, the beauty of the foliage, and the outline of the mountains, are the elements which determine the total impression produced by the aspect of any particular region. To apprehend these characteristics, and to reproduce them visibly, is the province of landscape painting ; while it is permitted to the artist, by analyzing the various groups, to resolve beneath his touch the great enchantment of nature—if I may venture on so metaphorical an expression—as the written words of men are resolved into a few simple characters.

But, even in the present imperfect condition of pictorial delineations of landscapes, the engravings which accompany, and too often disfigure, our books of travels, have, however, contributed considerably toward a knowledge of the physiognomy of distant regions, to the taste for voyages in the tropical zones, and to a more active study of nature. The improvements in landscape painting on a large scale (as decorative paintings, panoramas, dioramas, and neoramas) have also increased the generality and force of these impressions. The representations satirically described by Vitruvius and the Egyptian, Julius Pollux, as " exaggerated representations of rural adornments of the stage," and which, in the sixteenth century, were contrived by Serlio's arrangement of Coulisses to increase the delusion, may now, since the discoveries of Prevost and Daguerre, be made, in Barker's panoramas, to serve, in some degree, as a substitute for traveling through different regions. Panoramas are more productive of effect than scenic decorations, since the spectator, inclosed, as it were, within a magical circle, and wholly removed from all the disturbing influences of reality, may the more easily fancy that he is actually surrounded by a foreign scene. These compositions give rise to impressions which, after many years, often become wonderfully interwoven with the feelings awakened by the aspect of the scenes when actually beheld. Hitherto panoramas, which are alone effective when of considerable diameter, have been applied more frequently to the representation of cities and inhabited districts than to that of scenes in which nature revels in wild luxuriance and richness of life. An enchanting effect might be produced by a characteristic delineation of nature, sketched on the rugged declivities of the Himalaya and the Cordilleras, or in the midst of the Indian or South American river valleys, and much aid might be further derived by taking photographic pictures, which, although they certainly can not give the leafy canopy of trees, would present the most perfect representation of the form of colossal trunks, and the characteristic ramification of the different branches.

All these means, the enumeration of which is specially comprised within the limits of the present work, are calculated to raise the feeling of admiration for nature ; and I am of opinion that the knowledge of the works of creation, and an appreciation of their exalted grandeur, would be powerfully increased if, besides museums, and thrown open, like them, to the public, a number of panoramic buildings, containing alternating pictures of landscapes of different geographical latitudes and from different zones of elevation, should be erected in our large cities. The conception of the natural unity and the feeling of the harmonious accord pervading the universe can not fail to increase in vividness among men, in proportion as the means are multiplied by which the phenomena of nature may be more characteristically and visibly manifested.

CULTIVATION OF TROPICAL PLANTS—CONTRAST AND ASSEMBLAGE OF VEGETABLE FORMS.—IMPRESSIONS INDUCED BY THE PHYSIOGNOMY AND CHARACTER OF THE VEGETATION.

LANDSCAPE painting, notwithstanding the multiplication of its productions by engravings, and by the recent improvements in lithography, is still productive of a less powerful effect than that excited in minds susceptible of natural beauty by the immediate aspect of groups of exotic plants in hot-houses or in gardens. I have already alluded to the subject of my own youthful experience, and mentioned that the sight of a colossal dragon-tree and of a fan palm in an old tower of the botanical garden at Berlin implanted in my mind the seeds of an irresistible desire to undertake distant travels. He who is able to trace through the whole course of his impressions that which gave the first leading direction to his whole career, will not deny the influence of such a power.

I would here consider the different impression produced by the picturesque arrangement of plants, and their association for the purposes of botanical exposition ; in the first place, by groups distinguished for their size and mass, as Musaceæ and Helleoniæ, growing in thick clumps, and alternating with Corypha palms, Araucariæ, and Mimosæ, and moss-covered trunks, from which shoot forth Dracontia, delicately-leaved Ferns, and richly-blossoming Orchideæ ; and, in the next, by an abundance of separate lowly plants, classed and cultivated in rows for the purpose of affording instruction in descriptive and systematic botany. In the first case, our attention is challenged by the luxuriant development of vegetation in Cecropiæ, Coriniæ, and light, feathery Bamboos ; by the picturesque association of the grand and noble forms which embellish the shores of the Upper Orinoco, the wooded banks of the Amazon, or of the Huallaga, so vividly and admirably described by Martius and Edward Pöppig ; and by the sentiment of longing for the lands in which the current of life flows more abundantly and richly, and of whose beauty a faint but still pleasing image is reflected to the mind by means of our hot-houses, which originally served as mere nurseries for sickly plants.

It undoubtedly enters within the compass of landscape painting to afford a richer and more complete picture of nature than the most skillfully-arranged grouping of cultivated plants is able to present, since this branch of art exercises an almost magical command over masses and forms. Almost

unlimited in space, it traces the skirts of the forest till they
are wholly lost in the aërial distance, dashes the mountain
torrent from cliff to cliff, and spreads the deep azure of the
tropical sky alike over the summits of the lofty palms and
over the waving grass of the plain that bounds the horizon.
The luminous and colored effects imparted to all terrestrial
objects by the light of the thinly-vailed or pure tropical sky,
gives a peculiar and mysterious power to landscape painting,
when the artist succeeds in reproducing this mild effect of light.
The sky in the landscape has, from a profound appreciation
for the nature of Greek tragedy, been ingeniously compared to
the charm of the *chorus* in its general and mediative effect.[*]

The multiplication of means at the command of painting
for exciting the fancy, and concentrating the grandest phenomena of sea and land on a small space, is denied to our
plantations and gardens, but this deficiency in the total effect
is compensated for by the sway which reality every where
exercises over the senses. When, in the Messrs. Loddiges'
palm-house, or in the *Pfauen-Insel*, near Potsdam (a monument of the simple love of nature of my noble and departed
sovereign), we look down from the high gallery in the bright
noonday sun on the luxuriant reed and tree-like palms below,
we feel, for a moment, in a state of complete delusion as to
the locality to which we are transported, and we may even
believe ourselves to be actually in a tropical climate, looking
from the summit of a hill on a small grove of palms. It is
true that the aspect of the deep azure of the sky, and the impression produced by a greater intensity of light, are wanting,
but, notwithstanding, the illusion is more perfect, and exercises a stronger effect on the imagination than is excited by
the most perfect painting. Fancy associates with every plant
the wonders of some distant region, as we listen to the rustling of the fan-like leaves, and see the changing and flitting
effect of the light, when the tops of the palms, gently moved
by currents of air, come in contact as they wave to and fro.
So great is the charm produced by reality, although the recollection of the artificial care bestowed on the plants certainly
exercises a disturbing influence. Perfect development and
freedom are inseparably connected with nature, and in the
eyes of the zealous and botanical traveler, the dried plants of
an herbarium, collected on the Cordilleras of South America
or in the plains of India, are often more precious than the aspect of the same species of plants within a European hot-

* Wilh. von Humboldt, in his *Briefwechsel mit Schiller*, 1830, s. 470.

house. Cultivation blots out some of the original characters of nature, and checks the free development of the several parts of the exotic organization.

The physiognomy and arrangement of plants and their contrasted apposition must not be regarded as mere objects of natural science, or incitements toward its cultivation ; for the attention devoted to the physiognomy of plants is likewise of the greatest importance with reference to the art of landscape gardening. I will not yield to the temptation here held out to me of entering more fully into this subject, merely limiting myself to a reference to the beginning of this section of the present work, where, as we found occasion to praise the more frequent manifestation of a profound sentiment of nature noticed among nations of Semitic, Indian, and Iranian descent, so also we find from history that the cultivation of parks originated in Central and Southern Asia. Semiramis caused gardens to be laid out at the foot of the Mountain Bagistanos, which have been described by Diodorus,* and whose fame induced Alexander, on his progress from Kelone to the horse pastures of Nysæa, to deviate from the direct road. The parks of the Persian kings were adorned with cypresses, whose obelisk-like forms resembled the flame of fire, and were, on that account, after the appearance of Zerduscht (Zoroaster), first planted by Gushtasp around the sacred precincts of the Temple of Fire. It is thus that the form of the tree itself has led to the myth of the origin of the cypress in Paradise. †

* Diodor., ii., 13. He, however, ascribes to the celebrated gardens of Semiramis a circumference of only twelve stadia. The district near the pass of Bagistanos is still called the "bow or circuit of the gardens"— Tauk-i-bostan (Droysen, *Gesch. Alexanders des Grossen*, 1833, s. 553).

† In the *Schahnameh* of Firdusi it is said, "a slender cypress, reared in Paradise, did Zerdusht plant before the gate of the temple of fire" (at Kishmeer in Khorassan). "He had written on this tall cypress that Gushtasp had adopted the genuine faith, of which the slender tree was a testimony, and thus did God diffuse righteousness. When many years had passed away, the tall cypress spread and became so large that the hunter's cord could not gird its circumference. When its top was surrounded by many branches, he encompassed it with a palace of pure gold . . . . . . and caused it to be published abroad, Where is there on the earth a cypress like that of Kishmeer? From Paradise God sent it me, and said, Bow thyself from thence to Paradise." When the Calif Motewekkil caused the cypresses, sacred to the Magians, to be cut down, the age ascribed to this one was said to be 1450 years. See Vuller's *Fragmente über die Religion des Zoroaster*, 1831, s. 71 und 114; and Ritter, *Erdkunde*, th. vi., i., s. 242. The original native place of the cypress (in Arabic *arar*, wood, in Persian *serw kohi*) appears to be the mountains of Busih, west of Herat (*Géographie d'Edrisi*, trad. par Jaubert, 1836, t. i., p. 464).

The gardens of the Asiatic terrestrial paradises (παραδεισοι) excited the early admiration of the inhabitants of the West ;* and the worship of trees may be traced among the Iranians to the remote date of the prescripts of Hom, named, in the Zend-Avesta, the promulgator of the old law. We learn from Herodotus the delight taken by Xerxes in the great plane-tree in Lydia, on which he bestowed decorations of gold, appointing one of the " immortal ten thousand" as its special guard.† The ancient adoration of trees was connected, owing to the refreshing and humid shadow of the leafy canopy, with the worship of the sacred springs.

To this consideration of the primitive worship of nature belongs a notice of the fame attached among the Hellenic races to the remarkably large palm-tree in the island of Delos, and to an ancient palm-tree in Arcadia. The Buddhists of Ceylon venerate the colossal Indian fig-tree, the Banyan of Anurahdepura, which is supposed to have sprung from the branches of the original tree under which Buddha, as the inhabitant of the ancient Magadha, fell into a state of beatitude, spontaneous extinction, *nirwâna*.‡ As separate trees became objects of adoration from the beauty of their forms, so likewise groups of trees were venerated as groves of the gods. Pausanias speaks in high terms of admiration of a grove round the Temple of Apollo at Grynion Æolis,§ while the grove of Colonus is likewise celebrated in the famous chorus of Sophocles.

* Achill. Tat., i., 25 ; Longus, *Past.*, iv., p. 108 ; Schäfer. " Gesenius (*Thes. Linguæ Hebr.*, t. ii., p. 1124) very justly advances the view that the word Paradise belonged originally to the ancient Persian language, but that its use has been lost in the modern Persian. Firdusi, although his own name was taken from it, usually employs only the word *behischt;* the ancient Persian origin of the word is, however, expressly corroborated by Pollux, in the *Onomast.*, ix., 3 ; and by Xenophon (*Œcon.*, 4, 13, and 21 ; *Anab.*, i., 2, 7, and i., 4, 10 ; *Cyrop.*, i., 4, 5). In its signification of pleasure-garden, or garden, the word has probably passed from the Persian into the Hebrew (*pardês, Cant.*, iv., 13 ; *Nehem.*, ii., 8 ; and *Eccl.*, ii., 5); into the Arabic (*firdaus*, plur. *faradisu*, compare Alcoran, 23, 11, and Luc., 23, 43); into the Syrian and Armenian (*partês*, see Ciakciak, *Dizionario Armeno*, 1837, p. 1194; and Schröder, *Thes. Ling. Armen.*, 1711, Præf., p. 56). The derivation of the Persian word from the Sanscrit (*pradêsa*, or *paradêsa*, circuit, or district, or foreign land), which was noticed by Benfey (*Griech. Wurzellexikon*, bd. i., 1839, s. 138), and previously by Bohlen and Gesenius, suits perfectly in form, but not so well in sense."—Buschmann.

† Herod., vii., 31 (between Kallatebus and Sardes).

‡ Ritter, *Erdkunde*, th. iv., 2, s. 237, 251, und 681 ; Lassen, *Indische Alterthumskunde*, bd. i., s. 260.

§ Pausanias, i., 21, 9. Compare, also, *Arboretum Sacrum, in Meursii Op. ex recensione* Joann. Lami, vol. x., Florent., 1753, p. 777–844.

The feeling for nature manifested by the early cultivated East Asiatic nations, in the choice and the careful attention of sacred objects chosen from the vegetable kingdom, was most strongly and variously exhibited in their cultivation of parks. In the remotest parts of the Old Continent the Chinese gardens appear to have approached most nearly to what we are now accustomed to regard as English parks. Under the victorious dynasty of Han, gardens were so frequently extended over a circuit of many miles that agriculture was injured by them, and the people excited to revolt.* " What is it that we seek in the possession of a pleasure garden ?" asks an ancient Chinese writer, Lieu-tscheu. It has been universally admitted, throughout all ages, that plantations should compensate to man for the loss of those charms of which he is deprived by his removal from a free communion with nature, his proper and most delightful place of abode. " The art of laying out gardens consists in an endeavor to combine cheerfulness of aspect, luxuriance of growth, shade, solitude, and repose, in such a manner that the senses may be deluded by an imitation of rural nature. Diversity, which is the main advantage of free landscape, must therefore be sought in a judicious choice of soil, an alternation of chains of hills and valleys, gorges, brooks, and lakes covered with aquatic plants. Symmetry is wearying, and ennui and disgust will soon be excited in a garden where every part betrays constraint and art."† The description given by Sir George Staunton of the great imperial garden of Zhe-hol,‡ north of the Chinese wall, corresponds with these precepts of Lieu-tscheu—precepts to which our ingenious cotemporary, who formed the charming park of Muskau,§ will not refuse his approval.

In the great descriptive poem written in the middle of the last century by the Emperor Kien-long, in praise of the former Mantchou capital, Mukden, and of the graves of his ancestors, the most ardent admiration is expressed for free nature, when but little embellished by art. The poetic prince shows a happy power in fusing the cheerful images of the luxuriant

* *Notice Historique sur les Jardins des Chinois,* in the *Mémoires concernant les Chinois,* t. viii., p. 309.

† See the work last quoted, p. 318–320.

‡ Sir George Staunton, *Account of the Embassy of the Earl of Macartney to China,* vol. ii., p. 245.

§ Prince Pückler-Muskau, *Andeutungen über Landschaftsgärtnerei,* 1834. Compare, also, his Picturesque Descriptions of the Old and New English Parks, as well as that of the Egyptian Gardens of Schubra.

freshness of the meadows, of the forest-crowned hills, and the peaceful dwellings of men, with the somber picture of the tombs of his forefathers. The sacrifices which he offers in obedience to the rites prescribed by Confucius, and the pious remembrance of the departed monarchs and warriors, form the principal objects of this remarkable poem. A long enumeration of the wild plants and animals that are natives of the region is wearisome, like every other didactic work ; but the blending of the visible impressions produced by the land-scape, which serves, as it were, for the back-ground of the picture, with the exalted objects of the ideal world, with the fulfillment of religious duties, together with the mention of great historical events, gives a peculiar character to the whole composition. The feeling of adoration for mountains, which was so deeply rooted among the Chinese, leads Kien-long to give a careful delineation of the physiognomy of inanimate nature, for which the Greeks and Romans evinced so little feeling. The form of the separate trees, the character of their ramification, the direction of the branches, and the form of the foliage, are all dwelt on with special predilection.*

If I have not yielded to the distaste for Chinese literature, which is, unfortunately, disappearing too slowly from among us, and if I have dwelt too long on the consideration of the delineations of nature met with in the works of a cotemporary of Frederic the Great, I am so much the more bound to ascend seven and a half centuries further back into the annals of time, in order to refer to the poem of the *Garden*, by See-ma-kuang, a celebrated statesman. The pleasure grounds described in this poem are certainly much crowded by buildings in the fashion of the old Italian villas, but the minister likewise celebrates a hermitage, which is situated among rocks and surrounded by high fir-trees. He extols the open view over the broad river Kiang, crowded with vessels, and expects, with contentment, the arrival of friends, who will read their verses to him, since they will also listen to his compositions.† See-ma-kuang wrote about the year 1086, when, in Germany, poetry was in the hands of a rude clergy, and was not even clothed in the garb of the national tongue.

At this period, and probably five hundred years earlier, the inhabitants of China, of Eastern India, and Japan, were al-

* *Eloge de la Ville de Moukden*, Poême composé par l'Empereur Kien-long, traduit par le P. Amiot, 1770, p. 18, 22–25, 37, 63–68, 73–87, 104, and 120.

† *Mémoires concernant les Chinois*, t. ii., p. 643–650.

ready acquainted with a great variety of vegetable forms. The intimate connection which existed among the different Buddhist sacerdotal establishments contributed its influence in this respect. Temples, cloisters, and burying-places were surrounded by gardens, adorned with exotic trees, and covered by variegated flowers of different forms. Indian plants were early diffused over China, Corea, and Nipon. Siebold, whose writings give a comprehensive view of all matters referring to Japan, was the first to draw attention to the cause of the mixture of the floras of remotely-separated Buddhist lands.*

The rich abundance of characteristic vegetable forms presented by the present age to scientific observation and to landscape painting, must act as a powerful incentive to trace the sources which have yielded us this increased knowledge and enjoyment of nature. The enumeration of these sources must be reserved for the history of the contemplation of nature in the succeeding portion of this work. Here my object has been to depict, in the reflection of the external world on the mental activity and the feelings of mankind, those means which, in the progress of civilization, have exercised so marked and animated an influence on the study of nature. Notwithstanding a certain freedom of development of the several parts, the primitive force of organization binds all animal and vegetable forms to fixed and constantly-recurring types, determining, in every zone, the character that peculiarly appertains to it, or *the physiognomy of nature*. We may therefore regard it as one of the most precious fruits of European civilization, that it is almost every where permitted to man, by the cultivation and arrangement of exotic plants, by the charm of landscape painting, and by the inspired power of language, to procure a substitute for familiar scenes during the period of absence, or to receive a portion of that enjoyment from nature which is yielded by actual contemplation during long and not unfrequently dangerous journeys through the interior of distant continents.

* Ph. Fr. von Siebold, *Kruidkundige Naamlijst van Japansche en Chineesche Planten*, 1844, p. 4. What a difference do we not find on comparing the variety of vegetable forms cultivated for so many centuries past in Eastern Asia, with those enumerated by Columella, in his meager poem *De Cultu Hortorum* (v. 95–105, 174–176, 225–271, 295–306), and to which the celebrated garland-weavers of Athens were confined! It was not until the time of the Ptolemies that in Egypt, and especially in Alexandria, the more skillful gardeners appear to have devoted any great attention to variety, particularly for winter cultivation. (Compare *Athen.*, v., p. 196.)

PART II.

The history of the physical contemplation of the universe is the history of the recognition of the unity of nature, the representation of the efforts made by man to comprehend the combined action of natural forces on the earth and in the regions of space, and hence it designates the epochs of advancement in the generalization of views, being a portion of the history of our world of thought, in as far as it refers to objects manifested by the senses, to the form of conglomerated matter, and the forces inherent in it.

In the section of the first portion of this work, relating to the limitation and scientific treatment of a physical description of the universe, I hope I may have succeeded in developing with clearness the relation existing between the separate natural sciences and the description of the universe (the science of the Cosmos), and the manner in which this science simply draws from these various branches of study the materials for its scientific foundation. The history of the knowledge of the universe, of which I here present the leading ideas, and which, for the sake of brevity, I name either simply the history of the Cosmos, or the *history of the physical contemplation of the universe*, must not, therefore, be confounded with the history of the natural sciences, as given in many of our leading elementary works on physics and physiology, or on the morphology of plants and animals.

In order to give some idea of what has been collected at separate epochs under this point of view, it appears most desirable to adduce separate instances illustrative of the subjects which must either be treated of or discarded in the succeeding portions of this work. The discoveries of the compound microscope, of the telescope, and of colored polarization, belong to the history of the Cosmos, since they have afforded the means of discovering that which is common to all organisms ; of penetrating into the remotest regions of space ; of distinguishing between reflected or borrowed light, and the light of self-luminous bodies, or, in other words, determining whether solar light be radiated from a solid mass or from a gaseous envelope. The enumeration of the experiments which, since Huygens's time, have gradually led to Arago's discovery of colored polarization, must be reserved for the history of optics. The consideration of the development of the principles, in accordance with which variously-formed plants admit of being classified in families, falls, in like manner, within the domain of the history of phytognosy, or botany ; while the geography of plants, or a study of the local and climatic distribution of vegetation over the whole earth—alike over the solid portions and in the basins of the sea—constitutes an important section in the history of the physical contemplation of the universe.

The intellectual consideration of that which has led man to an insight into the unity of nature is, as we have already observed, as little entitled to the appellation of the complete history of the cultivation of mankind as to that of a history of the natural sciences. An insight into the connection of the vital forces of the universe must certainly be regarded as the noblest fruit of human civilization, and as the tendency to arrive at the highest point to which the most perfect development of the intellect can attain ; but the subject at present under consideration must still constitute only a part of the history of human civilization, embracing all that has been attained by the advance of different nations in the pursuit of every branch of mental and moral culture. By assuming a more limited physical point of view, we necessarily become restricted to one section of the history of human knowledge, and our attention is specially directed to the relation existing between the knowledge that has been gradually acquired and the whole extent of the domain of nature ; and we dwell less on the extension of separate branches of science than on the results capable of generalization, and the material aids contributed by different ages toward a more accurate observation of nature.

We must, above all, distinguish carefully between an early presentiment of knowledge and knowledge itself. With the increasing cultivation of the human race, much has passed from the former to the latter, and by this transition the history of discovery has been rendered indistinct. An intellectual and ideal combination of the facts already established often guides almost imperceptibly the course of presage, elevating it as by a power of inspiration. How much has been enounced among the Indians and Greeks, and during the Middle Ages, regarding the connection of natural phenomena, which, at first, either vague, or blended with the most unfounded hypotheses, has, at a subsequent epoch, been confirmed by sure experience, and then been recognized as a scientific truth ! The presentient fancy and the vivid activity of spirit which animated Plato, Columbus, and Kepler, must not be disregarded, as if they had effected nothing in the domain of science, or as if they tended, of necessity, to draw the mind from the investigation of the actual.

As we have defined the history of the physical contemplation of the universe to be the history of the recognition of nature in the unity of its phenomena, and of the connection of the forces of the universe, our mode of proceeding must consist in the enumeration of those subjects by which the idea of the unity of the phenomena has been gradually developed. We would here distinguish :

1. The independent efforts of reason to acquire a knowledge of natural laws, by a meditative consideration of the phenomena of nature.

2. Events in the history of the world which have suddenly enlarged the horizon of observation.

3. The discovery of new means of sensuous perception, as well as the discovery of new organs by which men have been brought into closer connection, both with terrestrial objects and with remote regions of space.

This three-fold view serves as a guide in defining the principal epochs that characterize the history of the science of the Cosmos. For the purpose of further illustration, I would again adduce some examples indicative of the diversity of the means by which mankind attained to the intellectual possession of a great portion of the universe. Under this head I include examples of an enlarged field of natural knowledge, great historical events, and the discovery of new organs.

The *knowledge of nature*, as it existed among the Hellenic nations under the most ancient forms of physics, was derived more from the depth of mental contemplation than from the sensuous consideration of phenomena. Thus the natural philosophy of the Ionian physiologists was directed to the fundamental ground of origin, and to the metamorphoses of one sole element, while the mathematical symbolicism of the Pythagoreans, and their consideration of numbers and forms, disclose a philosophy of measure and harmony. The Doric-Italian school, by its constant search for numerical elements, and by a certain predilection for the numerical relations of space and time, laid the foundation, as it were, of the subsequent development of our experimental sciences. The history of the contemplation of the universe, as I interpret its limits, designates not so much the frequently-recurring oscillations between truth and error, as the principal epochs of the gradual approximation to more accurate views regarding terrestrial forces and the planetary system. It shows us that the Pythagoreans, according to the report of Philolaüs of Croton, taught the progressive movement of the non-rotating Earth, its revolution round the focus of the world (the central fire, *hestia*), while Plato and Aristotle imagined that the Earth neither rotated nor advanced in space, but that, fixed to one central point, it merely oscillated from side to side. Hicetas of Syracuse, who must, at least, have preceded Theophrastus, Heraclides Ponticus, and Ecphantus, all appear to have had a knowledge of the rotation of the Earth on its axis ; but Aristarchus of Samos, and more particularly Seleucus of Babylon, who lived one hundred and fifty years after Alexander, first arrived at the knowledge that the Earth not only rotated on its own axis, but also moved round the Sun as the center of the whole planetary system. And if, in the dark period of the Middle Ages, Christian fanaticism, and the lingering influence of the Ptolemaic school, revived a belief in the immobility of the Earth, and if, in the hypothesis of the Alexandrian, Cosmas Indicopleustes, the globe again assumed the form of the disk of Thales, it must not be forgotten that a German cardinal, Nicholas de Cusa, was the first who had the courage and the independence of mind again to ascribe to our planet, almost a hundred years before Copernicus, both rotation on its axis and translation in space. After Copernicus, the doctrines of Tycho Brahe gave a retrograde movement to science, although this was only of short duration ; and when once a large mass of accurate observations had been collected, to which Tycho Brahe himself contributed largely, a correct view of the structure of the universe could not fail to be speedily established. We have already shown how a period of fluctuations between truth and error is especially one of presentiments and fanciful hypotheses regarding natural philosophy.

After treating of the extended knowledge of nature as a simultaneous consequence of direct observations and ideal combinations, we have proceeded to the consideration of those historical events which have materially extended the horizon of the physical contemplation of the universe. To these belong migrations of races, voyages of discovery, and military expeditions. Events of this nature have been the means of acquiring a knowledge of the natural character of the Earth's surface (as, for instance, the configuration of continents, the direction of mountain chains, and the relative height of elevated plateaux), and in the case of extended tracts of land, of presenting us with materials for expounding the general laws of nature. It is unnecessary, in this historical sketch, to give a connected tissue of events, and it will be sufficient, in the history of the recognition of nature as a whole, to refer merely to those events which, at early periods, have exercised a decided influence on the mental efforts of mankind, and on a more extended view of the universe. Considered in this light, the navigation of Colæus of Samos beyond the Pillars of Hercules; the expedition of Alexander to Western India; the dominion exercised by the Romans over the then discovered portions of the world; the extension of Arabian cultivation, and the discovery of the New Continent, must all be regarded as events of the greatest importance for the nations settled round the basin of the Mediterranean. My object is not so much to dwell on the relation of events that may have occurred, as to refer to the action exercised on the development of the idea of the Cosmos by events, whether it be a voyage of discovery, the establishment of the predominance of some highly-developed language rich in literary productions, or the sudden extension of the knowledge of the Indo-African monsoons.

As I have already incidentally mentioned the influence of language in my enumeration of heterogeneous inducements, I will draw attention generally to its immeasurable importance in two wholly different directions. Languages, when extensively diffused, act individually as means of communication between widely-separated nations, and collectively when several are compared together, and their internal structure and degrees of affinity are investigated, as means of promoting a more profound study of the history of mankind. The Greek language, which is so intimately connected with the national life of the Hellenic races, has exercised a magic power over all the foreign nations with which these races came in contact.* The Greek language appears in the interior of Asia, through the influence of the Bactrian empire, as a conveyer of knowledge, which, a thousand years afterward, was brought

* Niebuhr, *Röm. Geschichte*, th. i., s. 69; Droysen, *Gesch. der Bildung des Hellenistischen Staatensystems*, 1843, s. 31–34, 567–573; Fried. Cramer, *De Studiis quæ veteres ad aliarum Gentium contulerint Linguas*, 1844, p. 2–13.

back by the Arabs to the extreme west of Europe, blended with hypotheses of Indian origin. The ancient Indian and Malayan tongues furthered the advance of commerce and the intercourse of nations in the island-world of the southwest of Asia, in Madagascar, and on the eastern shores of Africa ; and it is also probable that tidings of the Indian commercial stations of the Banians may have given rise to the adventurous expedition of Vasco de Gama. The predominance of certain languages, although it unfortunately prepared a rapid destruction for the idioms displaced, has operated favorably, like Christianity and Buddhism, in bringing together and uniting mankind.

Languages compared together, and considered as objects of the natural history of the mind, and when separated into families according to the analogies existing in their internal structure, have become a rich source of historical knowledge ; and this is probably one of the most brilliant results of modern study in the last sixty or seventy years. From the very fact of their being products of the intellectual force of mankind, they lead us, by means of the elements of their organism, into an obscure distance, unreached by traditionary records. The comparative study of languages shows us that races now separated by vast tracts of land are allied together, and have migrated from one common primitive seat ; it indicates the course and direction of all migrations, and, in tracing the leading epochs of development, recognizes, by means of the more or less changed structure of the language, in the permanence of certain forms, or in the more or less advanced destruction of the formative system, *which* race has retained most nearly the language common to all who had migrated from the general seat of origin. The largest field for such investigations into the ancient condition of language, and, consequently, into the period when the whole family of mankind was, in the strict sense of the word, to be regarded as one living whole, presents itself in the long chain of Indo-Germanic languages, extending from the Ganges to the Iberian extremity of Europe, and from Sicily to the North Cape. The same comparative study of languages leads us also to the native country of certain products, which, from the earliest ages, have constituted important objects of trade and barter. The Sanscrit names of genuine Indian products, as those of rice, cotton, spikenard, and sugar, have, as we find, passed into the language of the Greeks, and, to a certain extent, even into those of Semitic origin.*

* In Sanscrit, rice is *vrihi,* cotton *karpâsa,* sugar *'sarkare,* and spike-

From the above considerations, and the examples by which
they have been illustrated, the comparative study of languages
appears as an important rational means of assistance, by which
scientific and genuinely philological investigations may lead to
a generalization of views regarding the affinity of races, and
their conjectural extension in various directions from one common point of radiation. The rational aids toward the gradual
development of the science of the Cosmos are, therefore, of
very different kinds, viz., investigations into the structure of
languages; the deciphering of ancient inscriptions and historical monuments in hieroglyphics and arrow-headed writing;
the greater perfection of mathematics, especially of that powerful analytic calculus by which the form of the earth, the ebb
and flow of the sea, and the regions of space are brought within
the compass of calculation. To these aids must be further
added the material inventions which have procured for us, as
it were, new organs, sharpened the power of our senses, and
enabled men to enter into a closer communication with terres-
frial forces, and even with the remote regions of space. In
order to enumerate only a few of the instruments whose invention characterizes great epochs in the history of civilization,
I would name the telescope, and its too long-delayed connection with instruments of measurement; the compound microscope, which furnishes us with the means of tracing the conditions of the process of development of organisms, which
Aristotle gracefully designates as "the formative activity, the
source of being;" the compass, and the different contrivances
invented for measuring terrestrial magnetism; the use of the

nard, *nanartha.* See Lassen, *Indische Alterthumskunde,* bd. i., 1843, s.
245, 250, 270, 289, und 538. On '*sarkara* and *kanda* (whence our sugar-
candy), consult my *Prolegomena de Distributione Geographicâ Plantarum,* 1817, p. 211. "Confudisse videntur veteres saccharum verum
cum Tebaschiro Bambusæ, cum quia utraque in arundinibus inveniun-
tur, tum etiam quia vox Sanscradana *scharkara,* quæ hodie (ut Pers.
schakar et Hindost. *schukur*) pro saccharo nostro adhibetur, observante
Boppio, ex auctoritate Amarasinhæ, proprie nil dulce (*madu*) significat,
sed quicquid lapidosum et arenaceum est, ac vel calculum vesicæ.
Verisimile igitur, vocem *scharkara* initio dumtaxat tebaschirum (*saccar
nombu*) indicasse, posterius in saccharum nostrum humilioris arundinis
(*ikschu, kandekschu, kanda*), ex similitudine aspectus translatam esse.
Vox Bambusa, ex *mambu* derivatur; ex *kanda* nostratium voces *candis*
zuckerkand. In *tebaschiro* agnoscitur Persarum *schir,* h. e. lac. Sanscr.
kschiram." The Sanscrit name for *tabaschir* is *tvakkschirâ,* bark-milk;
milk from the bark. See Lassen, bd. i., s. 271–274. Compare, also,
Pott, *Kurdische Studien* in the *Zeitschrift für die Kunde des Morgenlandes,* bd. vii., s. 163–166, and the masterly treatise by Carl Ritter, in
his *Erdkunde von Asien,* bd. vi., 2, s. 232–237.

pendulum as a measure of time; the barometer; the thermometer; hygrometric and electrometric apparatuses; and the polariscope, in its application to the phenomena of colored polarization, in the light of the stars, or in luminous regions of the atmosphere.

The history of the physical contemplation of the universe, which is based, as we have already remarked, on a meditative consideration of natural phenomena, on the connection of great events, and on inventions which enlarge the domain of sensuous perception, can only be presented in a fragmentary and superficial manner, and only in its leading features. I flatter myself with the hope that the brevity of this mode of treatment will enable the reader the more readily to apprehend the spirit in which a picture should be sketched, whose limits it is so difficult to define. Here, as in the picture of nature which is given in the former part of this work, it will be my object to treat the subject, not with the completeness of an individualizing enumeration, but merely by the development of leading ideas, that indicate some of the paths which must be pursued by the physicist in his historical investigations. The knowledge of the connection of events and their causal relations is assumed to be possessed by the reader, and it will consequently be sufficient merely to indicate these events, and determine the influence which they have exercised on the gradual increase of the knowledge of nature as a whole. Completeness, I must again repeat, is neither to be attained, nor is it to be regarded as the object of such an undertaking. In the announcement of the mode in which I propose treating my subject, in order to preserve for the present work its peculiar character, I shall, no doubt, expose myself again to the animadversions of those who think less of what a book contains than of that which, according to their individual views, ought to be found in it. I have purposely been much more circumstantial with reference to the more ancient than the modern portions of history. Where the sources of information are less copious, the difficulty of a proper combination is increased, and the opinions advanced then require to be supported by the testimony of facts less generally known. I would also observe that I have permitted myself to treat my subject with inequality, where the enumeration of individual facts afforded the advantage of imparting greater interest to the narrative.

As the recognition of the unity of the Cosmos began in an intuitive presentiment, and with merely a few actual observations on isolated portions of the domain of nature, it seems in-

cumbent that we should begin our historical representation of
the universe from some definite point of our terrestrial planet.
We will select for this purpose that sea basin around which
have dwelt those nations whose knowledge has formed the
basis of our western civilization, which alone has made an
almost uninterrupted progress. We may indicate the main
streams from which Western Europe has received the elements
of the cultivation and extended views of nature, but amid the
diversity of these streams we are unable to trace one primitive
source. A deep insight into the forces of nature and a recognition of the unity of the Cosmos does not appertain to a so-
called *primitive race :* a term that has been applied, amid
the alternations of historical views, sometimes to a Semitic
race in Northern Chaldea—Arpaxad (the Arrapachitis of
Ptolemy)*—and sometimes to a race of Indians and Iranians,
in the ancient Zend, in the district surrounding the sources
of the Oxus and the Jaxartes.† History, as far as it is based
on human testimony, knows of no *primitive race,* no one primitive seat of civilization, and no primitive physical natural
science whose glory has been dimmed by the destructive barbarism of later ages. The historical inquirer must penetrate
through many superimposed misty strata of symbolical myths
before he can reach that solid foundation where the earliest
germ of human culture has been developed in accordance
with natural laws. In the dimness of antiquity, which constitutes, as it were, the extreme horizon of true historical
knowledge, we see many luminous points, or centers of civilization, simultaneously blending their rays. Among these we
may reckon Egypt at least five thousand years before our
era,‡ Babylon, Nineveh, Kashmir, Iran, and also China, after

---

* Ewald, *Geschichte des Volkes Israel,* bd. i., 1843, s. 332–334; Lassen,
Ind. Alterthumskunde, bd. i., s. 528. Compare Rödiger, in the *Zeit-
schrift für die Kunde des Morgenlandes,* bd. iii., s. 4, on Chaldeans and
Kurds, the latter of whom Strabo terms Kyrti.

† *Bordj,* the water-shed of the Ormuzd, nearly where the chain of the
Thian-schan (or Celestial Mountains), at its western termination, abuts
in veins against the Bolor (Belur-tagh), or rather intersects it, under the
name of the Asferah chain, north of the highland of Pamer (Upa-Mêru,
or country above Meru). Compare Burnouf, *Commentaire sur le Yaçna,*
t. i., p. 239, and Addit., p. clxxxv., with Humboldt, *Asie Centrale,* t. i.,
p. 163 ; t. ii., p. 16, 377–390.

‡ The principal chronological data for Egypt are as follows : " Menes,
3900 B.C. at least, and probably tolerably correct ; 3430, commencement of the fourth dynasty, which included the pyramid builders, Che-
phren-Schafra, Cheops-Chufu, and Mykerinos or Menkera ; 2200, invasion of the Hyksos under the twelfth dynasty, to which belongs Ame-
nemha III., the builder of the original Labyrinth. A thousand years,

the first colony migrated from the northeastern declivity of
the Kuen-lun into the lower river valley of the Hoang-ho.

at least, and probably still more, must be conjectured for the gradual
growth of a civilization which had been completed, and had in part
begun to degenerate, at least 3430 years B.C." (Lepsius, in several
letters to myself, dated March, 1846, and therefore after his return from
his memorable expedition.) Compare, also, Bunsen's *Considerations
on the Commencement of Universal History*, which, strictly defined, is
only a history of recent times, in his ingenious and learned work,
Ægyptens Stelle in der Weltgeschichte, 1845, erstes Buch, s. 11–13.
The historical existence and regular chronology of the Chinese go back
to 2400, and even to 2700 before our era, far beyond Ju to Hoang-ty.
Many literary monuments of the thirteenth century B.C. are extant,
and in the twelfth century B.C. Thscheu-li records the measurement
of the length of the solstitial shadow taken with such exactness by
Tscheu-kung, in the town of Lo-yang, south of the Yellow River, that
Laplace found that it accorded perfectly with the theory of the alteration of the obliquity of the ecliptic, which was only established at the
close of the last century. All suspicion of a measurement of the Earth's
direction derived by calćulating back, falls therefore to the ground of
itself. See Edouard Biot, *Sur la Constitution Politique de la Chine au
12ème Siècle avant notre ère* (1845), p. 3 and 9. The building of Tyre
and of the original temple of Melkarth (the Tyrian Hercules) would,
according to the account which Herodotus received from the priests
(II., 44), reach back 2760 years before our era. Compare, also, Heeren, *Ideen über Politik und Verkehr der Völker*, th. i., 2, 1824, s. 12.
Simplicius calculates, from a notice transmitted by Porphyry, that the
date of the earliest Babylonian astronomical observations which were
known to Aristotle was 1903 years before Alexander the Great; and
Ideler, who is so profound and cautious as a chronologist, considers this
estimate in no way improbable. See his *Handbuch der Chronologie*,
bd. i., s. 207 ; the *Abhandlungen der Berliner Akad. auf das Jahr* 1814, s.
217 ; and Böckh, *Metrol. Untersuchungen über die Masse des Alterthums*,
1838, s. 36. Whether safe historic ground is to be found in India earlier
than 1200 B.C., according to the chronicles of Kashmeer (*Radjataran-
gini*, trad. par Troyer), is a question still involved in obscurity ; while
Megasthenes (*Indica*, ed. Schwanbeck, 1846, p. 50) reckons for 153
kings of the dynasty of Magadha, from Manu to Kandragupta, from
sixty to sixty-four centuries, and the astronomer Aryabhatta places the
beginning of his chronology 3102 B.C. (Lassen, *Ind. Alterthumsk.*, bd.
i., s. 473–505, 507, und 510). In order to give the numbers contained
in this note a higher significance in respect to the history of human
civilization, it will not be superfluous to recall the fact that the destruction of Troy is placed by the Greeks 1184, by Homer 1000 or 950, and
by Cadmus the Milesian, the first historical writer among the Greeks,
524 years before our era. This comparison of epochs proves at what
different periods the desire for an exact record of events and enterprises was awakened among the nations most highly susceptible of culture, and we are involuntarily reminded of the exclamation which
Plato, in the *Timæus*, puts in the mouth of the priests of Sais : " O Solon, O Solon ! ye Greeks still remain ever children ; nowhere in Hellas
is there an aged man. Your souls are ever youthful ; ye have in them
no knowledge of antiquity, no ancient belief, no wisdom grown venerable by age."

These central points involuntarily remind us of the largest among the sparkling stars of the firmament, those eternal suns in the regions of space, the intensity of whose brightness we certainly know, although it is only in the case of a few that we have been able to arrive at any certain knowledge regarding the relative distances which separate them from our planet.

The hypothesis regarding the physical knowledge supposed to have been revealed to the primitive races of men—the natural philosophy ascribed to savage nations, and since obscured by civilization—belongs to a sphere of science, or, rather, of belief, which is foreign to the object of the present work. We find this belief deeply rooted in the most ancient Indian doctrine of Krischna.* "Truth was originally implanted in mankind, but, having been suffered gradually to slumber, it was finally forgotten, knowledge returning to us since that period as a recollection." We will not attempt to decide the question whether the races, which we at present term savage, are all in a condition of original wildness, or whether, as the structure of their languages often allows of our conjecturing, many among them may not be tribes that have degenerated into a wild state, remaining as the scattered fragments saved from the wreck of a civilization that was early lost. A more intimate acquaintance with these so-called children of nature reveals no traces of that superiority of knowledge regarding terrestrial forces which a love of the marvelous has led men to ascribe to these rude nations. A vague and terror-stricken feeling of the unity of natural forces is no doubt awakened in the breast of the savage, but such a feeling has nothing in common with the attempt to prove, by the power of thought, the connection that exists among all phenomena. True cosmical views are the result of observation and ideal combination, and of a long-continued communion with the external world ; nor are they a work of a single people, but the fruits yielded by reciprocal communication, and by a great, if not general, intercourse between different nations.

As, in the considerations on the reflection of the external world on the powers of the imagination at the beginning of this section of the present work, I selected from the general history of literature examples illustrative of the expression of an animated feeling for nature, so, in *the history of the contemplation of the universe*, I would likewise bring forward from the general history of civilization whatever may serve to

* Wilhelm von Humboldt, *Ueber eine Episode des Maha-Bharata*, in his *Gesammelte Werke*, bd. i., s. 73.

indicate the progress that has been made toward the recognition of the unity of nature. Both portions—not separated arbitrarily, but by determined principles—have the same relations to one another as the studies from which they have been borrowed. The history of the civilization of mankind comprises in itself the history of the fundamental powers of the human mind, and also, therefore, of the works in which these powers have been variously displayed in the different departments of literature and art. In a similar manner, we recognize in the depth and animation of the sentiment of love for nature, which we have delineated according to its various manifestations at different epochs and among different races of men, active means of inducement toward a more careful observation of phenomena, and a more earnest investigation of their cosmical connection.

Owing to the diversity of the streams which have in the course of ages so unequally diffused the elements of a more extended knowledge of nature over the whole earth, it will be most expedient, as we have already observed, to start in the history of the contemplation of the external world from a single group of nations, and for this object I select the one from which our present scientific cultivation, and, indeed, that of the whole of Western Europe, has originated. The mental cultivation of the Greeks and Romans must certainly be regarded as very recent in comparison with that of the Egyptians, Chinese, and Indians ; but all that the Greeks and Romans received from the east and south, blended with what they themselves produced and developed, has been uninterruptedly propagated on our European soil, notwithstanding the continual alternation of historical events, and the admixture of foreign immigrating races. In those regions in which a much greater degree of knowledge existed thousands of years earlier, a destructive barbarism has either wholly darkened the pre-existing enlightenment, or, where a stable and complex system of government has been preserved, together with a maintenance of ancient customs, as is the case in China, advancement in science or the industrial arts has been very inconsiderable, while the almost total absence of a free intercourse with the rest of the world has interposed an insuperable barrier to the generalization of views. The cultivated nations of Europe, and their descendants who have been transplanted to other continents, may be said, by the gigantic extension of their maritime expeditions to the remotest seas, to be familiarized with the most distant shores ; and those countries which they do not already possess, they may threaten. In the almost uninterrupted course of the knowledge transmitted to them, and in their ancient scientific nomenclature, we may trace, as the guiding points of the history of the human race, recollections of the various channels through which important inventions, or, at any rate, their germs, have been conveyed to the nations of Europe; thus from Eastern Asia has flowed the knowledge of the direction and declination of a freely-suspended magnetic rod; from Phœnicia and Egypt the knowledge of chemical preparations, as glass, animal and vegetable dyes, and metallic oxyds; and from India the general use of *position* in determining the increased values of a few numerical signs.

Since civilization has left its most ancient seat within the tropics or the sub-tropical zone, it has remained permanently settled in the portion of the earth whose northern regions are less cold than those of Asia and America under the same latitude. The continent of Europe may be regarded as a western peninsula of Asia, and I have already observed how much general civilization is favored by the mildness of its climate, and how much it owes to the circumstances of its variously articulated form, first noticed by Strabo; to its position in respect to Africa, which extends so far into the equatorial zone, and to the prevalence of the west winds, which are warm winds in winter, owing to their passing over the surface of the ocean. The physical character of Europe has opposed fewer obstacles to the diffusion of civilization than are presented in Asia and Africa, where far-extending parallel ranges of mountain chains, elevated plateaux, and sandy deserts interpose almost impassable barriers between different nations.

We will therefore start in our enumeration of the principal momenta that characterize the history of the physical consideration of the universe from a portion of the earth which is, perhaps, more highly favored than any other, owing to its geographical position, and its constant intercourse with other countries, by means of which the cosmical views of nations experience so marked a degree of enlargement.

PRINCIPAL MOMENTA THAT HAVE INFLUENCED THE HIS-
TORY OF THE PHYSICAL CONTEMPLATION OF THE UNI-
VERSE.

THE MEDITERRANEAN CONSIDERED AS THE STARTING-POINT FOR
THE REPRESENTATION OF THE RELATIONS WHICH HAVE LAID THE
FOUNDATION OF THE GRADUAL EXTENSION OF THE IDEA OF THE
COSMOS.—SUCCESSION OF THIS RELATION TO THE EARLIEST CUL-
TIVATION AMONG HELLENIC NATIONS.—ATTEMPTS AT DISTANT
MARITIME NAVIGATION TOWARD THE NORTHEAST (BY THE ARGO-
NAUTS); TOWARD THE SOUTH (TO OPHIR); TOWARD THE WEST
(BY COLÆUS OF SAMOS).

PLATO, in his *Phædo*, describes the narrow limits of the
Mediterranean in a manner that accords with the spirit of enlarged cosmical views.* "We, who inhabit the region extending from Phasis to the Pillars of Hercules, occupy only a small
portion of the earth," he writes, "where we have settled ourselves round the inner sea like ants or frogs round a swamp."
This narrow basin, on the borders of which Egyptian, Phœnician, and Hellenic nations flourished and attained to a high
degree of civilization, is the point from which the most important historical events have proceeded, no less than the colonization of vast territories in Africa and Asia, and those
maritime expeditions which have led to the discovery of the
whole western hemisphere of the globe.

The Mediterranean shows in its present configuration the
traces of an earlier subdivision into three contiguous smaller
closed basins.†

The Ægean is bounded to the south by the curved line
formed by the Carian coast of Asia Minor, and the islands of
Rhodes, Crete, and Cerigo, and terminating at the Pelopon-

* Plato, *Phædo*, p. 109, B. (Compare *Herod.*, ii., 21.) Cleomedes
supposed that the surface of the earth was depressed in the middle, in
order to receive the Mediterranean (Voss, *Crit. Blätter*, bd. ii., 1828,
s. 144 und 150).
† I first developed this idea in my *Rel. Hist. du Voyage aux Régions
Equinoxiales*, t. iii., p. 236, and in the *Examen Crit. de l'Hist. de la
Géogr. au 15ème Siècle*, t. i., p. 36–38. See, also, Otfried Müller, in
the *Göttingische gelehrte Anzeigen*, 1838, bd. i., s. 376. The most western basin, which I name generally the Tyrrhenian, includes, according
to Strabo, the Iberian, Ligurian, and Sardinian Seas. The Syrtic basin,
east of Sicily, includes the Ausonian or Siculian, the Libyan, and the
Ionian Seas. The southern and southwestern part of the Ægean Sea
was called Cretic, Saronic, and Myrtoic. The remarkable passage in
Aristot.; *De Mundo*, cap. iii. (p. 393, Bekk.), refers only to the bay-like
configuration of the coasts of the Mediterranean, and its effect on the
ocean flowing into it.

nesus, not far from the Promontory of Malea. Further westward is the Ionian Sea, the Syrtic basin, in which lies Malta.
The western extremity of Sicily here approaches within forty-
eight geographical miles of the coast of Africa. The sudden
appearance and short continuance of the upheaved volcanic
island of Ferdinandea in 1831, to the southwest of the calcareous rocks of Sciacca, seem to indicate an effort of nature to
reclose the Syrtic basin between Cape Grantola, Adventure
Bank, examined by Captain Smyth, Pantellaria, and the African Cape Bon, and thus to divide it from the third western
basin, the Tyrrhenian. This last sea receives the ocean
which enters the Pillars of Hercules from the west, and surrounds Sardinia, the Balearic Islands, and the small volcanic
group of the Spanish Columbratæ.

This triple constriction of the Mediterranean has exercised
a great influence on the earliest limitations, and the subsequent extension of Phœnician and Greek voyages of discovery.
The latter were long limited to the Ægean and Syrtic Seas.
In the Homeric times the continent of Italy was still an "unknown land." The Phocæans opened the Tyrrhenian basin
west of Sicily, and Tartessian mariners reached the Pillars
of Hercules. It must not be forgotten that Carthage was
founded at the boundary of the Tyrrhenian and Syrtic basins.
The physical configuration of the coast-line influenced the
course of events, the direction of nautical undertakings, and
the changes in the dominion of the sea ; and the latter reacted
again on the enlargement of the sphere of ideas.

The northern shore of the Mediterranean possesses the advantage of being more richly and variously articulated than
the southern or Libyan shore, and this was, according to
Strabo, noticed already by Eratosthenes.* Here we find
three peninsulas, the Iberian, the Italian, and the Hellenic,
which, owing to their various and deeply-indented contour,
form, together with the neighboring islands and the opposite

* Humboldt, *Asie Centrale*, t. i., p. 67. The two remarkable passages of Strabo are as follows : "Eratosthenes enumerates three, and
Polybius five points of land in which Europe terminates. The first-
mentioned of these writers names the projecting point which extends
to the Pillars of Hercules, on which Iberia is situated ; next, that which
terminates at the Sicilian Straits, to which Italy belongs ; and, thirdly,
that which extends to Malea, and comprises all the nations between
the Adriatic, the Euxine, and the Tanais" (lib. ii., p. 109). "We begin with Europe because it is of irregular form, and is the quarter most
favorable to the mental and social ennoblement of men. It is habitable
in all parts except some districts near the Tanais, which are not peopled
on account of the cold" (lib. ii., p. 126).

coasts, many straits and isthmuses. Such a configuration of continents and of islands that have been partly severed and partly upheaved by volcanic agency in rows or in far projecting fissures, early led to geognostic views regarding eruptions, terrestrial revolutions, and outpourings of the swollen higher seas into those below them. The Euxine, the Dardanelles, the Straits of Gades, and the Mediterranean, with its numerous islands, were well fitted to draw attention to such a system of sluices. The Orphic Argonaut, who probably lived in Christian times, has interwoven old mythical narrations in his composition. He sings of the division of the ancient Lyktonia into separate islands, "when the dark-haired Poseidon, in anger with Father Kronion, struck Lyktonia with the golden trident." Similar fancies, which may often certainly have sprung from an imperfect knowledge of geographical relations, were frequently elaborated in the erudite Alexandrian school, which was so partial to every thing connected with antiquity. Whether the myth of the breaking up of Atlantis be a vague and western reflection of that of Lyktonia, as I have elsewhere shown to be probable, or whether, according to Otfried Müller, "the destruction of Lyktonia (Leukonia) refers to the Samothracian legend of a great flood which changed the form of that district,"* is a question that it is unnecessary here to decide.

* Ukert, *Geogr. der Griechen und Römer*, th. 1., abth. 2, s. 345–348, and th. ii., abth. 1, s. 194; Johannes v. Müller, *Werke*, bd. i., s. 38; Humboldt, *Examen Critique*, t. i., p. 112 and 171; Otfried Müller, *Minyer*, s. 64; and the latter, again, in a too favorable critique of my memoir on the *Mythische Geographie der Griechen* (*Gött. gelehrte Anzeigen*, 1838). I expressed myself as follows: "In raising questions which are of so great importance with respect to philological studies, I can not wholly pass over all mention of that which belongs less to the description of the actual world than to the cycle of mythical geography. It is the same with space as with time. History can not be treated from a philosophical point of view, if the heroic ages be wholly lost sight of. National myths, when blended with history and geography, can not be regarded as appertaining wholly to the domain of the ideal world. Although vagueness is one of its distinctive attributes, and symbols cover reality by a more or less thick vail, myths, when intimately connected together, nevertheless reveal the ancient source from which the earliest glimpses of cosmography and physical science have been derived. The facts recorded in primitive history and geography are not mere ingenious fables, but rather the reflection of the opinion generally admitted regarding the actual world." The great investigator of antiquity (whose opinion is so favorable to me, and whose early death in the land of Greece, on which he had bestowed such profound and varied research, has been universally lamented) considered, on the contrary, that "the chief part of the poetic idea of the earth, as it oc-

But that which, as has already been frequently remarked, has rendered the geographical position of the Mediterranean most beneficial in its influence on the intercourse of nations, is the proximity of the eastern continent, where it projects into the peninsula of Asia Minor; the number of islands in the Ægean Sea, which have served as a means for facilitating the spread of civilization;* and the fissure between Arabia, Egypt, and Abyssinia, through which the great Indian Ocean penetrates under the name of the Arabian Gulf or the Red Sea, and which is separated by a narrow isthmus from the Delta of the Nile and the southeastern coasts of the Mediterranean. By means of all these geographical relations, the influence of the sea as a connecting element was speedily manifested in the growing power of the Phœnicians, and subsequently in that of the Hellenic nations, and in the rapid extension of the sphere of general ideas. Civilization, in its early seats in Egypt, on the Euphrates, and the Tigris, in Indian Pentapotamia and China, had been limited to lands rich in navigable rivers; the case was different, however, in Phœnicia and Hellas. The active life of the Greeks, especially of the Ionian race, and their early predilection for maritime expeditions, found a rich field for its development in the remarkable configuration of the Mediterranean, and in its relative position to the oceans situated to the south and west.

curs in Greek poetry, is by no means to be ascribed to actual experience, which may have been invested, from credulity and love of the marvelous, with a fabulous character, as has been conjectured especially with respect to the Phœnician maritime legends, but rather that it was to be traced to the roots of the images which lie in certain ideal presuppositions and requirements of the feelings, on which a *true geographical knowledge has only gradually begun to work.* From this fact there has often resulted the interesting phenomenon that purely subjective creations of a fancy guided by certain ideas become almost imperceptibly blended with actual countries and well-known objects of scientific geography. From these considerations, it may be inferred that all genuine or artificially mythical pictures of the imagination belong, in their proper ground-work, to an ideal world, and have no original connection with the actual extension of the knowledge of the earth, or of navigation beyond the Pillars of Hercules." The opinion expressed by me in the French work agreed more fully with the earlier views of Ötfried Müller, for, in the *Prolegomenon zu einer wissenschaftlichen Mythologie*, s. 68 und 109, he said very distinctly that, "in mythical narratives of that which is done and that which is imagined, the real and the ideal are most closely connected together." See, also, on the Atlantis and Lyktonia, Martin, *Etudes sur le Timée de Platon*, t. i., p. 293–326.

* *Naxos*, by Ernst Curtius, 1846, s. 11; Droysen, *Geschichte der Bildung des Hellenistischen Staatensystems*, 1843, s. 4–9.

The existence of the Arabian Gulf as the result of the ir-ruption of the Indian Ocean through the Straits of Bab-el-Mandeb belongs to a series of great physical phenomena, which could alone have been revealed to us by modern geognosy. The European continent has its main axis directed from northeast to southwest; but almost at right angles to this direction there is a system of fissures, which have given occasion partly to a penetration of sea-water, and partly to the elevation of parallel mountain chains. This inverse line of strike, directed from the southeast to the northwest, is discernible from the Indian Ocean to the efflux of the Elbe in Northern Germany; in the Red Sea, the southern part of which is inclosed on both sides by volcanic rocks; in the Persian Gulf, with the deep valleys of the double streams of the Euphrates and the Tigris; in the Zagros chain in Luristan; in the mountain chains of Hellas, and in the neighboring islands of the Archipelago; and, lastly, in the Adriatic Sea, and the Dalmatian calcareous Alps. The intersection* of these two systems of geodetic lines directed from N.E. to S.W., and from S.E. to N.W. (the latter of which I consider to be the more recent of the two), and whose cause must undoubt-edly be traced to disturbances in the interior of our planet, has exercised the most important influence on the destiny of man-kind, and in facilitating intercourse among different nations. This relative position, and the unequal degrees of heat experi-enced by Eastern Africa, Arabia, and the peninsula of West-ern India at different periods of the year, occasion a regular alternation of currents of air (monsoons), favoring navigation to the Myrrhifera Regio of the Adramites in Southern Arabia, to the Persian Gulf, India, and Ceylon; for, at the season of the year (from April and May to October) when north winds are prevailing in the Red Sea, the southwest monsoon is blowing from Eastern Africa to the coast of Malabar, while the northeast monsoon (from October to April), which favors the return passage, corresponds with the period of the south winds between the Straits of Bab-el-Mandeb and the Isthmus of Suez.

After having sketched that portion of the earth to which foreign elements of civilization and geographical knowledge might have been conveyed to the Greeks from so many different directions, we will first turn to the consideration of those na-tions inhabiting the coasts of the Mediterranean who enjoyed

* Leopold von Buch, *Ueber die Geognostischen Systeme von Deutsch-land*, s. xi.; Humboldt, *Asie Centrale*, t. i., p. 284–286.

an early and distinguished degree of civilization, viz., the
Egyptians, the Phœnicians, with their north and west African
colonies, and the Etrurians. Immigration and commercial
intercourse have here exercised the most powerful influence.
The more our historical horizon has been extended in modern
times by the discovery of monuments and inscriptions, as well
as by philosophical investigation of languages, the more varied
does the influence appear which the Greeks in the earliest
ages experienced from Lycia and the district surrounding the
Euphrates, and from the Phrygians allied to Thracian races.

In the Valley of the Nile, which plays so conspicuous a part
in the history of mankind, " there are well-authenticated car-
touches of the kings as far back as the beginning of the fourth
dynasty of Manetho, in which are included the builders of the
Pyramids of Giseh (Chephren or Schafra, Cheops-Chufu, and
Menkera or Mencheres)." I here avail myself of the account
of the most recent investigations of Lepsius,* whose expedition has resulted in throwing much important light on the
whole of antiquity. " The dynasty of Manetho began more
than thirty-four centuries before our Christian era, and twenty-
three centuries before the Doric immigration of the Heraclidæ
into the Peloponnesus.† The great stone pyramids of Daschur,
somewhat to the south of Giseh and Sakara, are considered by
Lepsius to be the work of the third dynasty. Sculptural inscriptions have been discovered on the blocks of which they
are composed, but as yet no names of kings. The last dynasty
of the ancient kingdom, which terminated at the invasion of
the Hyksos, and probably 1200 years before Homer, was the
twelfth of Manetho, and the one to which belonged Ame-
nemha III., the prince who caused the original labyrinth to
be constructed, and who formed Lake Mœris artificially by
means of excavations and large dikes of earth running north
and west. After the expulsion of the Hyksos, the new kingdom began under the eighteenth dynasty (1600 years B.C.).
Rameses Miamoun the Great (Rameses II.) was the second
ruler of the nineteenth dynasty. The sculptured delineations
which perpetuate his victories were explained to Germanicus

* All that relates to Egyptian chronology and history, and which is
distinguished in the text by marks of quotation, is based on manuscript
communications which I received from my friend Professor Lepsius,
in March, 1846.

† I place the Doric immigration into the Peloponnesus 328 years
before the first Olympiad, agreeing in this respect with Otfried Müller
(*Dorier*, abth. ii., s. 436).

by the priests of Thebes.* He is noticed by Herodotus under the name of Sesostris, which is probably owing to a confusion with the almost equally victorious and powerful conqueror Seti (Setos), who was the father of Rameses II."

I have deemed it necessary to mention these few points of chronology, in order that where we meet with solid historical ground, we may pause to determine the relative ages of great events in Egypt, Phœnicia, and Greece. As I have already briefly described the geographical relations of the Mediterranean, I would now also call attention to the number of centuries that intervened between the epoch of human civilization in the Valley of the Nile and its subsequent transmission to Greece; for, without such simultaneous reference to space and time, it would be impossible, from the nature of our mental faculties, to form to ourselves any clear and satisfactory picture of history.

Civilization, which was early awakened and arbitrarily modeled in the Valley of the Nile, owing to the mental requirements of the people, the peculiar physical charaċter of the country, and its hierarchical and political institutions, excited there, as in every other portion of the earth, an impulse toward increased intercourse with other nations, and a tendency to undertake distant expeditions and establish colonies. But the records preserved to us by history and monumental representations testify only to transitory conquests on land, and to few extensive voyages of the Egyptians themselves. This anciently and highly civilized race appears to have exercised a less permanent influence on foreigners than many other smaller nations less stationary in their habits. The national cultivation of the Egyptians, which, from the long course of its development, was more favorable to masses than to individuals, appears isolated in space, and has, on that account, probably remained devoid of any beneficial result for the extension of cosmical views. Rameses Miamoun (who lived from 1388 to 1322 B.C., and therefore 600 years before the first Olympiad of Corœbus) undertook distant expeditions, having, according to the testimony of Herodotus, penetrated into Ethiopia (where Lepsius believed that he found his most southern architectural works at Mount Barkal) through Palestinian Syria, and crossed from Asia Minor to Europe, through the

* Tac., Annal., ii., 59. In the Papyrus of Sallier (Campagnes de Sésostris) Champollion found the names of the Javani or Iouni, and that of the Luki (Ionians and Lycians?). See Bunsen, Ægypten, buch. i., s. 60.

lands of the Scythians and Thracians, to Colchis and the
River Phasis, where those of his soldiers who were weary of
their wanderings remained as settlers. Rameses was also the
first, according to the priests, " who, by means of his long ships,
subjected to his dominion the people who inhabited the coasts
of the Erythrean Sea. After this achievement, he continued
his course until he came to a sea which was not navigable,
owing to its shallowness."* Diodorus expressly says that Sesostris (Rameses the Great) penetrated into India beyond the
Ganges, and that he brought captives back with him from
Babylon. " The only certain fact with reference to Egyptian
navigation is, that, from the earliest ages, not only the Nile,
but the Arabian Gulf, was navigated. The celebrated copper mines near Wadi-Magaha, on the peninsula of Sinai, were
worked as early as the fourth dynasty, under Cheops-Chufu.
The sculptural inscriptions of Hamamat on the Cosseir road,
which connected the Valley of the Nile with the western
coasts of the Red Sea, go back as far as the sixth dynasty.
Attempts were made under Rameses the Great† to form the

---

* Herod., ii., 102 and 103; Diod. Sic., i., 55 and 56. Of the memorial pillars (στῆλαι) which Rameses Miamoun set up as tokens of victory
in the countries through which he passed, Herodotus expressly names
three (ii., 106): "one in Palestinian Syria, and two in Ionia, on the
road from the Ephesian territory to Phocæa, and from Sardis to Smyrna." A rock inscription, in which the name of Rameses is frequently
met with, has been found near the Lycus in Syria, not far from Beirut
(Berytus), as well as another ruder one in the Valley of Karabel, near
Nymphio, and, according to Lepsius, on the road from the Ephesian
territory to Phocæa. Lepsius, in the *Ann. dell' Institute Archeol.*, vol.
x., 1838, p. 12; and in his letter from Smyrna, Dec., 1845, published
in the *Archäologische Zeitung*, Mai, 1846, No. 41, s. 271–280. Kiepert,
in the same periodical, 1843, No. 3, s. 35. Whether, as Heeren believes (see in his *Geschichte der Staaten des Alterthums*, 1828, s. 76),
the great conqueror penetrated as far as Persia and Western India, " as
Western Asia did not then contain any great empire" (the building of
Assyrian Nineveh is placed only 1230 B.C.), is a question that will undoubtedly soon be settled from the rapidly advancing discoveries now
made in archæology and phonetic languages. Strabo (lib. xvi., p. 760)
speaks of a memorial pillar of Sesostris near the Strait of Deire, now
known as Bab-el-Mandeb. It is, moreover, also very probable, that
even in " the Old Kingdom," above 900 years before Rameses Miamoun,
Egyptian kings may have undertaken similar military expeditions into
Asia. It was under Setos II., the Pharaoh belonging to the nineteenth
dynasty, and the second successor of the great Rameses Miamoun, that
Moses went out of Egypt, and this, according to the researches of Lepsius, was about 1300 years before our era.

† According to Aristotle, Strabo, and Pliny, but not according to
Herodotus. See Letronne, in the *Révue des deux Mondes*, 1841, t.
xxvii., p. 219; and Droysen, *Bildung des Hellenist. Staatensystems*, s. 735.

canal from Suez, probably for the purpose of facilitating intercourse with the land of the Arabian copper mines." More considerable maritime expeditions, as, for instance, the frequently contested, but not, I think, improbable* circumnavigation of Africa under Neku II. (611–595 B.C.), were confided to Phœnician vessels. About the same period or a little earlier, under Neku's father, Psammitich (Psemetek), and somewhat later, after the termination of the civil war under Amasis (Aahmes), Greek mercenaries, by their settlement at Naucratia, laid the foundation of a permanent foreign commerce, and by the admission of new elements, opened the way for the gradual penetration of Hellenism into Lower Egypt. Thus was introduced a germ of mental freedom and of greater independence of local influences—a germ which was rapidly

* To the important opinions of Rennell, Heeren, and Sprengel, who are inclined to believe in the reality of the circumnavigation of Libya, we must now add that of a most profound philologist, Etienne Quatremère (*Mémoires de l'Acad. des Inscriptions*, t. xv., Part ii., 1845, p. 380–388). The most convincing argument for the truth of the report of Herodotus (iv., 42) appears to me to be the observation which seems to him so incredible, viz., " that the mariners who sailed round Libya (from east to west) had the sun on their right hand." In the Mediterranean, in sailing from east to west, from Tyre to Gadeira, the sun at noon was seen to the left only. A knowledge of the possibility of such a navigation must have existed in Egypt previous to the time of Neku II. (Nechos), as Herodotus makes him distinctly command the Phœnicians "to return to Egypt through the passage of the Pillars of Hercules." It is singular that Strabo, who (lib. ii., p. 98) discusses at such length the attempted circumnavigation of Eudoxus of Cyzicus under Cleopatra, and mentions fragments of a ship from Gadeira which were found on the Ethiopian (eastern) shore, considers the accounts given of the circumnavigations actually accomplished as Bergaic fables (lib. ii., p. 100); but he does not deny the possibility of the circumnavigation itself (lib. i., p. 38), and declares that from the east to the west there is but little that remains to its completion (lib. i., p. 4). Strabo by no means agreed to the extraordinary isthmus hypothesis of Hipparchus and Marinus of Tyre, according to which Eastern Africa is joined to the southeast end of Asia, and the Indian Ocean converted into a *Mediterranean Sea*. (Humboldt, *Examen Crit. de l'Hist. de la Géographie*, t. i., p. 139–142, 145, 161, and 229; t. ii., p. 370–373). Strabo quotes Herodotus, but does not name Nechos, whose expedition he confounds with one sent by Darius round Southern Persia and Arabia (Herod., iv., 44). Gosselin even proposed, somewhat too boldly, to change the reading from Darius to Nechos. A counterpart for the horse's head of the ship of Gadeira, which Eudoxus is said to have exhibited in a market-place in Egypt, occurs in the remains of a ship of the Red Sea, which was brought to the coast of Crete by westerly currents, according to the account of a very trustworthy Arabian historian (Masudi, in the *Morudj-al-dzeheb*, Quatremère, p. 389, and Reinaud, *Relation des Voyages dans l'Inde*, 1845, t. i., p. xvi., and t. ii., p. 46).

and powerfully developed during the period of the new cosmical views that succeeded the Macedonian conquest. The opening of the Egyptian ports under Psammitich is an event of very great importance, as the country up to that period, at least at its northern extremity, had for a long time been completely closed to strangers, as Japan is at the present day.*

In our enumeration of the non-Hellenic civilized nations who dwelt around the basin of the Mediterranean—the most ancient seat and the starting point of our mental cultivation—we must rank the Phœnicians next to the Egyptians. This race is to be regarded as the most active in maintaining intercourse between the nations from the Indian Ocean to the west and north of the Old Continent. Although circumscribed in many spheres of mental cultivation, and less familiar with the fine arts than with mechanics, and not endowed with the grand form of creative genius common to the more highly-gifted inhabitants of the Valley of the Nile, the Phœnicians, as an adventurous and commercial race, and especially by the establishment of colonies (one of which far surpassed the parent city in political power), exerted an influence on the course of ideas, and on the diversity and number of cosmical views, earlier than all the other nations inhabiting the coasts of the Mediterranean. The Phœnicians made use of Babylonian weights and measures,† and, at least since the Persian dominion, employed stamped metallic coinage as a monetary currency, which, strangely enough, was not known in the artificially-arranged political institutions of the highly-cultivated Egyptians. But that by which the Phœnicians contributed most powerfully to the civilization of the nations with which they came in contact was the general spread of alphabetical writing, which they had themselves employed for a long period. Although the whole mythical relation of the colony of Cadmus in Bœotia remains buried in obscurity, it is not the less certain that the Hellenes obtained the alphabetical characters long known as Phœnician symbols by means of the commercial in-

---

* Diod., lib. i., cap. 67, 10; Herod., ii., 154, 178, and 182. On the probability of the existence of intercourse between Egypt and Greece, before the time of Psammetichus, see the ingenious observations of Ludwig Ross, in *Hellenika*, where he expresses himself as follows, bd. i., 1846, s. v. and x. "In the times immediately preceding Psammetichus, there was in both countries a period of internal disturbance, which must necessarily have brought about a diminution and partial interruption of intercourse."

† Böckh, *Meterologische Untersuchungen über Gewichte, Münzfüsse und Masse des Alterthums in ihrem Zusammenhang*, 1838, s. 12 und 273.

tercourse subsisting between the Ionians and the Phœnicians.*
According to the views which, since Champollion's great discovery, have been generally adopted regarding the earlier condition of the development of alphabetical writing, the Phœnician as well as the Semitic characters are to be regarded as a
phonetic alphabet, that has originated from pictorial writing,
and as one in which the ideal signification of the symbols is
wholly disregarded, and the characters are considered as mere
signs of sounds. Such a phonetic alphabet was, from its very
nature and fundamental character, *syllabic*, and perfectly able
to satisfy all requirements of a graphical representation of
the phonetic system of a language. "As the Semitic written
characters," says Lepsius, in his treatise on alphabets, "passed into Europe to Indo-Germanic nations, who showed throughout a much stronger tendency to define strictly between vowels
and consonants, and were by that means led to ascribe a higher significance to the vowels in their languages, important and
lasting modifications were effected in these syllabic alphabets."†
The endeavor to do away with syllabic characters was very
strikingly manifested among the Greeks. The transmission
of Phœnician signs not only facilitated commercial intercourse
among the races inhabiting almost all the coasts of the Mediterranean, and even the northwest coast of Africa, by forming a bond of union that embraced many civilized nations,
but these alphabetical characters, when generalized by their
graphical flexibility, were destined to be attended by even
higher results. They became the means of conveying, as an
imperishable treasure, to the latest posterity, those noble fruits
developed by the Hellenic races in the different departments
of the intellect, the feelings, and the inquiring and creative
faculties of the imagination.

The share taken by the Phœnicians in increasing the elements of cosmical contemplation was not, however, limited
to the excitement of indirect inducements, for they widened
the domain of knowledge in several directions by independent
inventions of their own. A state of industrial prosperity, based
on an extensive maritime commerce, and on the enterprise
manifested at Sidon in the manufacture of white and colored

* See the passages collected in Otfried Müller's *Minyer*, s. 115, and
in his *Dorier*, abth. i., s. 129; Franz, *Elementa Epigraphices Græcæ*,
1840, p. 13, 32, and 34.

† Lepsius, in his memoir, *Ueber die Anordnung und Verwandtschaft
des Semitischen, Indischen, Alt-Persischen, Alt-Ægyptischen und Æthio-
pischen Alphabets*, 1836, s. 23, 28, und 57; Gesenius, *Scripturæ Phœ-
niciæ Monumenta*, 1837, p. 17.

glass-wares, tissues, and purple dyes, necessarily led to advancement in mathematical and chemical knowledge, and more particularly in the technical arts. "The Sidonians," writes Strabo, "are described as industrious inquirers in astronomy, as well as in the science of numbers, to which they have been led by their skill in arithmetical calculation, and in navigating their vessels by night, both of which are indispensable to commerce and maritime intercourse."* In order to give some idea of the extent of the globe opened by the navigation and caravan trade of the Phœnicians, we will mention the colonies in the Euxine, on the Bithynian shore (Pronectus and Bithynium), which were probably settled at a very early age; the Cyclades, and several islands of the Ægean Sea, first known at the time of the Homeric bard; the south of Spain, rich in silver (Tartessus and Gades); the north of Africa, west of the Lesser Syrtis (Utica, Hadrumetum, and Carthage); the tin and amber lands of the north of Europe;†

* Strabo, lib. xvi., p. 757.

† The locality of the "land of tin" (Britain and the Scilly Islands) is more easily determined than that of the "amber coast;" for it appears very improbable that the old Greek denomination κασσιτερος, which was already in use in the Homeric times, is to be derived from a mountain in the southwest of Spain, called Mount Cassius, celebrated for its tin ore, and which Avienus, who was well acquainted with the country, placed between Gaddir and the mouth of a small southern Iberus (Ukert, *Geogr. der Griechen und Romer*, theil ii., abth. i., s. 479). Kassiteros is the ancient Indian Sanscrit word *kastíra*. *Dän* in Icelandic; *zinn* in German; *tin* in English and Danish; and *tenn* in Swedish, are rendered, in the Malay and Javanese language, by *timah*; a similarity of sound which calls to mind that of the old German word *glessum* (the name applied to transparent amber), with the modern German *glas*, glass. The names of wares and articles of commerce pass from one nation to another, and into the most different families of languages. Through the intercourse which the Phœnicians maintained with the eastern coast of India, by means of their factories in the Persian Gulf, the Sanscrit word *kastira*, which expressed so useful a product of Further India, and still exists among the old Aramæic idioms in the Arabian word *kasdir*, may have become known to the Greeks even before Albion and the British Cassiterides had been visited (Aug. Wilh. v. Schlegel, in the *Indische Bibliothek*, bd. ii., s. 393; Benfey, *Indien*, s. 307; Pott, *Etymol. Forschungen*, th. ii., s. 414; Lassen, *Indische Alterthumskunde*, bd. i., s. 239). A name often becomes a historical monument, and the etymological analysis of languages, however it may be derided, is attended by valuable results. The ancients were also acquainted with the existence of tin—one of the rarest metals—in the country of the Artabri and the Callaici, in the northwest part of the Iberian continent (Strabo, lib. iii., p. 147; Plin., xxxiv., c. 16), which was nearer of access than the Cassiterides (Œstrymnides of Avienus), from the Mediterranean. When, before embarking for the Canaries, I was in Galicia in 1799, mining operations, although of very inferior

and two commercial factories in the Persian Gulf* (the Baharian islands, Tylos and Aradus).

The amber trade, which was probably directed first to the west Cimbrian shores,† and subsequently to the land of the

---

nature, were still carried on in the granitic mountains (see my *Rel. Hist.*, t. i., p. 51 and 53). The occurrence of tin is of some geognostic importance, on account of the former connection of Galicia, the peninsula of Brittany, and Cornwall.

* Etienne Quatremère, op. cit., p. 363–370.

† The opinion early expressed (see Heinzen's *Neue Kielisches Magazin*, th. ii., 1787, s. 339; Sprengel, *Gesch. der Geogr. Entdeckungen*, 1792, s. 51; Voss, *Krit. Blätter*, bd. ii., s. 392–403) that amber was brought by sea at first only from the west Cimbrian coast, and that it reached the Mediterranean chiefly by land, being brought across the intervening countries by means of inland barter, continues to gain in validity. The most thorough and acute investigation of this subject is contained in Ukert's memoir *Ueber das Electrum*, in *Die Zeitschrift für Alterthumswissenschaft*, Jahr 1838, No. 52–55, s. 425–452. (Compare with it the same author's *Geographie der Griechen und Römer*, th. ii., abth. 2, 1832, s. 25–36; th. iii., i., 1843, s. 86, 175, 182, 320, und 349.) The Massilians, who, under Pytheas, advanced, according to Heeren, after the Phœnicians, as far as the Baltic, hardly penetrated beyond the mouths of the Weser and the Elbe. Pliny (iv., 16) placed the amber islands (Glessaria, also called Austrania) decidedly west of the Cimbrian promontory, in the German Sea; and the connection with the expedition of Germanicus sufficiently teaches us that the island signified is not in the Baltic. The great effect of the ebb and flood tides in the estuaries which throw up amber, where, according to the expression of Servius, "mare vicissim tum accedit, tum recedit," applies to the coasts between the Helder and the Cimbrian Peninsula, but not to the Baltic, in which the island of Baltia is placed by Timæus (Plin., xxxvii., 2). Abalus, a day's journey from an æstuarium, can not, therefore, be the Kurish Nehrung. See, also, on the voyage of Pytheas to the west shores of Jutland, and on the amber trade along the whole coast of Skage as far as the Netherlands, Werlauff, *Bidrag til den Nordiske Ravhandels Historie* (Kopenh., 1835). In Tacitus, and not in Pliny, we find the first acquaintance with the glessum of the shores of the Baltic, in the land of the Æstui (Æstuorum gentium) and of the Venedi, concerning whom the great philologist Shaffarik (*Slawische Alterthümer*, th. i., s. 151–165) is uncertain whether they were Slaves or Germani. The more active direct connection with the Samland coast of the Baltic, and with the Esthonians, by means of the over-land route through Pannonia, by Carnuntum, which was first followed by a Roman knight under Nero, appears to me to have belonged to the later times of the Roman Cæsars (Voigt, *Gesch. Preussen's*, bd. i., s. 85). The relations between the Prussian coasts and the Greek colonies on the Black Sea are proved by fine coins, struck probably before the eighty-fifth Olympiad, which have been recently found in the Netz district (Lewezow, in the *Abhandl. der Berl. Akad. der Wiss. aus dem Jahr* 1833, s. 181–224). The electron, the sun-stone of the very ancient mythus of the Eridanus (Plin., xxxvii., cap. 2), the amber stranded or buried on the coast, was, no doubt, frequently brought to the south, both by land and by sea, from very different districts. The "amber which was found buried at two places

Æstii on the Baltic, owed its origin to the daring perseverance of Phœnician coasting traders. Its subsequent extension affords a remarkable example in the history of the contemplation of the universe, of the influence which may be exercised on the establishment of international intercourse, and on the extension of the knowledge of large tracts of land, by a predilection for even a single product. In the same manner as the Phocæan Massilians conveyed British tin through the whole extent of Gaul to the shores of the Rhone, amber passed from people to people through Germany and the territory of the Celts, on both sides of the Alps, to the Padus, and through Pannonia to the Borysthenes. This inland trade thus first connected the inhabitants of the coasts of the North Sea with those living on the shores of the Adriatic and the Euxine.

The Phœnicians of Carthage, and probably those inhabiting the cities of Tartessus and Gades, which had been colonized two hundred years earlier, visited a considerable portion of the northwest coast of Africa, even beyond Cape Bojador, although the Chretes of Hanno is neither the Chremetes of the *Meteorologica* of Aristotle, nor yet our Gambia.* Here were situated the numerous Tyrian cities, whose numbers were estimated by Strabo at 300, which were destroyed by Pharusians and Nigritians. Among these was Cerne (Dicuil's Gaulea according to Letronne), the principal station for ships, as well as the chief emporium of the colonies on the coast. The Canary Islands and the Azores (which latter were regarded by Don Fernando, the son of Columbus, as the Cassiterides

in Scythia was, in part, very dark colored." Amber is still collected near Kaltschedansk, not far from Kamensk, on the Ural; and we have obtained at Katharinenburg fragments imbedded in lignite. See G. Rose, *Reise nach dem Ural*, bd. i., s. 481; and Sir Roderic Murchison, in the *Geology of Russia*, vol. i., p. 366. The petrified wood which frequently surrounds the amber had early attracted the attention of the ancients. This resin, which was, at that time, regarded as so precious a product, was ascribed either to the black poplar (according to the Chian Scymnus, v. 396, p. 367, Letronne), or to a tree of the cedar or pine genus (according to Mithridates, in Plin., xxxvii., cap. 2 and 3). The recent admirable investigations of Prof. Göppert, at Breslau, have shown that the conjecture of the Roman collector was the more correct. Respecting the petrified amber-tree (Pinites succifer) belonging to an extinct vegetation, see Berendt, *Organische Reste im Bernstein*, bd. i., abth. 1, 1845, s. 89.

* On the Chremetes, see Aristot., *Meteor.*, lib. i., p. 350 (Bekk.); and on the most southern points of which Hanno makes mention in his ship's journal, see my *Rel. Hist.*, t. i., p. 172; and *Examen Crit. de la Géog.*, t. i., p. 39, 180, and 288; t. iii., p. 135. Gosselin, *Recherches sur la Géog. System. des Anciens*, t. i., p. 94 and 98; Ukert, th. i., 1, s. 61–66.

discovered by the Carthaginians), and the Orkneys, Färoë Islands, and Iceland, became the respective western and northern intermediate stations for passing to the New Continent. They indicate the two directions by which the European portion of the human race first became acquainted with the natives of North and Central America. This consideration gives a great, and, I might almost say, a cosmical importance to the question whether and how early the Phœnicians of the mother country, or those of the Iberian and African settlements (Gadeira, Carthage, and Cerne), were acquainted with Porto Santo, Madeira, and the Canary Islands. In a long series of events, we willingly seek to trace the first and guiding link of the chain. It is probable that fully 2000 years elapsed from the foundation of Tartessus and Utica by Phœnicians, to the discovery of America by the northern course, that is to say, to Eric Randau's voyage to Greenland, which was followed by voyages to North Carolina ; and that 2500 years intervened before Christopher Columbus, starting from the old Phœnician settlement of Gadeira, made the passage by the southwest route.*

In accordance with the requirements for the generalization of ideas demanded by the present work, I have considered the discovery of a group of islands lying only 168 miles from the African shore as the first member of a long series of similarly-directed efforts, but I have made no allusion to the *Elysium, the Islands of the Blessed,* fabled by the poetic visions of fancy, as situated on the confines of the earth, in an ocean warmed by the rays of the near setting sun. All the enjoyments of life and the choicest products of nature were supposed to be placed at the remotest distance of the terrestrial globe.† The ideal land—the geographical myth of the Elysion—was removed further to the west, even beyond the Pillars of Hercules, as the knowledge of the Mediterranean was extended among the Hellenic races. True cosmical knowledge, and the earliest discoveries of the Phœnicians, regard-

---

* Strabo, lib. xvii., p. 826. The destruction of Phœnician colonies by Nigritians (lib. ii., p. 131) appears to indicate a very southern locality ; more so, perhaps, than the crocodiles and elephants mentioned by Hanno, since both these were certainly, at one period, found north of the desert of Sahara, in Maurusia, and in the whole western Atlas country, as is proved from Strabo, lib. xvii., p. 827 ; Ælian., *De Nat. Anim.,* vii., 2 ; Plin., v., 1, and from many occurrences in the wars between Rome and Carthage. See, on this important subject, referring to the geography of animals, Cuvier, *Ossemens Fossiles,* 2 éd., t. i., p. 74, and Quatremère, op. cit., p. 391–394.   † Herod., iii., 106.

ing whose precise period no certain tidings have come down to us, did not probably give rise to this myth of the "Islands of the Blessed," the application to which was made subsequently. Geographical discovery has merely embodied a phantom of the imagination, to which it served as a substratum.

Later writers (as an unknown compiler of the *Collection of Wonderful Relations* ascribed to Aristotle, who made use of Timæus, and more especially of Diodorus Siculus) have spoken of "Pleasant Islands," which must be supposed to be the Canaries, and of the great storms to which their accidental discovery is due. It is said that "Phœnician and Carthaginian vessels, which were sailing toward the settlements already then founded on the coast of Libya, were driven out to sea." This event is supposed to have occurred in the early period of the Tyrrhenian navigation, and in that of the contest between the Tyrrhenian Pelasgians and Phœnicians. Statius Sebosus and the Numidian king Juba first gave names to the separate islands, but, unfortunately, not Punic names, although undoubtedly in accordance with notices taken from Punic works. As Plutarch says that Sertorius, when driven away from Spain, wished to save himself and his attendants, after the loss of his fleet, on a group of two Atlantic islands, ten thousand stadia to the west of the mouth of the Bætis, it has been supposed that he meant to designate the two islands of Porto Santo and Madeira,* which were clearly indicated by Pliny as the Pur-

* I have treated in detail this often-contested subject, as well as the passages of Diodorus (v. 19 and 20), and of the Pseudo-Aristot. (*Mirab. Auscult.*, cap. 85, p. 172, Bekk.), in another work (*Examen Crit.*, t. i., p. 130–139; t. ii., p. 158 and 169; t. iii., p. 137–140). The compilation of the *Mirab. Auscult.* appears to have been of a date prior to the end of the first Punic war, since, in cap. 105, p. 211, it describes Sardinia as under the dominion of the Carthaginians. It is also worthy of notice that the wood-clad island, which is mentioned in this work, is described as uninhabited (therefore not peopled by Guanches). The whole group of the Canary Islands was inhabited by Guanches, but not the island of Madeira, in which no inhabitants were found either by John Gonzalves and Tristan Vaz in 1519, or, still earlier, by Robert Masham and Anna Dorset (supposing their Crusoe-like narrative to possess a character of veracity). Heeren applies the description of Diodorus to Madeira alone; yet he thinks that in the account of Festus Avienus (v. 164), who is so conversant with Punic writings, he can recognize the frequent volcanic earthquakes of the Peak of Teneriffe. (See *Ideen über Politik und Handel*, th. ii., abth. i., 1826, s. 106.) To judge from the geographical connection, the description of Avienus would appear to indicate a more northern locality, perhaps even the Kronic Sea. (*Examen Crit.*, t. iii., p. 138.) Ammianus Marcellinus (xxii., 15) also notices the Punic sources of which Juba availed himself. Respecting the probability of the Semitic origin of the appellation of the Canary Islands (the dog

purariæ. The strong oceanic current, which is directed beyond the Pillars of Hercules from northwest to southeast, might long have prevented the coast navigators from discovering the islands most remote from the continent, and of which only the smaller, Porto Santo, was found to be inhabited in the fifteenth century ; and, owing to the curvature of the Earth, the summit of the great volcano of Teneriffe could not be seen, even with a strong refraction, by Phœnician mariners sailing along the coast, although I found, from my own observations, that it was discernible from the slight elevations that surround Cape Bojador,* especially in cases of eruption, and by the reflection of a high cloud resting over the volcano. It is even asserted that eruptions of Mount Ætna have been seen, in recent times, from Mount Taygetos in Greece.†

island of Pliny's Latin etymology !), see Credner's *Biblische Vorstellung vom Paradiese*, in Illgen's *Zeitschr. für die Historische Theologie*, bd. vi., 1836, s. 166–186. Joaquim Jose da Costa de Macedo, in a work entitled *Memoria em que se pretende provar que os Arabes nao conhecerão as Canarias antes dos Portuguezes*, 1844, has recently collected all that has been written from the most ancient times to the Middle Ages respecting the Canary Islands. Where history, so far as it is founded on certain and distinctly-expressed evidence, is silent, there remain only different degrees of probability ; but an absolute denial of all facts in the world's history, of which the evidence is not distinct, appears to me no happy application of philological and historical criticism. The many indications which have come down to us from antiquity, and a careful consideration of the relations of geographical proximity to ancient undoubted settlements on the African shore, lead me to believe that the Canary Islands were known to the Phœnicians, Carthaginians, Greeks, and Romans, perhaps even to the Etruscans.

* Compare the calculations in my *Rel. Hist.*, t. i., p. 140 and 287. The Peak of Teneriffe is distant 2° 49′ of an arc from the nearest point of the African coast. In assuming a mean refraction of 0·08, the summit of the Peak may be seen from a height of 1291 feet, and, therefore, from the Montañas Negras, not far from Cape Bojador. In this calculation, the elevation of the Peak above the level of the sea has been taken at 12,175 feet; Captain Vidal has recently determined it trigonometrically at 12,405, and Messrs. Coupvent and Dumoulin, barometrically, at 12,150. (D'Urville, *Voyage au Pole Sud, Hist.*, t. i., 1842, p. 31, 32.) But Lancerote, with a volcano, la Corona, 1918 feet in height (Leop. v. Buch, *Canarische Inseln*, s. 104), and Fortaventura, lie much nearer to the main land than Teneriffe ; the distance of the first-named island being 1° 15′, and that of the second 1° 2′.

† Ross has only mentioned this assertion as a report (*Hellenika*, bd. i., s. xi.). May the observation not have rested on a mere deception ? If we take the elevation of Ætna above the sea at 10,874 feet (lat. 37° 45′, long. from Paris 12° 41′), and that of the place of observation, on the Taygetos (Mount Elias), at 7904 feet (lat. 36° 57′, long. from Paris 20° 1′), and the distance between the two at 352 geographical miles, we have for the point from which light was emitted above Ætna, and was visible on Taygetos, fully 48,675 feet, which is four and a half

In the enumeration of the elements of an extended knowledge of the universe, which were early brought to the Greeks
from other parts of the Mediterranean basin, we have hitherto followed the Phœnicians and Carthaginians in their intercourse with the northern tin and amber lands, as well as in
their settlements near the tropics, on the west coast of Africa.
It now, therefore, only remains for us to refer to a voyage of
the Phœnicians to the south, when they proceeded 4000 geographical miles east of Cerne and Hanno's Western Horn, far
within the tropics, to the Prasodic and Indian Seas. Whatever doubt may exist regarding the localization of the distant
gold lands (Ophir and Supara), and whether these gold lands
are the western coasts of the Indian Peninsula or the eastern
shores of Africa, it is, at any rate, certain that this active,
enterprising Semitic race, who so early employed alphabetical
writing, had a direct acquaintance with the products of the
most different climates, from the Cassiterides to the south of
the Straits of Bab-el-Mandeb, far within the tropics. The
Tyrian flag floated simultaneously in the British and Indian
Seas. The Phœnicians had commercial settlements in the
northern parts of the Arabian Gulf, in the ports of Elath and
Ezion-Geber, as well as on the Persian Gulf at Aradus and
Tylos, where, according to Strabo, temples had been erected,
which, in their style of architecture, resembled those on the
Mediterranean.* The caravan trade, which was carried on
by the Phœnicians in seeking spices and incense, was directed
to Arabia Felix, through Palmyra, and to the Chaldean or
Nabathæic Gerrha, on the western or Arabian side of the Persian Gulf.

The expeditions sent by Hiram and Solomon, and which
were undertaken conjointly by Tyrians and Israelites, sailed
from Ezion-Geber through the Straits of Bab-el-Mandeb to
Ophir (Opheir, Sophir, Sophora, the Sanscrit Supara of Ptolemy).† Solomon, who loved pomp, caused a fleet to be con-

times greater than the elevation of Ætna. If, however, we might assume, as my friend Professor Encke has remarked, the reflecting surface to be 184 miles from Ætna and 168 miles from Taygetos, its height
above the sea would only require to be 1829 feet.

* Strabo, lib. xvi., p. 767, Casaub. According to Polybius, it would
seem that the Euxine and the Adriatic Sea were discernible from
Mount Aimon—an assertion ridiculed by Strabo (lib. vii., p. 313).
Compare Scymnus, p. 93.

† On the synonym of Ophir, see my *Examen Crit. de l'Hist. de la
Géographie*, t. ii., p. 42. Ptolemy, in lib. vi., cap. 7, p. 156, speaks of
a Sapphara, the metropolis of Arabia; and in lib. vii., cap. 1, p. 168, of

structed at the Red Sea, and Hiram supplied him with experienced Phœnician seamen, and Tyrian vessels, "ships of
Tarshish."* The articles of commerce which were brought
from Ophir were gold, silver, sandal-wood (*algummin*), precious stones, ivory, apes (*kophim*), and peacocks (*thukkiim*).
These are not Hebrew, but Indian names.† It would appear
highly probable, from the careful investigations of Gesenius,
Benfey, and Lassen, that the Phœnicians, who had been early

Supara, in the Gulf of Camboya (Barigazenus Sinus, according to Hesy-
chius), as "a district rich in gold!" Supara signifies in Indian a *fair
shore* (Lassen, *Diss. de Taprobane*, p. 18, and *Indische Alterthumskunde*,
bd. i., s. 107; also Professor Keil, of Dorpat, *Ueber die Hiram-Salomo-
nische Schiffahrt nach Ophir und Tarsis*, s. 40–45).

* On the question whether ships of Tarshish mean ocean ships, or
whether, as Michaelis contends, they have their name from the Phœnician Tarsus, in Cilicia, see Keil, op. cit., s. 7, 15–22, and 71–84.

† Gesenius, *Thesaurus Linguæ Hebr.*, t. i., p. 141; and the same in
the *Encycl.* of Ersch and Gruber, sect. iii., th. iv., s. 401; Lassen, *Ind.
Alterthumskunde*, bd. i., s. 538; Reinaud, *Relation des Voyages faits par
les Arabes. dans l'Inde et en Chine*, t. i., 1845, p. xxviii. The learned
Quatremère, who, in a very recently-published treatise (*Mém. de l'Acad.
des Inscriptions*, t. xvi., Part ii., 1845, p. 349–402), still maintains, with
Heeren, that Ophir is the east coast of Africa, has explained the word
thukkiim (thukkiyyim) as parrots, or Guinea-fowls, and not peacocks (p.
375). Regarding Dahotora, compare Dohlen, *Das alte Indien*, th. ii., s.
139, with Benfey, *Indien*, s. 30–32. Sofala is described by Edrisi (in
Amédée Jaubert's translation, t. i., p. 67), and subsequently by the
Portuguese, after Gama's voyage of discovery (*Barros*, Dec. i., liv. x.,
cap. i.; Part ii., p. 375; Külb, *Geschichte der Entdeckungsreisen*, th i ,
1841, s. 236), as a country rich in gold. I have elsewhere drawn attention to the fact that Edrisi, in the middle of the twelfth century, speaks
of the application of quicksilver in the gold-washings of the negroes of
this district, as a long-known process of amalgamation. When we bear
in mind the great frequency of the interchange of *r* and *l*, we find that
the name of the East African Sofala is perfectly represented by that of
Sophara, which is used, with several other forms, in the version of the
Septuagint, for the Ophir of Solomon and Hiram. Ptolemy also, as has
been already noticed, was acquainted with a Sapphara, in Arabia (Ritter, *Asien*, bd. viii., 1, 1846, s. 252), and a Supara in India. The significant (Sanscrit) names of the mother country had been conferred on
neighboring or opposite coasts, as we find, under similar relations in
the present day, in the Spanish and English parts of America. The
trade to Ophir might thus, according to my view, be extended in the
same manner as a Phœnician expedition to Tartessus might touch at
Cyrene and Carthage, Gadeira and Cerne, and as one to the Cassiterides
might touch at the Artabrian, British, and East Cimbrian coasts. It is
nevertheless remarkable that incense, spices, silk, and cotton cloth are
not named among the wares from Ophir, together with ivory, apes, and
peacocks. The latter are exclusively Indian, although, on account of
their gradual extension to the west, they were frequently termed by
the Greeks "Median and Persian birds;" the Samians even supposed
them to have belonged originally to Samos, on account of their being

made acquainted with the periodic prevalence of the monsoons through their colonies on the Persian Gulf, and their intercourse with the inhabitants of Gerrha, must have visited the western coasts of the Indian Peninsula. Christopher Columbus was even persuaded that Ophir (the El Dorado of Solomon) and Mount Sopora were a portion of Eastern Asia, the *Chersonesus Aurea* of Ptolemy.* As it appears difficult to form an idea of Western India as a fruitful source of gold, it will, I think, scarcely be necessary to refer to the " gold-seeking ants" (or to the unmistakable account given by Ctesias of a foundery in which, however, gold and iron were said, according to his account, to be fused together),† it being sufficient to direct attention to the geographical proximity of Southern Arabia, of the island of Dioscorides (the Diu Zokotora of the moderns, a corruption of the Sanscrit Dvipa Sukhatara), cultivated by Indian colonists, and to the auriferous coast of Sofala in Eastern Africa. Arabia and the island last referred to, to the southeast of the Straits of Bab-el-Mandeb, may be regarded as affording intermediate links of connection between the Indian Peninsula and Eastern Africa for the combined commerce of the Hebrews and Phœnicians. The Indians had,

reared by the priests in the sanctuary of Hera. From a passage in Eustathius (*Comm. in Iliad*, t. iv., p. 225, ed. Lips., 1827) on the sacredness of peacocks in Libya, it has been unjustly inferred that the ταώς also belonged to Africa.

* See the remarks of Columbus on Ophir and el Monte Sopora, " which Solomon's fleet could not reach within a term of three years," in Navarrete, *Viages y Descubrimientos que hiciéron los Españoles*, t. i., p. 103. In another work, the great discoverer says, still in the hope of reaching Ophir, " the excellence and power of the gold of Ophir can not be described ; he who possesses it does what he will in this world ; nay, it even enables him to draw souls from purgatory to paradise" (" llega a que echa las animas al paraiso"), *Carta del Almirante, escrita en la Jamaica,* 1503 ; Navarrete, t. i., p. 309. (Compare my *Examen Critique,* t. i., p. 70 and 109 ; t. ii., p. 38–44 ; and on the proper duration of the Tarshish voyage, see Keil, op. cit., s. 106.)

† *Ctesiæ Cnidii Operum Reliquiæ*, ed. Felix Baehr, 1824, cap. iv. and xii., p. 248, 271, and 300. But the accounts collected by the physician at the Persian court from native sources, which are not, therefore, altogether to be rejected, refer to districts in the north of India, and from these the gold of the Daradas must have come by many circuitous routes to Abhira, the mouth of the Indus, and the coast of Malabar. (Compare my *Asie Centrale*, t. i., p. 157, and Lassen, *Ind. Alterthumskunde*, bd. i., s. 5.) May not the wonderful story related by Ctesias of an Indian spring, at the bottom of which iron was found, which was very malleable when the fluid gold had run off, have been based on a misunderstood account of a foundery ? The molten iron was probably taken for gold owing to its color, and when the yellow color had disappeared in cooling, the black mass of iron was found below it.

from the earliest time, made settlements in the eastern part of Africa, and on the coasts immediately opposite their native country ; and the traders to Ophir might have found, in the basin of the Erythreian and Indian Seas, other sources of gold besides India itself.

Less influential than the Phœnicians in extending the geographical sphere of our views, and early affected by the Greek influence of a band of Pelasgian Tyrrhenians, who invaded their country from the sea, the Etruscans present themselves to our observation as a gloomy and stern race. They carried on no inconsiderable inland trade to distant amber countries, through Northern Italy and across the Alps, where a *via sacra** was protected by all the neighboring tribes. The primitive Tuscan race of the Rasenæ appears to have followed almost the same road on their way from Rhætia to the Padus, and even further southward. In accordance with our object, which is always to seize on the most general and permanent features, we would here consider the influence which the general character of the Etruscans exercised on the most ancient political institutions of Rome, and through these on the whole of Roman life. It may be said that the reflex action of this influence still persists in its secondary and remote political effects, inasmuch as, for ages, Rome stamped her character, with more or less permanence, on the civilization and mental culture of mankind.†

A peculiar characteristic of the Tuscans, which demands our special notice in the present work, was their inclination for cultivating an intimate connection with certain natural phenomena. Divination, which was the occupation of their equestrian hierarchical caste, gave occasion for a daily observation of the meteorological processes of the atmosphere. The *Fulguratores*, observers of lightning, occupied themselves in investigating the direction of the lightning, with " drawing it down," and " turning it aside."‡ They carefully distinguished

* Aristot., *Mirab. Auscult.*, cap. 86 and 111, p. 175 and 225, Bekk.

† *Die Etrusker*, by Otfried Müller, abth. ii., s. 350 ; Niebuhr, *Römische Geschichte*, th. ii., s. 380.

‡ The story formerly current in Germany, and reported on the testimony of Father Angelo Cortenovis, that the tomb described by Varro of the hero of Clusium, Lars Porsena, ornamented with a bronze hat and bronze pendant chains, was an apparatus for collecting atmospherical electricity, or for conducting lightning (as were also, according to Michaelis, the metal points on Solomon's temple), was related at a time when men were inclined to attribute to the ancients the remains of a supernaturally-revealed primitive knowledge of physics, which was, however, soon again obscured. The most important notice of the rela-

between flashes of lightning from the higher regions of the
clouds, and those which Saturn, an earth god,[*] caused to
ascend from below, and which were called Saturnine-terres-
trial lightning, a distinction which modern physicists have
thought worthy of especial attention. Thus were established
regular official notices of the occurrence of storms.[†] The
Aquælicium, the art of discovering springs of waters, which
was much practiced by the Etruscans, and the drawing forth
of water by their *Aquileges*, indicate a careful investigation
of the natural stratification of rocks and of the inequalities of
the ground. Diodorus, on this account, extols the Etruscans
as industrious inquirers of nature. We may add to this com-
mendation that the patrician and powerful hierarchical caste
of the Tarquinii offered the rare example of favoring physical
science.

We have spoken of the ancient seats of human civilization
in Egypt, Phœnicia, and Etruria, before proceeding to the
highly-gifted Hellenic races, with whose culture our own civ-

tions between lightning and conducting metals (which it was not difficult to discover) appears to me to be that of Ctesias (*Indica*, cap. 4, p.
169, ed. Lion; p. 248, ed. Baehr). " He had possessed, it is said, two
iron swords, presents from the King Artaxerxes Mnemon, and from
Parysatis, the mother of the latter, which, when planted in the earth,
averted clouds, hail, and strokes of lightning. He had himself seen the
results of this operation, for the king had twice made the experiment
before his eyes." The great attention paid by the Etruscans to the
meteorological processes of the atmosphere in all that differed from the
ordinary course of natural phenomena, makes it certainly a cause for
regret that nothing has come down to us from the books of the Fulguratores. The epochs of the appearance of great comets, of the fall of
meteoric stones, and of showers of falling stars, were no doubt recorded
in them, as in the more ancient Chinese annals made use of by Edouard
Biot. Creuzer (*Symbolik und Mythologie der alten Völker*, th. iii., 1842,
s. 659) has endeavored to prove that the natural features of Etruria acted
on the peculiar direction of mind of its inhabitants. A " calling forth"
of the lightning, which is ascribed to Prometheus, calls to mind the
strange pretended "drawing down" of lightning by the Fulguratores.
This operation consisted, however, in a mere conjuration, which was
probably not more efficacious than the skinned ass's head, supposed, in
accordance with Etruscan religious usages, to have the faculty of preserving against the danger of thunder-storms.
* Otfr. Müller, *Etrusker*, abth. ii., s. 162–178. It would appear that,
in accordance with the very complicated Etruscan augur-theory, a distinction was made between the "soft reminding lightnings propelled
by Jupiter by his own independent power, and the violent electrical
means of chastisement which he could only send forth in obedience to
established constitutional prescriptions, after consulting with the other
twelve gods" (Seneca, *Nat. Quæst.*, ii., p. 41).
† Joh. Lydus, *De Ostentis*, ed. Hase, p. 18, in præfat.

ilization is most deeply rooted, and from whom we have derived a considerable portion of our early knowledge of other
nations, and of our views regarding the universe. We have
considered the basin of the Mediterranean in its characteristic
configuration and position, and the influence of these relations
on the commercial intercourse established with the western
coasts of Africa, the extreme north, and the Indo-Arabian
Sea. No portion of the earth has been the theater of greater
changes of power, or of greater or more animated activity under the influence of mental guidance. This movement was
transmitted far and enduringly by the Greeks and Romans,
especially after the latter had destroyed the Phœnicio-Cartha-
ginian power. That which we term the beginning of history
is, therefore, only the period when later generations awoke to
self-consciousness. It is one of the advantages of the present
age that, by the brilliant progress that has been made in general and comparative philology, by the careful investigation
of monuments and their more certain interpretation, the views
of the historical inquirer are daily enlarged, and the strata of
remote antiquity gradually opened, as it were, before our eyes.
Besides the civilized nations of the Mediterranean which we
have just enumerated, there are many others who show traces
of ancient cultivation ; among these we may mention the
Phrygians and Lycians in Western Asia, and the Turduli and
Turdetani in the extreme West.* Of the latter, Strabo observes, " they are the most cultivated of all the Iberians ; they
employ the art of writing, and have written books containing
memorials of ancient times, and also poems and laws set in
verse, for which they claim an antiquity of six thousand years."
I have dwelt on these separate examples in order to show
how much of ancient cultivation, even among European nations, has been lost without our being able to discover any
trace of its existence, and how the history of the earliest contemplation of the universe must continue to be limited to a
very narrow compass.

Beyond the forty-eighth degree of latitude, north of the Sea
of Azof and of the Caspian, between the Don, the Wolga, and
the Jaik, where the latter flows from the southern auriferous

* Strabo, lib. iii., p. 139, Casaub. Compare Wilhelm von Humboldt,
Ueber die Urbewohner Hispaniens, 1821, s. 123 und 131–136. The Iberian
alphabet has been successfully investigated in our own times by M. de
Saulcy ; the Phrygian, by the ingenious discoverer of arrow-headed
writing, Grotefend ; and the Lycian, by Sir Charles Fellowes. (Compare Ross, *Hellenika*, bd. i., s. xvi.)

Uralian Mountains, Europe and Asia are, as it were, fused together by flat steppes. Herodotus, in the same manner as
Pherecydes of Syros had previously done, regarded the whole
of northern Scythian Asia (Siberia) as belonging to Sarmatian
Europe, and even as forming a portion of Europe itself.* Toward the south, our quarter of the globe is sharply separated
from Asia, but the far-projecting peninsula of Asia Minor and
the richly-varied Ægean Archipelago (serving as a bridge between the two separate continents) have afforded an easy
passage for different races, languages, customs, and manners.
Western Asia has, from the earliest ages, been the great thoroughfare for races migrating from the east, as was the northwest of Greece for the Illyric races. The Ægean Archipelago,
which was in turn subject to Phœnician, Persian, and Greek
dominion, was the intermediate link between Greece and the
far East.

When Phrygia was incorporated with Lydia, and both
merged into the Persian empire, the contact led to the general extension of the sphere of ideas among Asiatic and European Greeks. The Persian rule was extended by the warlike expeditions of Cambyses and Darius Hystaspes from Cyrene and the Nile to the fruitful lands of the Euphrates and
the Indus. A Greek, Scylax of Karyanda, was employed to
explore the course of the Indus, from the then-existing territory of Caschmeer (Kaspapyrus)† to its mouth. An active
intercourse was carried on between Greece and Egypt (with
Naucratis and the Pelusian arm of the Nile) before the Persian conquest, and even under Psammitichus and Amasis.‡
These extensive relations of intercourse with other nations
drew many Greeks from their native land, not only for the
purpose of establishing those distant colonies which we shall
consider in a subsequent part of the present work, but also as
hired soldiers, who formed the nucleus of foreign armies in
Carthage,§ Egypt, Babylon, Persia, and in the Bactrian district of the Oxus.

A deeper insight into the individuality and national character of the different Greek races has shown that, if a grave

* Herod., iv., 42 (Schweighäuser ad Herod., t. v., p. 204). Compare Humboldt, *Asie Centrale*, t. i., p. 54 and 577.

† Regarding the most probable etymology of *Kaspapyrus* of Hecatæus (*Fragm.*, ed. Klausen, No. 179, v. 94), and the *Kaspatyrus* of
Herodotus (iii., 102, and iv., 44), see my *Asie Centrale*, t. i., p. 101–104.

‡ Regarding Psammitichus and Aahmes, see *ante*, p. 127.

§ Droysen, *Geschichte der Bildung des Hellenistischen Staatensystems*,
1843, s. 23.

and reserved exclusiveness prevailed among the Dorians, and in part, also, among the Æolians, we must, on the other hand, ascribe to the gayer Ionic race a mobility of mind, which, under the stimulus of an eager spirit of inquiry, and an ever-wakeful activity, was alike manifested in a faculty for mental contemplation and sensuous perception. Directed by the objective bent of their mode of thought, and adorned by a luxuriance of fancy in poetry and in art, the Ionians scattered the beneficent germs of progressive cultivation wherever they established their colonies in other countries.

As the landscape of Greece was so strikingly characterized by the peculiar charm of an intimate blending of land and sea, the configuration of the coast-line to which this character was owing could not fail early to awaken in the minds of the Greeks a taste for navigation, and to excite them to an active commercial intercourse and contact with foreign nations.* The maritime dominion of the Cretans and Rhodians was followed by the expeditions of the Samians, Phocæans, Taphians, and Thesprotians, which were, it must be owned, originally directed to plunder and to the capture of slaves. Hesiod's disinclination to a sea-faring life is probably to be regarded merely as the expression of an individual opinion, or as the result of a timid ignorance of nautical affairs, which may have prevailed on the main land of Greece at the early dawn of civilization. On the other hand, the most ancient legends and myths abound in reference to distant expeditions by land and sea, as if the youthful imagination of mankind delighted in the contrast between its own ideal creations and a limited reality. In illustration of this sentiment we may mention the expeditions of Dionysus and Hercules (Melkarth in the temple at Gadeira) ; the wanderings of Io ;† of the often-resuscitated Aristeas ; and of the Hyperborean magician Abaris, in whose "guiding arrow"‡ some commentators have supposed that they recognized the compass. In these narratives we trace

* See *ante*, p. 25.

† Völker, *Mythische Geographie der Griechen und Römer*, th. i., 1832, s. 1–10 ; Klausen, *Ueber die Wanderungen der Io und des Herakles*, in Niebuhr and Brandis *Rheinische Museen für Philologie, Geschichte und Griech. Philosophie*, Jahrg., iii., 1829, s. 293–323.

‡ In the myth of Abaris (Herod., iv., 36), the magician does not travel through the air on an arrow, but he carries the arrow, "which Pythagoras gave him (Jambl., *De Vita Pythag.*, xxix., p. 194, Kiessling), in order that it may be useful to him in all difficulties on his long journey." Creuzer, *Symbolik*, th. ii., 1841, s. 660–664. On the repeatedly disappearing and reappearing Arimaspian bard, Aristeas of Proconnesus, see Herod., iv., 13–15.

the reciprocal reflection of passing events and ancient cosmical views, and the progressive modification which the latter effected in these mythical representations of history. In the wanderings of the heroes returning from Troy, Aristonicus makes Menelaus circumnavigate Africa more than five hundred years before Neco sailed from Gadeira to India.*

At the period which we are here considering, of the history of Greece before the Macedonian expeditions into Asia, there occurred three events which exercised a special influence in extending the views of the Greeks regarding the universe. These events were the attempts to penetrate beyond the basin of the Mediterranean toward the east; the attempts toward the west; and the establishment of numerous colonies from the Pillars of Hercules to the northeastern extremity of the Euxine, which, by the more varied form of their political constitution, and by their furtherance of mental cultivation, were more influential than those of the Phœnicians and Carthaginians in the Ægean Sea, Sicily, Theria, and on the north and west coasts of Africa.

The advance toward the East, about twelve centuries before our era, or one hundred and fifty years after Rameses Miamoun (Sesostris), is known in history as the expedition of the Argonauts to Colchis. The true version of this event, which is clothed in a mythical garb, and concealed under a blending of ideal images, is simply the fulfillment of a national desire to open the inhospitable Euxine. The myth of Prometheus, and the unbinding of the fire-kindling Titan on the Caucasus by Hercules, during his expedition to the East; the ascent of Io from the Valley of the Hybrites† to the heights of the Caucasus; the myth of Phryxus and Helle, all indicate

* Strabo, lib. i., p. 38, Casaub.

† Probably the Valley of the Don or of the Kuban. See my *Asie Centrale*, t. ii., p. 164. Pherecydes expressly says (*Fragm.* 37, *ex Schol. Apollon.*, ii., 1214) that the Caucasus burned, and that, therefore, Typhon fled to Italy; a notice from which Klausen, in the work already mentioned, s. 298, explains the ideal relation of the "fire-kindler" (πυρκαεύς), Prometheus, to the burning mountain. Although the geognostical constitution of the Caucasus (which has been recently so ably investigated by Abich), and its connection with the volcanic chain of the Thianschan, in the interior of Asia (which I think I have shown in my *Asie Centrale*, t. ii., p. 55–59), render it in no way improbable that reminiscences of great volcanic eruptions may have been preserved in the most ancient traditions of men, yet we may rather assume that a bold and somewhat hazardous spirit of etymological conjecture may have led the Greeks to the hypothesis of the burning. On the Sanscrit etymologies of Graucasus (or shining mountain), see Bohlen's and Burnouf's statements, in my *Asie Centrale*, t. i., p. 109.

the same direction of the course on which the early Phœnician navigators had adventured.

Before the migrations of the Dorians and Æolians, the Bœotian Orchomenus, near the eastern extremity of the Lake of Copais, was already a rich commercial city of the Minyans. The Argonautic expedition began at Iolcus, the principal seat of the Thessalian Minyans, on the Pagasæan Gulf. The locality of the myth, considered with respect to the aim of the undertaking, after having been variously modified* at different times, was finally associated with the mouth of the Phasis (Rion), and with Colchis, a seat of ancient civilization, instead of with the uncertain and remote land of Æa. The expeditions of the Milesians and their numerous colonial cities on the Euxine enabled them to obtain a more exact knowledge of the eastern and northern limits of that sea, and thus gave a more definite outline to the geographical portion of the myth. A number of important new views was thus simultaneously opened. The Caspian had long been known only on its western coast; and even Hecatæus regarded this shore as the western boundary of the encircling Eastern Ocean.† The father of history was the first who taught that the Caspian Sea was a basin closed on all sides, a fact which, after him, was again contested, for six centuries, until the time of Ptolemy.

* Otfried Müller, *Minyer*, s. 247, 254, und 274. Homer was not acquainted with the Phasis, or with Colchis, or with the Pillars of Hercules; but the Phasis is named by Hesiod. The mythical traditions concerning the return of the Argonauts through the Phasis into the Eastern Ocean, and across the "double" Triton Lake, formed either by the conjectured bifurcation of the Ister, or by volcanic earthquakes (*Asie Centrale*, t. i., p. 179; t. iii., p. 135–137; Otfr. Müller, *Minyer*, s. 357), are especially important in arriving at a knowledge of the earliest views regarding the form of the continents. The geographical phantasies of Peisandros, Timagetus, and Apollonius of Rhodes were continued until late in the Middle Ages, and showed themselves sometimes as bewildering and deterring obstacles, and sometimes as stimulating incitements to actual discoveries. This reaction of antiquity on later times, when men suffered themselves to be led more by opinions than by actual observations, has not been hitherto sufficiently considered in the history of geography. My object is not merely to present bibliographical sources from the literature of different nations for the elucidation of the facts advanced in the text, but also to introduce into these notes, which permit of greater freedom, such abundant materials for reflection as I have been able to derive from my own experience and from long-continued literary studies.

† *Hecatæi, Fragm.*, ed. Klausen, p. 39, 92, 98, and 119. See, also, my investigations on the history of the geography of the Caspian Sea, from Herodotus down to the Arabian El-Istâchri, Edrisi, and Ibn-el-Vardi, on the Sea of Aral, and on the bifurcation of the Oxus and the Araxes, in my *Asie Centrale*, t. ii., p. 162–297.

At the northeastern extremity of the Black Sea a wide field was also opened to ethnology. Astonishment was felt at the multiplicity of languages among the different races,* and the necessity for skillful interpreters (the first aids and rough instruments in a comparative study of languages) was keenly felt. The intercourse established by barter and trade was carried from the Mæotic Gulf, then supposed to be of very vast extent, over the Steppe where the central Kirghis horde now pasture their flocks, through a chain of the Scythio-Scolotic tribes of the Argippæans and Issedones,† whom I regard as of Indo-Germanic origin, to the Arimaspes on the northern declivity of the Altai Mountains, who possessed large treasures in gold.‡ Here, therefore, we have the ancient realm of the

---

* Cramer, *De Studiis quæ veteres ad aliarum gentium contulerint Linguas*, 1844, p. 8 and 17. The ancient Colchians appear to have been identical with the tribe of the Lazi (Lazi, gentes Colchorum, Plin., vi., 4; the Λαζοί of Byzantine writers); see Vater (Professor in Kasan), *Der Argonautenzug aus den Quellen dargestellt*, 1845, Heft. i., s. 24; Heft. ii., s. 45, 57, und 103. In the Caucasus, the names Alani (Alanethi, for the land of the Alani), Ossi, and Ass may still be heard. According to the investigations begun with a truly philosophic and philological spirit by George Rosen in the Valleys of the Caucasus, the language spoken by the Lazi possesses remains of the ancient Colchian idiom. The Iberian and Grussic family of languages includes the Lazian, Georgian, Suanian, and Mingrelian, all belonging to the group of the Indo-Germanic languages. The language of the Osseti bears a greater affinity to the Gothic than to the Lithuanian.

† On the relationship of the Scythians (Scolotes or Sacæ), Alani, Goths, Massagetæ, and the Yueti of the Chinese historians, see Klaproth, in the commentary to the *Voyage du Comte Potocki*, t. i., p. 129, as well as my *Asie Centrale*, t. i., p. 400; t. ii., p. 252. Procopius himself says very definitely (*De Bello Gothico*, iv., 5, ed. Bonn, 1833, vol. ii., p. 476), that the Goths were formerly called Scythians. Jacob Grimm, in his recently-published work, *Ueber Jornandes*, 1846, s. 21, has shown the identity of the Getæ and the Goths. The opinion of Niebuhr (see his *Untersuchungen über die Geten und Sarmaten*, in his *Kleine Historische und Philologische Schriften*, 1te Sammlung, 1828, s. 362, 364, und 395), that the Scythians of Herodotus belong to the family of the Mongolian tribes, appears the less probable, since these tribes, partly under the yoke of the Chinese, and partly under that of the Hakas or Kirghis (Χερχίς of Menander), still lived, far in the east of Asia, round Lake Baikal, in the beginning of the thirteenth century. Herodotus distinguishes also the bald-headed Argippæans (iv., 23) from the Scythians; and if the first-named are characterized as "flat-nosed," they have, at the same time, a "long chin," which, according to my experience, is by no means a physiognomical characteristic of the Calmucs, or of other Mongolian races, but rather of the blonde (Germanizing?) Usun and Tingling, to whom the Chinese historians ascribe 'long horse faces."

‡ On the dwelling-place of the Arimaspes, and on the gold trade of

Griffins, the seat of the meteorological myth of the Hyperboreans,* which has wandered with Hercules far to the West.

We may conjecture that the portion of Northern Asia above alluded to, which has again, in our days, become celebrated by the Siberian gold washings, as well as the large quantity of gold accumulated, in the time of Herodotus, by the Gothic tribe of the Massagetæ, must have become an important source of wealth and luxury to the Greeks, by means of the intercourse opened with the Euxine. I place the locality of this source of wealth between the 53d and 55th degrees of latitude. The region of the gold-sand, of which the travelers were informed by the Daradas (Darder or Derder), mentioned in the Mahabharata, and in the fragments collected by Megasthenes, and which, owing to the accidental double meaning of the names of some animals,† has been associated with the often-

Northwestern Asia in the time of Herodotus, see my *Asie Centrale*, t. i., p. 389–407.

* " The story of the Hyperboreans is a meteorological myth. The wind of the mountains (B'Oreas) is believed to issue from the Rhipean Mountains, while beyond these mountains there prevail a calm air and a genial climate, as on the Alpine summits, beyond the region of clouds. In this we trace the dawn of a physical science, which explains the distribution of heat and the difference of climates by local causes, by the direction of predominating winds, the vicinity of the sun, and the action of a saline or humid principle. The consequence of these systematic ideas was the assumption of a certain independence supposed to exist between the climate and the latitude of the place; thus the myth of the Hyperboreans, connected by its origin with the Dorian worship of Apollo, which was primitively Boreal, may have proceeded from the north toward the west, thus following Hercules in his progress toward the sources of the Ister, to the island of Erythia, and to the gardens of the Hesperides. The Rhipes, or Rhipean Mountains, have also a meteorological meaning, as the word indicates. They are the mountains of impulsion, or of the glacial souffle (ῥιπή), the place from which the Boreal tempests are unloosened."—*Asie Centrale*, t. i., p. 392, 403.

† In Hindostanee there are two words which might easily be confounded, as Wilford has already remarked, one of which is *tschiúntá*, a kind of large black ant (whence the diminutive *tschiunti, tschinti*, the small common ant); the other *tschitá*, a spotted panther, the little hunting leopard (the Felis jubata, Schreb.). This word (*tschitá*) is the Sanscrit *tschitra*, variegated or spotted, as is shown by the Bengalee name for the animal (*tschitábágh* and *tschitibágh*, from *bágh*, Sanscrit *wyaghra*, tiger).—(Buschmann.) In the *Mahabharata* (ii., 1860) there is a passage recently discovered, in which the ant-gold is mentioned. " Wilso invenit (*Journ. of the Asiat. Soc.*, vii., 1843, p. 143), mentionem fieri etiam in Indicis litteris bestiarum aurum effodientium, quas, quum terram effodiant, eodem nomine (pipilica) atque formicas Indi nuncupant." Compare Schwanbeck, in *Megasth. Indicis*, 1846, p. 73. It struck me to see that, in the basaltic districts of the Mexican high-

repeated fable of the gigantic ants, is situated within a more southern latitude of 35° or 37°. This region must, according to one of two combinations, be situated either in the Thibetian highlands east of the Bolor chain, between the Himalaya and Kuen-lun, west of Iskardo, or north of the latter mountain chain toward the desert of Gobi, which has likewise been described as an auriferous district by the accurate Chinese observer and traveler Hiuen-thsang, who lived at the beginning of the seventh century of our era. How much more accessible must the gold of the Arimaspes and Massagetæ have been to the traders in the Milesian colonies on the northern shores of the Euxine! I have alluded to these sources of wealth for the purpose of not omitting to mention a fact which may be regarded as an important and still active result of the opening of the Euxine, and of the first advance of the Greeks toward the East.

The great event of the Doric migrations, and of the return of the Heraclidæ into Peloponnesus, which was productive of such important changes, falls about one hundred and fifty years after the demi-mythical expedition of the Argonauts, which is synonymous with the opening of the Euxine to Greek navigation and commercial intercourse. This navigation simultaneously gave occasion to the founding of new states and new governments, and to the establishment of a colonial system designating an important period in the life of the Hellenic races, and it has further been most influential in extending the sphere of cosmical views, based upon intellectual culture. Europe and Asia thus owed their more intimate connection to the establishment of the colonies, which formed a continuous chain from Sinope (Dioscurias) and the Tauric Panticapæum to Saguntum and Cyrene, the latter of which was founded by the inhabitants of the rainless island of Thera.

No nation of antiquity possessed more numerous, and, on the whole, more powerful colonial cities than the Greeks. It must, however, be remembered, that a period of four hundred or five hundred years intervened between the establishment of the most ancient Æolian colonies, among which Mytilene and Smyrna were pre-eminently distinguished, and the foundation of Syracuse, Croton, and Cyrene. The Indians and Malayans made only weak attempts to found colonies on the eastern coast of Africa, in Zokotora (Dioscorides), and in the South Asiatic Archipelago. Among the Phœnicians a highly-devel-

lands, the ants bring together heaps of shining grains of hyalite, which I was able to collect out of their hillocks.

oped colonial system had been extended over a larger space than that occupied by the Greeks, stretching, although with wide intervals between the stations, from the Persian Gulf to Cerne on the western coast of Africa. No mother country ever established a colony which was as powerful from conquests, and as famed for its commercial undertakings, as Carthage. But, notwithstanding this greatness, Carthage stood far below that degree of mental and artistical cultivation which has enabled the Greek colonial cities to transmit to us so many noble and lasting forms of art.

It must not be forgotten that many populous Greek cities flourished simultaneously in Asia Minor, the Ægean Sea, Lower Italy, and Sicily; and that, like Carthage, the colonial cities of Miletus and Massilia again founded other colonies; that Syracuse, when at the zenith of her power, fought against Athens, and the army of Hannibal and Hamilkar; and that Miletus was, for a long time, the first commercial city in the world after Tyre and Carthage. While a life so rich in enterprise was being developed externally by the activity of a people whose internal condition was frequently exposed to violent agitations, new germs of national intellectual development were continually called forth with the increase of prosperity and the transmission to other nations of native cultivation. One common language and religion bound together the most distant members of the whole body, and it was by this union that the small parent country was brought within the wider circle embraced by the life of other nations. Foreign elements were incorporated in the Hellenic world, without, on that account, depriving it of any portion of its great and characteristic independence. The influence of contact with the East, and with Egypt before it had been connected with Persia, and above one hundred years before the irruption of Cambyses, was, no doubt, from its very nature, more permanent than the influence of the colonies of Cecrops from Sais, of Cadmus from Phœnicia, and of Danaus and Chemmis, whose existence has so often been contested, and is, at any rate, wrapped in the deepest obscurity.

The characteristics by which the Greek colonies differed so widely from all others, especially from the less flexible Phœnicians, and which affected the whole organization of their system, arose from the individuality and the primitive differences existing in the tribes which constituted the whole mother country, and thus gave occasion to a mixture of connecting and separating forces in the colonies as well as in

Greece itself. These contrasts occasioned diversities in the direction of ideas and feelings, and in the form of poetry and harmonious art, and created a rich fullness of life, in which all the apparently hostile elements were dissolved, according to a higher law of universal order, into a gentle harmonious unison.

Notwithstanding that Miletus, Ephesus, and Colophon were Ionic ; Cos, Rhodes, and Halicarnassus Doric ; and Croton and Sybaris Achaic, the power and the inspired poetry of the Homeric song every where made their power appreciable in the midst of this diversity of cultivation, and even in Lower Italy, in the many contiguous colonial cities founded by different races. Amid the most firmly-rooted contrasts in manners and political institutions, and notwithstanding the fluctuations to which the latter were subject, Greece retained its nationality unbroken, and the wide domain of ideal and artistic creations achieved by the separate tribes was regarded as the common property of the whole nation.

It still remains for me to mention, in the present section, the third point, which we have already indicated as having, conjointly with the opening of the Euxine, and the establishment of colonies on the basin of the Mediterranean, exercised so marked an influence on the history of the contemplation of the universe. The foundation of Tartessus and Gades, where a temple was dedicated to the wandering divinity Melkart (a son of Baal), and of the colonial city of Utica, which was older than Carthage, remind us that the Phœnicians had already navigated the open sea for many centuries before the Greeks passed beyond the straits termed by Pindar the " Gadeirian Gate."* In the same manner as the Milesians in the East, by the way of the Euxine,† established relations of international contact which laid the foundation of an inland trade

---

* Strabo, lib. iii., p. 172 (Böckh, *Pind. Fragm.*, v., 155). The expedition of Colæus of Samos falls, according to Otfr. Müller (*Prolegomena zu einer wissenschaftlichen Mythologie*), in Olymp. 31, and according to Letronne's investigation (*Essai sur les Idées Cosmographiques qui se rattachent au nom d'Atlas*, p. 9), in Olymp. 35, 1, or in the year 640. The epoch depends, however, on the foundation of Cyrene, which is placed by Otfr. Müller between Olymp. 35 and 37 (*Minyer*, s. 344, *Prolegomena*, s. 63) ; for in the time of Colæus (Herod., iv., 152), the way from Thera to Libya was not as yet known. Zumpt places the foundation of Carthage in 878, and that of Gades in 1100 B.C.

† According to the manner of the ancients (Strabo, lib. ii., p. 126), I reckon the whole Euxine, together with the Mæotis (as required by physical and geological views), to be included in the common basin of the great " Inner Sea."

between the north of Europe and Asia, and subsequently with the Oxus and Indus, so the Samians* and Phocæans† were the first among the Greeks who endeavored to penetrate from the basin of the Mediterranean toward the west.

Colæus of Samos sailed for Egypt, where, at that time, an intercourse had begun, under Psammitichus, with the Greeks, which probably was only the renewal of a former connection. He was driven by easterly storms to the island of Platæa, and from thence Herodotus significantly adds, " not without divine direction," through the straits into the ocean. The accidental and unexpected commercial gain in Iberian Tartessus conduced less than the discovery of an entrance into an unknown world (whose existence was scarcely conjectured as a mythical creation of fancy) toward giving to this event importance and celebrity wherever the Greek language was understood on the shores of the Mediterranean. Beyond the Pillars of Hercules (earlier known as the Pillars of Briareus, of Ægæon, and of Cronos), at the western margin of the earth, on the road to Elysium and the Hesperides, the primeval waters of the circling Oceanus‡ were first seen, in which the source of all rivers was then sought.

At Phasis the navigators of the Euxine again found themselves on a coast beyond which a *Sun Lake* was supposed to be situated, and south of Gadeira and Tartessus their eyes for the first time ranged over a boundless waste of waters. It was this circumstance which, for fifteen hundred years, gave to the gate of the inner sea a peculiar character of importance. Ever striving to pass onward, Phœnicians, Greeks, Arabs, Catalans, Majorcans, Frenchmen from Dieppe and La Rochelle, Genoese, Venetians, Portuguese, and Spaniards in turn attempted to advance across the Atlantic Ocean, long held to be a miry, shallow, dark, and misty sea, *Mare tenebrosum;* until, proceeding from station to station, as it were, these southern nations, after gaining the Canaries and the

* Herod., iv., 152.

† Herod., i., 163, where even the discovery of Tartessus is ascribed to the Phocæans; but the commercial enterprise of the Phocæans was seventy years after the time of Colæus of Samos, according to Ukert (*Geogr. der Griechen und Römer*, th. 1, i., s. 40).

‡ According to a fragment of Phavorinus, ὠκεανός (and therefore ὠγήν also) are not Greek words, but merely borrowed from the barbarians (Spohn, *De Nicephor. Blemm. duobus Opusculis*, 1818, p. 23). My brother was of opinion that they were connected with the Sanscrit roots *ogha* and *ogh*. (See my *Examen Critique de l'Hist. de la Géogr.*, t. i., p. 33 and 182.)

Azores, finally came to the New Continent, which, however, had already been reached by the Northmen at an earlier period and from a different direction.

While Alexander was opening the far East, the great Stagirite* was led, by a consideration of the form of the earth, to conceive the idea of the proximity of India to the Pillars of Hercules ; while Strabo had even conjectured that there might be " *many other habitable tracts of land*† in the northern hemisphere, perhaps in the parallel which passes through those Pillars, the island of Rhodes and Thinæ, between the coasts of Western Europe and Eastern Asia." The hypothesis of the locality of such lands, in the prolongation of the major axis of the Mediterranean, was connected with a grand geographical view of Eratosthenes, current in antiquity, and in accordance with which the whole of the Old Continent, in its widest extension from west to east, and nearly in the thirty-sixth degree of latitude, was supposed to present an almost continuous line of elevation.‡

The expedition of Colæus of Samos does not, however, alone indicate an epoch in which the Hellenic races, and the nations to whom their cultivation was transmitted, developed new views that led to the extension of maritime expeditions, but it also immediately enlarged the sphere of ideas. The great natural phenomenon which, by the periodic elevation of the level of the sea, exhibits the connection existing between the earth, and the sun, and moon, now first permanently arrested the attention of men. In the African Syrtic Sea this phenomenon had appeared to the Greeks to be accidental, and had not unfrequently been attended by danger. Posidonius, who had observed the ebb and flow of the sea at Ilipa and

---

* Aristot., *De Cœlo*, ii., 14 (p. 298, b., Bekk.); *Meteor.*, ii., 5 (p. 362, Bekk.). Compare my *Examen Critique*, t. i., p. 125–130. Seneca ventures to say (*Nat. Quæst.*, in præfat., 11), " Contemnet curiosus spectator domicilii (terræ) angustias. Quantum enim est quod ab ultimis littoribus Hispaniæ usque ad Indos jacet ? Paucissimorum dierum spatium, si navem suus ventus implevit." (*Examen Critique*, t. i., p. 158.)

† Strabo, lib. i., p. 65 and 118, Casaub. (*Examen Critique*, t. i., p. 152.)

‡ In the Diaphragma of Dicæarchus, by which the earth is divided, the elevation passes through the Taurus, the chains of Demavend and Hindoo-Coosh, the Northern Thibetian Kuen-lun, and the mountains of the Chinese provinces Sse-tschuan and Kuang-si, which are perpetually covered with snow. See my orographical researches on these lines of elevation in my *Asie Centrale*, t. i., p. 104–114, 118–164; t. ii., p. 413 and 438.

Gadeira, compared his observations with the facts of which he
was informed by the experienced Phœnicians concerning the
influence supposed to be exercised by the moon.*

EXPEDITIONS OF THE MACEDONIANS UNDER ALEXANDER THE GREAT.
—CHANGES IN THE RELATIONS OF THE WORLD.—FUSION OF THE
WEST WITH THE EAST.—THE GREEKS PROMOTE THE INTERMIX-
TURE OF RACES FROM THE NILE TO THE EUPHRATES, THE JAX-
ARTES, AND THE INDUS.—SUDDEN EXTENSION OF COSMICAL VIEWS,
BOTH BY MEANS OF DIRECT OBSERVATION OF NATURE, AND BY THE
RECIPROCAL INTERCOURSE OF ANCIENT CIVILIZED AND INDUSTRIAL
NATIONS.

THE campaigns of the Macedonians under Alexander the
Great, the downfall of the Persian dominion, the rising intercourse with Western India, and the action of the Græco-Bactrian empire, which continued to prevail for one hundred and
sixteen years, may be regarded as among the most important
social epochs in the process of the development of the history
of mankind, as far as it indicates a closer connection of Southern Europe with the southwest of Asia, the Nile, and Libya.
Independently of the almost immeasurable extension opened
to the sphere of development by the advance of the Macedonians, their campaigns acquired a character of profound moral
greatness by the incessant efforts of the conqueror to amalgamate all races, and to establish, under the noble influence of
Hellenism, a unity throughout the world.† The foundation
of many new cities at points, the selection of which indicates
higher aims, the arrangement and classification of an independently responsible form of government for these cities, and
the tender forbearance evinced by Alexander for national customs and national forms of worship, all testify that the plan
of one great and organic whole had been laid. That which
was perhaps originally foreign to a scheme of this kind developed itself subsequently from the nature of the relations, as is
always the case under the influence of comprehensive events.
If we remember that only fifty-two Olympiads intervened
from the battle of the Granicus to the destructive irruption
into Bactria of the Sacæ and Tochi, we shall be astonished
at the permanence and the magical influence exercised by the

* Strabo, lib. iii., p. 173 (*Examen Crit.*, t. iii., p. 98).
† Droysen, *Gesch. Alexanders des Grossen*, s. 544; the same in his
Gesch. der Bildung des Hellenistischen Staatensystems, s. 23–34, 588–
592, 748–755.

introduction from the west of Hellenic cultivation. This cultivation, blended with the knowledge of the Arabians, the modern Persians and Indians, extended its influence in so great a degree even to the time of the Middle Ages, that it is often difficult to determine the elements which are due to Greek literature, and those which have originated, independently of all admixture, from the inventive spirit of the Asiatic races.

The principle of unity, or, rather, the feeling of the beneficent political influence incorporated in this principle, was deeply implanted in the breast of the great conqueror, as is testified by all the arrangements of his polity ; and its application to Greece itself was a subject that had already early been inculcated upon him by his great teacher. In the *Politica* of Aristotle we read as follows :* " The Asiatic nations are not deficient in activity of mind and artistic ingenuity, yet they live in subjection and servitude without evincing the courage necessary for resistance, while the Greeks, valiant and energetic, living in freedom, and, therefore, well governed, *might, if they were united into one state, exercise dominion over all barbarians.*" Thus wrote the Stagirite during his second stay at Athens,† before Alexander had passed the Granicus. These dogmas of the philosopher, however contrary to nature he may have professed to consider an unlimited dominion (the παμβασιλεία), no doubt made a more vivid impression on the conqueror than the fantastic narrations of Ctesias respecting India, to which August Wilhelm von Schlegel, and, prior to him, Ste. Croix, ascribed so important an influence.‡

In the preceding pages we have attempted to give a brief delineation of the sea as a means of furthering international contact and union, and of the influence exercised in this respect by the extended navigation of the Phœnicians, Carthaginians, Tyrrhenians, and Etruscans. We have further shown how the Greeks, whose maritime power was strengthened by numerous colonies, endeavored to penetrate beyond the basin of the Mediterranean toward the east and the west by the Argonautic expedition from Iolcus, and by the voyage of Colæus of Samos ; and, lastly, how the fleet of Solomon and Hiram visited distant gold lands in their voyages to Ophir through the Red Sea. The present section will lead us to the

---

* Aristot., *Polit.*, vii., 7, p. 1327, Bekker. (Compare, also, iii., 16, and the remarkable passage of Eratosthenes in Strabo, lib. i., p. 66 and 97, Casaub.)   † Stahr, *Aristotelia*, th. ii., s. 114.
‡ Ste. Croix, *Examen Critique des Historiens d'Alexandre*, p. 721. (Schlegel, *Ind. Bibliothek*, bd. i., s. 150.)

interior of a great continent, through different routes opened
to inland trade and river navigation.   In the short period of
twelve years are compressed the campaigns in Western Asia
and Syria, with the battles of the Granicus, and the passes of
Issus ; the taking of Tyre, and the easy conquest of Egypt ;
the Persico-Babylonian campaign, when the dominion of the
Achæmenidæ was annihilated at Arbela, in the plain of Gau-
gamela ; the expedition to Bactria and Sogdiana, between the
Hindoo-Coosh and the Jaxartes (Syr) ; and, lastly, the bold
advance into the country of the five rivers, the Pentapotamia
of Western India.   Alexander founded Greek colonies almost
every where, and diffused Greek manners and customs over
the vast tracts of land that extend from the Temple of Am-
mon in the Libyan Oasis, and from Alexandria on the Western Delta of the Nile to Alexandria on the Jaxartes, the present Khodjend in Fergana.

The extension of the sphere of new ideas—and this is the
point of view from which the Macedonian expeditions and the
prolonged duration of the Bactrian empire must be considered
—was owing to the magnitude of the space made known, and
to the variety of climates manifested from Cyropolis on the
Jaxartes (in the latitude of Tiflis and Rome) to the eastern
delta of the Indus at Tira, under the tropic of Cancer.   To
these we may further add the wonderful diversity in the configuration of the country, which alternated in luxurious and
fruitful districts, in arid plains and snow-crowned mountain
ranges ; the novelty and gigantic size of animal and vegetable
forms ; the aspect and geographical distribution of races of
men of various color ; the actual contact with Oriental nations in some respects so highly gifted and enjoying a civilization of almost primitive antiquity, and an acquaintance with
their religious myths, systems of philosophy, astronomical
knowledge, and astrological phantasies.   In no age, excepting only the epoch of the discovery and opening of tropical
America, eighteen centuries and a half later, has there been
revealed, at one time and to one race, a richer field of new
views of nature, or a greater mass of materials for laying the
foundation of a physical knowledge of the earth, and of comparative ethnological science.   The vividness of the impression thus produced is testified by the whole literature of the
West, and is also manifested by the doubts—such as accompany, in all cases, an appeal to the imagination in the description of natural scenery—which were excited in Greek, and
subsequently in Roman writers, by the narrations of Megas-

thenes, Nearchus, Aristobulus, and other companions of Alexander's campaigns. These narrators, influenced by the tone of feeling characteristic of their age, and closely connecting together facts and individual opinions, have experienced the varying fate of all travelers, meeting at first with bitter animadversion, and subsequently with a milder judgment. The latter has been more frequent in our own day, since a more profound study of Sanscrit, a more general knowledge of geographical names, the discovery of Bactrian coins in Topes, and, above all, an actual acquaintance with the country and its organic productions, have placed more correct elements of information at the disposal of the critic than those yielded to the partial knowledge of the caviling Eratosthenes, or of Strabo and Pliny.*

If we compare, according to differences in longitude, the length of the Mediterranean with the distance from west to east which separates Asia Minor from the shores of the Hyphasis (Beas), from *the Altars of Return*, we shall perceive that

* Compare Schwanbeck, "*De fide Megasthenis et pretio*," in his edition of that writer, p. 59–77. Megasthenes frequently visited Palibothra, the court of the King of Magadha. He was deeply initiated in the study of Indian chronology, and relates "how, in past times, the All had three times come to freedom ; how three ages of the world had run their course, and how the fourth had begun in his own time" (Lassen, *Indische Alterthumskunde*, bd. i., s. 510). Hesiod's doctrine of four ages of the world, as connected with four great elementary destructions, which together embrace a period of 18,028 years, is also to be met with among the Mexicans. (Humboldt, *Vues des Cordillères et Monumens des Peuples indigènes de l'Amérique*, t. ii., p. 119–129.) A remarkable proof of the exactness of Megasthenes has been discovered in modern times by the study of the *Rigveda* and of the *Mahabharata*. Consult what Megasthenes relates concerning "the land of the long-living blessed beings" in the most northern parts of India—the land of Uttarakuru (probably north of Kashmeer, toward Belurtagh), which, according to his Greek views, he associates with the supposed " thousand years of the life of the Hyperboreans." (Lassen, in the *Zeitschrift für die Kunde des Morgenlandes*, bd. ii., s. 62.) A tradition mentioned by Ctesias (who has been too long esteemed below his merits), of a sacred place in the northern desert, may be noticed in connection with this point. (*Ind.*, cap. viii., ed. Baehr, p. 249 and 285.) The martichoras mentioned by Aristotle (*Hist. de Animal.*, ii., 3, § 10 ; t. i., p. 51, Schneider), the griffin, half eagle and half lion, the kartazonon noticed by Ælian, and a one-horned wild ass, are certainly spoken of by Ctesias as real animals ; they were not, however, the creations of his inventive fancy, for he mistook, as Heeren and Cuvier have remarked, the pictured forms of symbolical animals, seen on Persian monuments, for representations of strange beasts still living in the remote parts of India. There is, however, as Guigniaut has well observed, much difficulty in identifying the martichoras with Persepolitan symbols. (Creuzer, *Religions de l'Antiquité : Notes et Eclaircissements*, p. 720.)

the geographical knowledge of the Greeks was doubled in extent in the course of a few years. In order to define more accurately that which we have termed the mass of materials added to the sciences of natural philosophy and physical geography by the different campaigns and by the colonial institutions of Alexander, I would first refer to the diversity in the form of the earth's crust, which has, however, only been more specially made known to us by the experiments and researches of recent times. In the countries through which he passed, low lands, deserts, and salt steppes devoid of vegetation (as on the north of the Asferah chain, which is a continuation of the Thian-schan, and the four large cultivated alluvial districts of the Euphrates, the Indus, Oxus, and Jaxartes), contrasted with snow-clad mountains, having an elevation of nearly 20,000 feet. The Hindoo-Coosh, or Indian Caucasus of the Macedonians, which is a continuation of the North Thibetian Kuen-lun, west of the south transverse chain of Bolor, is divided in its prolongation toward Herat into two great chains bounding Kafiristan,* the southern of which is the loftier of the two. Alexander passed over the plateau of Bamian, which lies at an elevation of about 8500 feet, and in which men supposed they had found the cave of Prometheus,† to the crest of the Kohibaba, and beyond Kabura along the Choes, crossing the Indus somewhat to the north of the present Attok. A comparison between the low Tauric chain, with which the Greeks were familiar, and the eternal snow surmounting the range of the Hindoo-Coosh, and which, according to Burnes, begins at an elevation of 13,000 feet, must have given occasion to a recognition, on a more colossal scale, of the superposition of different zones of climate and vegetation. In active minds direct contact with the elementary world produces the most vivid impression on the senses. And thus we find that Strabo has described, in the most perfectly truthful characters, the passage across the mountainous district of the Paropanisadæ, where the army with difficulty cleared a passage through the snow, and where arborescent vegetation had ceased.‡

* I have considered these intricate orographical relations in my *Asie Centrale*, t. ii., p. 429–434.
† Lassen, in the *Zeitschrift für die Kunde des Morgenl.*, bd. i., s. 230.
‡ The country between Bamian and Ghori. See Carl Zimmermann's excellent orographical work *Uebersichtsblatt von Afghanistan*, 1842. (Compare Strabo, lib. xv., p. 725; Diod. Sicul., xvii., 82; Menu, *Meletem. Hist.*, 1839, p. 25 and 31; Ritter, *Ueber Alexanders Feldzug am Indischen Kaukasus*, in the *Abhandl. der Berl. Akad.* of the year

More certain knowledge was now transmitted to the West from the Macedonian colonies respecting those Indian products of nature and art which had hitherto been only imperfectly known from commercial intercourse, or from the narrations of Ctesias of Cnidus, who lived seventeen years at the court of Persia as physician to Artaxerxes Mnemon. Among the objects thus made known we must reckon irrigated rice-fields, for whose cultivation Aristobulus gives special directions; the cotton-tree, and the fine tissues and the paper for which it* furnished the materials; spices and opium; wine made from rice and the juice of palms, whose Sanscrit name of *tala* has been preserved in the works of Arrian;† sugar from the sugar-cane,‡ which is certainly often confounded in the Greek and Roman writers with the *tabaschir* of the bamboo reed; wool from the great Bombax-tree;§ shawls made of the Thibetian goat's hair; silken (Seric) tissues;|| oil from the white sesamum (Sanscrit *tila*); attar of roses and other perfumes; lac (Sanscrit *lâkschâ*, in the vulgar tongue *lakkha*);¶ and, lastly, the hardened Indian wutz-steel.

Besides the knowledge of these products, which soon became objects of universal commerce, and many of which were transported by the Seleucidæ to Arabia,** the aspect of a rich-

1829, s. 150; Droysen, *Bildung des Hellenist. Staatensystems*, s. 614.) I write Paropanisus, as it occurs in all the good codices of Ptolemy, and not Paropamisus. I have explained the reasons in my *Asie Centrale*, t. i., p. 114–118. (See, also, Lassen, *zur Gesch. der Griechischen und Indoskythischen Könige*, s. 128). * Strabo, lib. xv., p. 717, Casaub.

† *Tala*, the name of the palm *Borassus flabelliformis*, which is very characteristically termed by Amarasinha "a king of the grasses." Arrian, *Ind.*, vii., 3.

‡ The word *tabaschir* is deduced from the Sanscrit *tvakkschîrâ* (bark milk). In 1817, in the historical additions to my work *De distributione Geographicâ Plantarum, secundum cœli, temperiem et altitudinem Montium*, p. 215, I drew attention to the fact that the companions of Alexander learned to know the true sugar of the sugar-cane of the Indians as well as the *tabaschir* of the bamboo. (Strabo, lib. xv., p. 693; *Peripl. Maris Erythr.*, p. 9.) Moses of Chorene, who lived in the middle of the fifth century, was the first (*Geogr.*, ed. Whiston, 1736, p. 364) who circumstantially described the preparation of sugar from the juice of the *Saccharum officinarum*, in the province of Chorasan.

§ Strabo, lib. xv., p. 694.

|| Ritter, *Erdkunde von Asien*, bd. iv., 1, s. 437; bd. vi., 1, s. 698; Lassen, *Ind. Alterthumskunde*, bd. i., s. 317–323. The passage in Aristotle's *Hist. de Animal.*, v. 17 (t. i., p. 209, ed. Schneider), relating to the web of a great horned caterpillar, refers to the island of Cos.

¶ Thus λάκκος χρωμάτινος in the *Peripl. Maris Erythr.*, p. 5 (Lassen, s. 316).

** Plin., *Hist. Nat.*, xvi., 32. (On the introduction of rare Asiatic plants into Egypt by the Ptolemies, see Pliny, xii., 14 and 17.)

ly-embellished tropical nature speedily yielded the Greeks enjoyments of another kind. The gigantic forms of hitherto
unknown animals and plants filled their imaginations with
the most exciting images. Writers, whose dry scientific style
is usually devoid of all animation, became poetic when they
described the characteristics of animals, as, for instance, elephants; or when they spoke of the height of trees, whose
summits can not be reached by the arrow in its flight, and
whose leaves are larger than the shields of the infantry; of
"the bamboo, a light, feathery, tree-like grass," "each of
whose jointed parts (internodia) may serve for a many-oared
keel;" or of the Indian fig-tree, that takes root by its branches,
and whose stem has a diameter of twenty-eight feet, and
which, as Onesicritus remarked, with much truth to nature,
forms "a leafy canopy similar to a tent, supported by numerous pillars." The tall, arborescent ferns, which, according
to my opinion, constitute the greatest ornament of tropical
scenery, are never mentioned by Alexander's companions,* although they speak of the noble, fan-like umbrella palm, and
the delicate and ever-fresh green of the cultivated banana.†

The knowledge of a great portion of the earth may now be
said to have been opened for the first time. The objective
world began to assume a preponderating force over that of
mere subjective creation; and while the fruitful seeds yielded
by the language and literature of the Greeks were scattered

---

* Humboldt, *De distrib. Geogr. Plantarum*, p. 178.

† I have often corresponded, since the year 1827, with Lassen on the
remarkable passage in Pliny, xii., 6: "Major alia (arbor) pomo et su-
avitate præcellentior, quo *sapientes* Indorum vivunt. Folium alas avi-
um imitatur, longitudine trium cubitorum, latitudine duûm. Fructum
córtice mittit, admirabilem succi dulcedine ut uno quaternos satiet.
Arbori nomen *palæ*, pomo *arienæ*." The following is the result of my
learned friend's investigation: "Amarasinha places the banana (*musa,
pisang*) at the head of all nutritive plants. Among the many Sanscrit
names which he adduces are *varanabuscha, bhanuphala* (sun fruit), and
moko, whence the Arabic *mauza*. *Phala* (*pala*) is fruit in general, and
it is therefore only by a misunderstanding that it has been taken for the
name of the plant. In Sanscrit, *varana* without *buscha* is not used as
the name of the banana, although the abbreviation may have been
characteristic of the popular language. *Varana* would be in Greek
ovαρενα, which is certainly not very far removed from *ariena*." (Compare Lassen, *Ind. Alterthumskunde*, bd. i., p. 262; my *Essai Politique
sur la Nouv. Espagne*, t. ii., 1827, p. 382; and *Relat. Hist.*, t. i. p. 491.)
The chemical connection of the nourishing amylum with sugar was de-
tected both by Prosper Alpinus and Abd-Allatif, and they sought to explain the origin of the banana by the insertion of the sugar-cane, or the
sweet date fruit, into the root of the colocasia (Abd-Allatif, *Relation de
l'Egypte*, trad. par Silvestre de Sacy, p. 28 and 105).

abroad by the conquests of Alexander, scientific observation
and the systematic arrangement of the knowledge already acquired were elucidated by the doctrines and expositions of
Aristotle.* We here indicate a happy coincidence of favoring
relations, for, at the very period when a vast amount of new
materials was revealed to the human mind, their intellectual
conception was at once facilitated and multiplied through the
direction given by the Stagirite to the empirical investigation
of facts in the domain of nature, to the profound consideration
of speculative hypothesis, and to the development of a language of science based on strict definition. Thus Aristotle
must still remain, for thousands of years to come, as Dante
has gracefully termed him,

" *Il maestro di color che sanno.*"†

The belief in the direct enrichment of Aristotle's zoological
knowledge by means of the Macedonian campaigns has, however, either wholly disappeared, or, at any rate, been rendered
extremely uncertain by recent and more carefully-conducted
researches. The wretched compilation of a life of the Stagirite, which was long ascribed to Ammonius, the son of Hermias, had contributed to the diffusion of many erroneous
views, and, among others, to the belief that‡ the philosopher
accompanied his pupil as far, at least, as the shores of the
Nile.§ The great work on Animals appears to have been
written only a short time after the *Meteorologica*, the date
of which would seem, from internal evidence,‖ to fall in the

* Compare, on this epoch, Wilhelm von Humboldt's work, *Ueber
die Kawi-Sprache und die Verschiedenheit des menschlichen Sprachbaues,*
bd. i., s. ccl. und ccliv. ; Droysen, *Gesch. Alexanders des Gr.,* s. 547 ; and
Hellenist. Staatensystem, s. 24.   † Dante, *Inf.,* iv., 130.

‡ Compare Cuvier's assertions in the *Biographie Universelle,* t. ii.,
1811, p. 458 (and unfortunately again repeated in the edition of 1843,
t. ii., p. 219), with Stahr's *Aristotelia,* th. i., s. 15 und 108.

§ Cuvier, when he was engaged on the *Life of Aristotle,* inclined to
the belief of the philosopher having accompanied Alexander to Egypt,
" whence," he says, " the Stagirite must have brought back to Athens
(Olymp. 112, 2) all the materials for the *Historia Animalium.*" Subsequently (1830) the distinguished French naturalist abandoned this opinion, because, after a more careful examination, he remarked "that the
descriptions of Egyptian animals were not sketched from life, but from
notices by Herodotus." (See, also, Cuvier, *Histoire des Sciences Naturelles,* publiée par Magdeleine de Saint Agy, t. i., 1841, p. 136.)

‖ To these internal indications belong the statement of the perfect
insulation of the Caspian Sea ; the notice of the great comet, which appeared under Nicomachus when holding the office of archon, Olymp.
109, 4 (according to Corsini), and which is not to be confounded with
that which Von Boguslawski has lately named the comet of Aristotle

106th, or, at the latest, in the 111th Olympiad, and, therefore, either fourteen years before Aristotle came to the court
of Philip, or, at the furthest, three years before the passage
across the Granicus. It must, however, be admitted, that
some few facts may be advanced as evidence against this assumption of an early completion of the nine books of Aristotle's History of Animals. Among these must be reckoned the
accurate knowledge possessed by Aristotle of the elephant, the
bearded horse-stag (hippelaphus), the Bactrian two-humped
camel, the hippardion, supposed to be the hunting-tiger (guepard), and the Indian buffalo, which does not appear to have
been introduced into Europe before the time of the Crusades.
But here it must be remarked that the native place of this
large and singular stag, having a horse's mane, which Diard
and Duvancel sent from Eastern India to Cuvier, who gave
to it the name of *Cervus Aristotelis*, is, according to Aristotle's own account, not the Indian Pentapotamia traversed by
Alexander, but Arachosia, west of Candahar, which, together
with Gedrosia, constituted one satrapy of ancient Persia.*

(under the Archon Asteus, Olymp. 101, 4; Aristot., *Meteor.*, lib. i., cap.
6, 10; vol. i., p. 395, Ideler; and which is probably identical with the
comets of 1695 and 1843?); and, lastly, the mention of the destruction
of the temple at Ephesus, as well as of a lunar rainbow, seen on two
occasions in the course of fifty years. (Compare Schneider, *ad Aristot.,*
Hist. de Animalibus, vol. i., p. xl., xlii., ciii., and exx.; Ideler, *ad Aris-*
tot. Meteor., vol. i., p. x.; and Humboldt, *Asie Cent.*, t. ii., p. 168.) We
know that the *Historia Animalium* "was written later than the *Meteor-*
ologica," from the fact that allusion is made in the last-named work
to the former as to a work about to follow (*Meteor.*, i., 1, 3, and iv.,
12, 13).
   * The five animals named in the text, and especially the hippelaphus
(horse-stag with a long mane), the hippardion, the Bactrian camel, and
the buffalo, are instanced by Cuvier as proofs of the later composition
of Aristotle's *Historia Animalium* (*Hist. des Sciences Nat.*, t. i., p. 154).
Cuvier, in the fourth volume of his admirable *Recherches sur les Osse-*
mens Fossiles, 1823, p. 40–43 and 502, distinguishes between two Asiatic
stags with manes, which he calls Cervus hippelaphus and Cervus Aristotelis. He originally regarded the first-named, of which he had seen
a living specimen in London, and of which Diard had sent him skins
and antlers from Sumatra, as Aristotle's hippelaphus from Arachosia
(*Hist. de Animal.*, ii., 2, § 3, and 4, t. i., p. 43, 44, Schneider); but he
afterward thought that a stag's head, sent to him from Bengal by Du-
vaucel, agreed still better, according to the drawing of the entire large
animal, with the Stagirite's description of the hippelaphus. This stag,
which is indigenous in the mountains of Sylhet in Bengal, in Nepaul,
and in the country east of the Indus, next received the name of Cervus
Aristotelis. If, in the same chapter in which Aristotle speaks generally
of animals with manes, the horse-stag (Equicervus), and the Indian
guepard, or hunting tiger (Felis jubata), are both understood, Schneider

May not the knowledge of the form and habits of the animals above referred to, and which, for the most part, was comprised in short notices, have been transmitted to Aristotle, independently of the Macedonian campaigns, either from Persia or from Babylon, which was the seat of a widely-extended foreign commercial intercourse? Owing to the utter ignorance that prevailed at this time of the preparation of alcohol,* nothing but

(t. iii., p. 66) considers the reading πάρδιον preferable to that of τὸ ἱππάρδιον. The latter reading would be best interpreted to mean the giraffe, as Pallas also conjectures (*Spicileg. Zool.*, fasc. i., p. 4). If Aristotle had himself seen the guepard, and not merely heard it described, how could he have failed to notice non-retractile claws in a feline animal? It is also surprising that Aristotle, who is always so accurate, if, as August Wilhelm von Schlegel maintains, he had a menagerie near his residence at Athens, and had himself dissected one of the elephants taken at Arbela, should have failed to describe the small opening near the temples of the animal, where, at the rutting season, a strong-smelling fluid is secreted, often alluded to by the Indian poets. (Schlegel's *Indische Bibliothek*, bd. i., s. 163–166.) I notice this apparently trifling circumstance thus particularly, because the above-mentioned small aperture was made known to us from the accounts of Megasthenes, to whom, nevertheless, no one would be led to ascribe anatomical knowledge. (Strabo, lib. xv., p. 704 and 705, Casaub.) I find nothing in the different zoological works of Aristotle which have come down to us that necessarily implies his having had the opportunity of making direct observations on elephants, or of his having dissected any. Although it is most probable that the *Historia Animalium* was completed before Alexander's campaigns in Asia Minor, there is undoubtedly a possibility that the work may, as Stahr supposes (*Aristotelia*, th. ii., s. 98), have continued to receive additions until the end of the author's life, Olymp. 114, 3, and therefore three years after the death of Alexander ; but we have no direct evidence on this subject. That which we possess of the correspondence of Aristotle is undoubtedly not genuine (Stahr, th. i., s. 194–208 ; th. ii., s. 169–234); and Schneider says very confidently (*Hist. de Animal.*, t. i., p. xl.), "hoc enim tanquam certissimum sumere mihi licebit, scriptas comitum Alexandri notitias post mortem demum regis fuisse vulgatas."
 * I have elsewhere shown that, although the decomposition of sulphuret of mercury by distillation is described in Dioscorides (*Mat. Med.*, v. 110, p. 667, Saracen.), the first description of the distillation of a fluid (the distillation of fresh water from sea water) is, however, to be found in the commentary of Alexander of Aphrodisias to Aristotle's work *De Meteorol.* See my *Examen Critique de l'Histoire de la Géographie*, t. ii., p. 308–316, and Joannis (Philoponi), *Grammatici in libro de Generat. et Alexandri Aphrod.*, *in Meteorol. Comm.*, Venet., 1527, p. 97, b. Alexander of Aphrodisias in Caria, the learned commentator of the *Meteorologica* of Aristotle, lived under Septimius Severus and Caracalla ; and although he calls chemical apparatuses χυικὰ ὄργανα, yet a passage in Plutarch (*De Iside et Osir.*, c. 33) proves that the word *Chemie*, applied by the Greeks to the Egyptian art, is not derived from χέω. Hoefer (*Histoire de la Chimie*, t. i., p. 91, 195, and 219 ; t. ii., p. 109).

the skins and bones of animals, and not the soft parts capable of dissection, could be sent from remote parts of Asia to Greece. However probable it may be that Aristotle received the most liberal aid from Philip and Alexander for the furtherance of his studies in physical science, for procuring an immense number of zoological specimens both from Greece and the neighboring seas, and for forming a collection of books unique in that age, and which passed successively into the hands, first of Theophrastes, and afterward of Neleus of Skepsis, we must nevertheless regard the accounts of " the presents of eight hundred talents, and the maintenance of so many thousand collectors, overseers of fish-ponds, and bird-keepers," as mere exaggerations of a later period, or as traditions misunderstood by Pliny, Athenæus, and Ælian.*

The Macedonian campaign, which opened so large and beautiful a portion of the earth to the influence of one sole highly-gifted race, may therefore certainly be regarded, in the strictest sense of the word, as a *scientific* expedition, and, moreover, as the first in which a conqueror had surrounded himself with men learned in all departments of science, as naturalists, geometricians, historians, philosophers, and artists. The results that we owe to Aristotle are not, however, solely to be referred to his own personal labors, for he acted also through the intelligent men of his school who accompanied the expedition. Among these shone pre-eminently Callisthenes of Olynthus, the near kinsman of the Stagirite, who had

---

* Compare Sainte-Croix, *Examen des Historiens d'Alexandre*, 1810, p. 207 ; and Cuvier, *Histoire des Sciences Naturelles*, t. i., p. 137, with Schneider, *ad Aristot. de Historiâ Animalium*, t. i., p. xlii., xlvi., and Stahr, *Aristotelia*, th. i., s. 116–118. If, therefore, the transmission of specimens from Egypt and the interior of Asia seems to be highly improbable, yet the latest writings of our great anatomist, Johannes Müller, show with what wonderful delicacy Aristotle dissected the fishes of the Greek seas. See the learned treatise of Johannes Müller, on the adherence of the ovum to the uterus, in one of the two species of the genus Mustelus living in the Mediterranean, which in its fœtal state possesses a placenta of the vitelline vesicle connected with the uterine placenta of the mother, and his researches on the γαλεός λεῖος of Aristotle, in the *Abhandl. der Berliner Akad. aus dem Jahr* 1840, s. 192–197. (Compare Aristot., *Hist. Anim.*, vi., 10, and *De Gener. Anim.*, iii., 3.) The distinction and detailed analysis of the species of cuttle-fish, the description of the teeth of snails, and the organs of other gasteropodes, all testify to the delicate nicety of Aristotle's own anatomical examinations. Compare *Hist. Anim.*, iv., 1 and 4, with Lebert, in Müller's *Archiv der Physiologie*, 1846, s. 463 und 467. I myself, in 1797, called the attention of modern naturalists to the form of snails' teeth. See my *Versuche über die gereizte Muskel und Nervenfaser*, bd. i., s. 261.

already, before the campaign, composed a work on Botany, and a treatise on the organs of vision. Owing to the rigid austerity of his morals, and the unchecked freedom of his speech, he was regarded with hatred by Alexander himself, who had already fallen from his noble and elevated mode of thought, and by the flatterers of the prince. Callisthenes undauntedly preferred liberty to life; and when, in Bactria, he was implicated, although guiltless, in the conspiracy of Hermolaus and the pages, he became the unhappy occasion of Alexander's exasperation against his former instructor. Theophrastes, the warm friend and fellow-disciple of Callisthenes, had the generosity to undertake his defense after his fall. Of Aristotle we only know that he recommended prudence to his friend before his departure; for being, as it would appear, familiar with a court life from his long sojourn with Philip of Macedon, he counseled him to " converse as little as possible with the king, and,· where necessity required that he should do so, always to coincide with the views of the sovereign."*

Aided by the co-operation of chosen men of the school of the Stagirite, Callisthenes, who was already conversant with nature before he left Greece, gave a higher direction to the investigations of his companions in the extended sphere of observation now first opened to them. The richness of vegetation and the diversity of animal forms, the configuration of the soil and the periodical rising of great rivers, no longer sufficed to engage exclusive attention, for the time was come when man and the different races of mankind, in their manifold gradations of color and of civilization, could not fail to be regarded, according to Aristotle's own expression,† " as the central point and the object of all creation, and as the beings in whom the divine nature of thought was first made manifest." From the little that remains to us of the narratives of Onesicritus, who was so much censured in antiquity, we find that the Macedonians were astonished, on penetrating far to the East, to meet with no African, curly-haired negroes, although they found the Indian races spoken of by Herodotus as " dark colored, and resembling Ethiopians."‡ The influence of the atmosphere on color, and the different effect produced by dry and moist winds, were carefully noticed. In the early Homeric

* *Valer. Maxim.*, vii., 2: " ut cum rege aut rarissime aut quam jucundissime loqueretur."
† Aristot., *Polit.*, i., 8, and *Eth. ad Eudemum*, vii., 14.
‡ Strabo, lib. xv., p. 690 and 695. Herod., iii., 101.

ages, and even long after that period, the dependence of the temperature of the air on latitude was wholly unknown, and the relations of east and west then constituted the whole thermic meteorology of the Greeks. The countries lying to the east were regarded as near the sun—*sun lands*, and the inhabitants as "colored by the near sun-god in his course with a sooty luster,* and their hair dried and crisped with the heat of his rays."

Alexander's campaigns first gave occasion to a comparison, on a grand scale, between the African races which predominated so much in Egypt with the Arian races beyond the Tigris and the ancient Indian Aborigines, who were very dark colored, but not woolly haired. The classification of mankind into varieties, and their distribution over the surface of the earth, which is to be regarded rather as a consequence of historical events than as the result of protracted climatic relations (when the types have been once firmly fixed), together with the apparent contradiction between color and places of abode, were subjects that could not fail to produce the most vivid impression on the mind of thoughtful observers. We still find, in the interior of the great Indian continent, an extensive territory, which is inhabited by a population of dark, almost black aborigines, totally different from the lighter-colored Arian races, who immigrated at a subsequent period. Among these we may reckon, as belonging to the Vindhya races, the Gonda, the Bhilla in the forest districts of Malava and Guzerat, and the Kola of Orissa. The acute observer Lassen regards it as probable that, at the time of Herodotus, the black Asiatic races, "the Ethiopians of the sun-rising," which resembled the Libyans in the color of their skin, but not in the character of their hair, were diffused much further toward the northwest than at present.† In like manner, in the ancient Egyptian empire, the actual woolly-hair-

---

* Thus says Theodectes of Phaselis: see vol. i., p. 353. Northern tracts of land were considered to lie more toward the west, and southern countries to the east. Consult Völcker, *Ueber Homerische Geographie und Weltkunde,* s. 43 und 87. The indefinite meaning of the word Indies, even at that age, as connected with ideas of position, of the complexion of the inhabitants, and of precious products, contributed to the extension of these meteorological hypotheses; for Western Arabia, the countries between Ceylon and the mouth of the Indus, Troglodytic Ethiopia, and the African myrrh and cinnamon lands south of Cape Aroma, were all termed India. (Humboldt, *Examen Crit.,* t. ii., p. 35.)

† Lassen, *Ind. Alterthumskunde,* bd. i., s. 369, 372–375, 379, und 389; Ritter, *Asien,* bd. iv., 1, s. 446.

ed negro races, which were so frequently conquered by other nations, moved their settlements far to the north of Nubia.*

The enlargement of the sphere of ideas, which arose from the contemplation of numerous hitherto unobserved physical phenomena, and from a contact with different races, and an acquaintance with their contrasted forms of government, was not, unfortunately, accompanied by the fruits of ethnological comparative philology, as far as the latter is of a philosophical nature depending on the fundamental relations of thought, or is simply historical.† This species of inquiry was wholly unknown to classical antiquity. But, on the other hand, Alexander's expedition added to the science of the Greeks those materials yielded by the long-accumulated knowledge of more anciently civilized nations. I would here especially refer to the fact that, with an increased knowledge of the earth and its productions, the Greeks likewise obtained from Babylon a considerable accession to their knowledge of the heavens, as we find from recent and carefully-conducted investigations. The conquest of Cyrus the Great had certainly greatly diminished the glory of the astronomical college of the priests in the Oriental capital. The terraced pyramid of Belus (at once a temple, a grave, and an observatory, from which the hours of the night were proclaimed) had been given over to destruction by Xerxes, and was in ruins at the time of the Macedonian campaign. But from the very fact of the dissolution of the close hierarchical caste, and owing to the formation of many schools of astronomy,‡ Callisthenes was enabled (and as Sim-

* The geographical distribution of mankind can no more be determined in entire continents by degrees of latitude than that of plants and animals. The axiom advanced by Ptolemy (*Geogr.*, lib. i., cap. 9), that north of the parallel of Agisymba there are no elephants, rhinoceroses, or negroes, is entirely unfounded (*Examen Critique*, t. i., p. 39). The doctrine of the universal influence of the soil and climate on the intellectual capacities and on the civilization of mankind, was peculiar to the Alexandrian school of Ammonius Sakkas, and more especially to Longinus. See Proclus, *Comment. in Tim.*, p. 50.

† See Georg. Curtius, *Die Sprachvergleichung in ihrem Verhältniss zur Classischen Philologie*, 1845, s. 5–7, and his *Bildung der Tempora und Modi*, 1846, s. 3–9. (Compare, also, Pott's Article, *Indogermanischer Sprachstamm*, in the *Allgem. Encyklopädie* of Ersch and Gruber, sect. ii., th. xviii., s. 1–112.) Investigations on language in general, in as far as they touch upon the fundamental relations of thought, are, however, to be found in Aristotle, where he develops the connection of categories with grammatical relations. See the luminous statement of this comparison in Adolf Trendelenburg's *Histor. Beiträge zur Philosophie*, 1846, th. i., s. 23–32.

‡ The schools of the Orchenes and Borsipenes (Strabo, lib. xvi., p.

plicius maintains, in accordance with the advice of Aristotle) to send to Greece observations of the stars for a very long period (Porphyrius says for 1903 years) before Alexander's entrance into Babylon, Ol. 112, 2. The earliest Chaldean observations mentioned by Almagest (probably the oldest which Ptolemy found available for his object) only go back 721 years before our era, that is to say, to the first Messenian war. It is certain "that the Chaldeans knew the mean motions of the moon with an exactness which induced the Greek astronomers to employ their calculations for the foundation of a lunar theory."* The planetary observations to which they were led by their ancient love of astrology appear also to have been used for the true construction of astronomical tables.

The present is not the place to decide how much of the Pythagorean views regarding the true structure of the heavens, the course of the planets, and of the comets which, according to Apollonius Myndius, return in long regulated orbits,† may be due to the Chaldeans. Strabo calls the mathematician Seleucus a Babylonian, and distinguishes him in this manner‡ from the Erythræan, who measured the tides of the sea. It is sufficient to remark that the Greek zodiac was most probably taken from "the Dodecatemoria of the Chaldeans, and that, according to Letronne's important investigations,§ it does

739). In this passage four Chaldean mathematicians are indicated by name, in conjunction with the Chaldean astronomers. This circumstance is so much the more important in an historical point of view, because Ptolemy always mentions the observers of the heavenly bodies under the collective name of Χαλδαῖοι, as if the observations at Babylon were only made collectively in collegiate bodies (Ideler, *Handbuch der Chronologie*, bd. i., 1825, s. 198).

* Ideler, op. cit., bd. i., s. 202, 206, und 218. When a doubt is advanced regarding the astronomical observations said to have been sent by Callisthenes from Babylon to Greece, on the ground that there is no trace of these observations of a Chaldean priestly caste to be found in the writings of Aristotle (Delambre, *Hist. de l'Astronomie Ancienne*, t. i., p. 308), it is forgotten that Aristotle, in speaking (*De Cœlo*, lib. ii., cap. 12) of an occultation of Mars by the Moon, observed by himself, expressly adds, that "similar observations had been made for many years on the other planets by the Egyptians and the Babylonians, many of which have come to our knowledge." On the probable use of astronomical tables by the Chaldeans, see Chasles, in the *Comptes Rendus de l'Académie des Sciences*, t. xxiii., 1846, p. 852–854.

† Seneca, *Nat. Quæst.*, vii., 17.

‡ Compare Strabo, lib. xvi., p. 739, with lib. iii., p. 174.

§ These investigations were made in the year 1824 (see Guigniaut, *Religions de l'Antiquité, ouvrage traduit de l'Allemand de F. Creuzer*, t. i., Part ii., p. 928). See a more recent notice by Letronne, in the *Journal des Savans*, 1839, p. 338 and 492, as well as the *Analyse Cri-*

not go further back than to the beginning of the sixth century before our era."

The direct result of the contact of the Hellenic races with nations of Indian origin at the time of the Macedonian expedition is wrapped in obscurity. In a scientific point of view the gain was probably inconsiderable, since Alexander did not advance beyond the Hyphasis, in the land of the five rivers (the Pantschanada), after he had traversed the kingdom of Porus between the Hydaspes (Jelum), skirted by cedars* and the Acesines (Tschinab) ; he reached the point of junction, however, between the Hyphasis and the Satadru, the Hesidrus of Pliny. Discontent among his troops, and the apprehension of a general revolt in the Persian and Syrian provinces, forced the hero to the great catastrophe of his return, notwithstanding his wish to advance to the Ganges. The countries traversed by the Macedonians were occupied by races who were but imperfectly civilized. In the territories intervening between the Satadru and the Yamuna (the district of the Indus and Ganges), an insignificant river, the sacred Sarasvati, constitutes an ancient classical boundary between the "pure, worthy, pious" worshipers of Brahma in the East, and the "impure kingless" tribes in the West, which are not divided into castes.† Alexander did not, therefore, reach the true seat

---

tique des Représentations Zodiacales en Egypte, 1846, p. 15 and 34. (Compare with these Ideler, *Ueber den Ursprung des Thierkreises,* in the *Abhandlungen der Akademie der Wissenschaften zu Berlin aus dem Jahr* 1838, s. 21.)

* The magnificent groves of Cedrus deodvara, which are most frequently met with at an elevation of from 8000 to nearly 12,000 feet on the Upper Hydaspes (Behut), which flows through the Pilgrim's Lake in the Alpine Valley of Kashmeer, supplied the materials for the fleet of Nearchus (Burnes's *Travels,* vol. i., p. 60). The trunk of this cedar is often forty feet in circumference, according to the observation of Dr. Hoffmeister, the companion of Prince Waldemar of Prussia, who was unhappily too early lost to science by his death on the battle field.

† Lassen, in his *Pentapotamia Indica,* p. 25, 29, 57–62, and 77 ; and also in his *Indische Alterthumskunde,* bd. i., s. 91. Between the Sarasvati in the northwest of Delhi, and the rocky Drischadvati, there lies, according to Menu's code of laws, Brahmavarta, a priestly district of Brahma, established by the gods themselves; on the other hand, in the wider sense of the word, Aryavarta, the land of the worthy (Arians), designates in the ancient Indian geography the whole country east of the Indus, between the Himalaya and the Vindhya chain, to the south of which the ancient non-Arian aboriginal population began. Madhya Desa, the middle land referred to in the text, see vol. i., p. 35, was only a portion of Aryavarta. Compare my *Asie Centrale,* t. i., p. 204, and Lassen, *Ind. Alterthumsk.,* bd. i., s. 5, 10, und 93. The ancient Indian free states, the territories of the "kingless" (condemned by orthodox

of higher Indian civilization. Seleucus Nicator, the founder of the great empire of the Seleucidæ, penetrated from Babylon toward the Ganges, and established political relations with the powerful Sandrocottus (Tschandraguptas) by means of the repeated missions of Megasthenes to Pataliputra.*

In this manner a more animated and lasting contact was established with the most civilized portions of Madhya Desa (the middle land). There were, indeed, learned Brahmins living as anchorites in the Pendschab (Pentapotamia), but we do not know whether those Brahmins and Gymnosophists were acquainted with the admirable Indian system of numbers, in which the value of a few signs is derived merely from position, or whether, as we may however conjecture, the value of position was already at that time known in the most civilized portions of India. What a revolution would have been effected in the more rapid development and the easier application of mathematical knowledge, if the Brahmin Sphines, who accompanied Alexander, and who was known in the army by the name of Calanos—or, at a later period, in the time of Augustus, the Brahmin Bargosa—before they voluntarily ascended the scaffold at Susa and Athens, could have imparted to the Greeks a knowledge of the Indian system of numbers in such a manner as to admit of its being brought into general use ! The ingenious and comprehensive investigations of Chasles have certainly shown that the method of the Abacus or Algorismus of Pythagoras, as we find it explained in the geometry of Boëthius, was nearly identical with the Indian numerical system based upon the value of position, but this method, which long continued devoid of practical utility among the Greeks and Romans, first obtained general application in the Middle Ages, and especially when the zero had been substituted for a vacant space. The most beneficent discoveries have often required centuries before they were recognized and fully developed.

Eastern poets), were situated between the Hydraotes and the Hyphasis (the present Ravi and Beas).

* Megasthenes, *Indica*, ed. Schwanbeck, 1846, p. 17.

EXTENSION OF THE CONTEMPLATION OF THE UNIVERSE UNDER THE
PTOLEMIES.—MUSEUM AT SERAPEUM.—PECULIAR CHARACTER OF
THE DIRECTION OF SCIENCE AT THIS PERIOD.—ENCYCLOPÆDIC
LEARNING.—GENERALIZATION OF THE VIEWS OF NATURE RESPECT-
ING THE EARTH AND THE REGIONS OF SPACE.

AFTER the dissolution of the Macedonian empire, which included territories in three continents, those germs were variously developed which the uniting and combining system of
government of the great conqueror had cast abroad in a fruitful soil. The more the national exclusiveness of the Hellenic
mode of thought vanished, and the more its creative force of
inspiration lost in depth and intensity, the greater was the increase in the knowledge acquired of the connection of phenomena by a more animated and extensive intercourse with other
nations, as well as by a rational mode of generalizing views
of nature. In the Syrian kingdom, under the Attalidæ of
Pergamus, and under the Seleucidæ and the Ptolemies, learning was universally favored by distinguished rulers. Grecian-
Egypt enjoyed the advantage of political unity, as well as that
of a geographical position, by which the traffic of the Indian
Ocean was brought within a few miles of the Mediterranean
by the influx of the Arabian Gulf from the Straits of Bab-el-
Mandeb to Suez and Akaba (running in the line of intersection that inclines from south-southeast to north-northwest).*

The kingdom of the Seleucidæ did not enjoy the same advantage of maritime trade as that afforded by the form and
configuration of the territories of the Lagides (the Ptolemies),
and its stability was endangered by the dissensions fomented
by the various nations occupying the different satrapies. The
traffic carried on in the Seleucidean kingdom was besides more
an inland one, limited to the course of rivers or to the caravan
routes, which defied all the natural obstacles presented by
snow-capped mountain chains, elevated plateaux, and extensive deserts. The great inland caravan trade, whose most
valuable articles of barter were silk, passed from the interior
of Asia, from the elevated plains of the Seres, north of Uttara
Kuru, by the stony tower† (probably a fortified caravansery),
south of the sources of the Jaxartes, through the Valley of the
Oxus to the Caspian and Black Seas. On the other hand,
the principal traffic of the Ptolemaic empire was, in the strict-

* See *ante*, p. 123.
† Compare my geographical researches, in *Asie Centrale*, t. i., p. 145
and 151–157; t. ii., p. 179.

est sense of the word, a sea trade, notwithstanding the animation of the navigation on the Nile, and the communication between the banks of the river, and the artificially constructed roads along the shores of the Red Sea. According to the grand views of Alexander, the newly-founded Egyptian city of Alexandria and the ancient Babylon were to have constituted the respective eastern and western capitals of the Macedonian empire ; Babylon never, at any subsequent period, realized these hopes, and the prosperity of Seleucia, which was built by Seleucus Nicator on the Lower Tigris, and had been connected by canals with the Euphrates,* contributed to its entire downfall.

Three great rulers, the three first Ptolemies, whose reigns occupied a whole century, gave occasion, by their love of science, their brilliant institutions for the promotion of mental culture, and their unremitting endeavors for the extension of maritime trade, to an increase of knowledge regarding distant nations and external nature hitherto unattained by any people. This treasure of genuine, scientific cultivation passed from the Greek settlers in Egypt to the Romans. Under Ptolemæus Philadelphus, scarcely half a century after the death of Alexander, and even before the first Punic war had shaken the aristocratic republic of the Carthaginians, Alexandria was the greatest commercial port in the world, forming the nearest and most commodious route from the basin of the Mediterranean to the southeastern parts of Africa, Arabia, and India. The Ptolemies availed themselves with unprecedented success of the advantages held out to them by a route which nature had marked, as it were, for a means of universal intercourse with the rest of the world by the direction of the Arabian Gulf,† and whose importance can not even now be duly appreciated until the savage violence of Eastern nations, and the injurious jealousies of Western powers, shall simultaneously diminish. Even after it had become a Roman province, Egypt continued to be the seat of immense wealth, for the increased luxury of Rome, under the Cæsars, reached to the territory of the Nile, and turned to the universal commerce of Alexandria for the chief means of its satisfaction.

The important extension of the sphere of knowledge regarding external nature and different countries under the Ptolemies was mainly owing to the caravan trade in the interior

* Plin., vi., 26 (?).
† See Droysen. *Gesch. des Hellenistischen Staatensystems*, s. 749.

of Africa by Cyrene and the Oases ; to the conquest in Ethiopia and Arabia Felix under Ptolemy Euergetes ; to the maritime trade with the whole of the western peninsula of India, from the Gulf of Barygaza (Guzerat and Cambay), along the shores of Canara and Malabar (Malayavara, a territory of Malaya), to the Brahminical sanctuaries of the promontory of Comorin (Kumari),* and to the large island of Ceylon (Lanka in the Ramayana, and known to the cotemporaries of Alexander as Taprobane, a corruption of the native name).† Nearchus had already materially contributed to the advance of nautical knowledge by his laborious five months' voyage along the coasts of Gedrosia and Caramania (between Pattala, at the mouths of the Indus, and the Euphrates).

Alexander's companions were not ignorant of the existence of the monsoons, by which navigation was so greatly favored between the eastern coasts of Africa and the northern and western parts of India. After having spent ten months in navigating the Indus, between Nicæa on the Hydaspes and Pattala, with a view of opening the river to a universal traffic, Nearchus hastened, at the beginning of October (Ol. 113, 3), to sail from Stura, at the mouth of the Indus, since he knew that his passage would be favored by the northeast and east monsoons to the Persian Gulf along the coasts running in the same parallel of latitude. The knowledge of this remarkable local law of the direction of the winds subsequently imboldened navigators to attempt to sail from Ocelis, on the Straits of Bab-el-Mandeb, across the open sea to Muzeris (south of Mangolar), the great Malabar emporium of trade, to which products from the eastern shores of the Indian peninsula, and even gold from the distant Chryse (Borneo ?), were brought by inland trade. The honor of having first applied the new system of Indian navigation is ascribed to an otherwise unknown seaman named Hippalus, but considerable doubt is attached to the age in which he lived.‡

* See Lassen, *Indische Alterthumskunde*, bd. i., s. 107, 153–158.
† A corruption of *Támbapanni.* This Pali form sounds in Sanscrit *Támraparni.* The Greek form *Taprobane* gives half the Sanscrit (*Támbra, Tabro*) and half the Pali. (Lassen, op. cit., s. 201. Compare Lassen, *Diss. de Taprobane Insula,* p. 19.) The Laccadives (*lakke* for *lakscha,* and *dive* for *dwípa,* one hundred thousand islands), as well as the Maldives (*Malayadiba,* islands of Malabar), were known to Alexandrian mariners.
‡ Hippalus is not generally supposed to have lived earlier than the time of Claudius ; but this view is improbable, even though under the first Lagides, a great portion of the Indian products were only procured in Arabian markets. The southwest monsoon was, moreover,

The history of the contemplation of the universe embraces the enumeration of all the means which have brought nations into closer contact with one another, rendered larger portions of the earth more accessible, and thus extended the sphere of human knowledge. One of the most important of these means was the opening of a road of communication from the Red Sea to the Mediterranean by means of the Nile. At the point where the scarcely-connected continents present a line of bay-like indentations, the excavation of a canal was begun, if not by Sesostris (Rameses Miamoun), to whom Aristotle and Strabo ascribe the undertaking, at any rate by Neku, although the work was relinquished in consequence of the threatening oracular denunciations directed against it by the priests. Herodotus saw and described a canal completed by Darius Hystaspes, one of the Achæmenidæ, which entered the Nile somewhat above Bubastus. This canal, after having fallen into decay, was restored by Ptolemy Philadelphus in so perfect a manner that, although (notwithstanding the skillful arrangement of sluices) it was not navigable at all seasons of the year, it nevertheless contributed to facilitate Ethiopian, Arabian, and Indian commerce at the time of the Roman dominion under Marcus Aurelius, or even as late as Septimius Severus, and, therefore, a century and a half after its construction. A similar object of furthering international communication through the Red Sea led to a zealous prosecution of the works necessary for forming a harbor in Myos Hormos and Berenice, which was connected with Coptos by means of an admirably made artificial road.*

All these various mercantile and scientific enterprises of the Lagides were based on an irrepressible striving to acquire new territories and penetrate to distant regions, on an idea of connection and unity, and on a desire to open a wider field of action by their commercial and political relations. This direction of the Hellenic mind, so fruitful in results, and which had been long preparing in silence, was manifested, under its

itself called Hippalus, and a portion of the Erythrean or Indian Ocean was known as the Sea of Hippalus. Letronne, in the *Journal des Savans*, 1818, p. 405; Reinaud, *Relation des Voyages dans l'Inde*, t. i., p. xxx.

* See the researches of Letronne on the construction of the canal between the Nile and the Red Sea, from the time of Neku to the Calif Omar, or during an interval of more than 1300 years, in the *Revue des deux Mondes*, t. xxvii., 1841, p. 215–235. Compare, also, Letronne, *De la Civilisation Égyptienne depuis Psammitichus jusqu'à la conquête d'Alexandre*, 1845, p. 16–19.

noblest type, in the efforts made by Alexander in his campaign
to fuse together the eastern and western worlds. Its extension under the Lagides characterizes the epoch which I would
here portray, and must be regarded as an important advance
toward the attainment of a knowledge of the universe in its
character of unity.

As far as abundance and variety in the objects presented to
the contemplation are conducive to an increased amount of
knowledge, we might certainly regard the intercourse existing
between Egypt and distant countries ; the scientific exploring
expeditions into Ethiopia at the expense of the government ;*
distant ostricht and elephant hunts ; and the establishment of
menageries of wild and rare animals in the " king's houses of
Bruchium" as means of incitement toward the study of natural history,‡ and as amply sufficient to furnish empirical
science with the materials requisite for its further development ; but the peculiar character of the Ptolemaic period, as
well as of the whole Alexandrian school, which retained the
same individuality of type until the third and fourth centuries,
manifested itself in a different direction, inclining less to an
immediate observation of particulars than to a laborious accumulation of the results of that which had already been noted
by others, and to a careful classification, comparison, and men-

* Meteorological speculations on the remote causes of the swelling
of the Nile gave occasion to some of these journeys, since, as Strabo
expresses it (lib. xvii., p. 789 and 790), " Philadelphus was constantly
seeking new diversions and new objects of interest from a desire for
knowledge and from bodily weakness."

† Two hunting inscriptions, " one of which principally records the
elephant hunts of Ptolemy Philadelphus," were discovered and copied
from the colossi of Abusimbel (Ibsambul) by Lepsius during his
Egyptian journey (compare, on this subject, Strabo, lib. xvi., p. 769
and 770; Ælian, De Nat. Anim., iii., 34, and xvii., 3 ; Athenæus, v., p.
196). Although Indian ivory was an article of export from Barygaza,
according to the Periplus Maris Erythræi, yet, from the statement of
Cosmas, ivory would also appear to have been exported from Ethiopia
to the western peninsula of India. Elephants have withdrawn more
to the south in Eastern Africa, also, since ancient times. According to
the testimony of Polybius (v., 84), when African and Indian elephants
were opposed to each other on fields of battle, the sight, smell, and
cries of the larger and stronger Indian elephants drove the African ones
to flight. The latter were probably never employed as war elephants
in such large numbers as in Asiatic expeditions, where Kandragupta
had assembled 9000, the powerful King of the Prasii 6000, and Akbar
an equally large number. (Lassen, Ind. Alterthumskunde, bd. i., s.
305–307.)

‡ Athen., xiv., p. 654. Compare Parthey, Das Alexandrinische Museum, eine Preisschrift, s. 55 und 171.

tal elaboration of these results. During a period of many centuries, and until the powerful mind of Aristotle was revealed, the phenomena of nature, not regarded as objects of acute observation, were subjected to the sole control of ideal interpretation, and to the arbitrary sway of vague presentiments and vacillating hypotheses, but from the time of the Stagirite a higher appreciation for empirical science was manifested. The facts already known were now first critically examined. As natural philosophy, by pursuing the certain path of induction, gradually approached nearer to the scrutinizing character of empiricism, it became less bold in its speculations and less fanciful in its images. A laborious tendency to accumulate materials enforced the necessity for a certain amount of polymathic learning ; and although the works of different distinguished thinkers occasionally exhibited precious fruits, these were unfortunately too often accompanied, in the decline of creative conception among the Greeks, by a mere barren erudition devoid of animation. The absence of a careful attention to the form as well as to animation and grace of diction, has likewise contributed to expose Alexandrinian learning to the severe animadversions of posterity.

The present section would be incomplete if it were to omit a notice of the accession yielded to general knowledge by the epoch of the Ptolemies, both by the combined action of external relations, the foundation and proper endowment of several large institutions (the Alexandrian Museum and two libraries at Bruchium and Rhakotis),* and by the collegiate association of so many learned men actuated by practical views. This encyclopedic species of knowledge facilitated the comparison of observations and the generalization of natural views.† The

* The library in the Bruchium, which was destroyed in the burning of the fleet under Julius Cæsar, was the more ancient. The library at Rhakotis formed a part of the " Serapeum," where it was connected with the museum. By the liberality of Antoninus, the collection of books at Pergamus was joined to the library of Rhakotis.

† Vacherot, *Histoire Critique de l'Ecole d'Alexandrie*, 1846, t. i., p. v. and 103. The institute of Alexandria, like all academical corporations, together with the good arising from the concurrence of many laborers, and from the acquisition of material aids, exercised also some narrowing and restraining influence, as we find from numerous facts furnished by antiquity. Adrian appointed his tutor, Vestinus, high-priest of Alexandria (a sort of minister presiding over the management of public worship), and at the same time head of the museum (or president of the academy). (Letronne, *Recherches pour servir à l'Histoire de l'Egypte pendant la Domination des Grecs et des Romains*, 1823, p. 251.)

great scientific institution which owes its origin to the first of the Ptolemies long enjoyed, among other advantages, that of being able to give a free scope to the differently directed pursuits of its members, and thus, although founded in a foreign country, and surrounded by nations of different races, it could still preserve the characteristics of the Greek acuteness of mind and a Greek mode of thought.

A few examples must suffice, in accordance with the spirit and form of the present work, to show how experiments and observations, under the protecting influence of the Ptolemies, acquired their appropriate recognition as the true sources of knowledge regarding celestial and terrestrial phenomena, and how, in the Alexandrian period, a felicitous generalization of views manifested itself conjointly with a laborious accumulation of knowledge. Although the different Greek schools of philosophy, when transplanted to Lower Egypt, gave occasion, by their Oriental degeneration, to many mythical hypotheses regarding nature and natural phenomena, mathematics still constituted the firmest foundation of the Platonic doctrines inculcated in the Alexandrian Museum ;* and this science comprehended, in the advanced stages of its development, pure mathematics, mechanics, and astronomy. In Plato's high appreciation of mathematical development of thought, as well as in Aristotle's morphological views, which embraced all organisms, we discover the germs of the subsequent advances of physical science. They became the guiding stars which led the human mind through the bewildering fanaticism of the Dark Ages, and prevented the utter destruction of a sound and scientific manifestation of mental vigor.

The mathematician and astronomer, Eratosthenes of Cyrene, the most celebrated of the Alexandrian librarians, employed the materials at his command to compose a system of universal geography. He freed geography from mythical legends, and, although himself occupied with chronology and history, separated geographical descriptions from that admixture of historical elements with which it had previously been not ungracefully embodied. The absence of these elements was, however, satisfactorily compensated for by the introduc-

---

* Fries, *Geschichte der Philosophie*, bd. ii., s. 5 ; and the same author's *Lehrbuch der Naturlehre*, th. i., s. 42. Compare, also, the considerations on the influence which Plato exercised on the foundation of the experimental sciences by the application of mathematics, in Brandis, *Geschichte der Griechisch-Römischen Philosophie*, th. ii., abth. i., s. 276.

tion of mathematical considerations on the articulation and expansion of continents ; by geognostic conjectures regarding the connection of mountain chains, the action of clouds, and the former submersion of lands, which still bear all the traces of having constituted a dried portion of the sea's bottom. Favorable to the oceanic sluice-theory of Strabo of Lampsacus, the Alexandrian librarian was led, by the belief of the former swelling of the Euxine, the penetration of the Dardanelles, and the consequent opening of the Pillars of Hercules, to an important investigation of the problem of the equal level of the whole "*external sea**" surrounding all continents." An additional proof of this philosopher's power of generalizing views is afforded by his assertions that the whole continent of Asia is traversed by a continually-connected mountain chain, running from west to east in the parallel of Rhodes (in the diaphragm of Dicæarchus).†

An animated desire to arrive at a generalization of views— the consequence of the intellectual movement of the age— gave rise to the first Greek measurement of degrees between Syene and Alexandria, and this experiment may be regarded as an attempt on the part of Eratosthenes to arrive at an approximative determination of the circumference of the Earth. In this case, it is not the result at which he arrived from the imperfect premises afforded by the *Bematists* which excites our interest, but rather the attempt to rise from the narrow limits of one circumscribed land to a knowledge of the magnitude of the whole earth.

A similar tendency toward generalization may be traced in the splendid progress made in the scientific knowledge of the heavens in the epoch of the Ptolemies. I allude here to the determination of the places of the fixed stars by the earliest Alexandrian astronomers, Aristyllus and Timochares ; to Aristarchus of Samos, the cotemporary of Cleanthes, who, conversant with ancient Pythagorean views, ventured upon an in-

---

* On the physical and geognostical opinions of Eratosthenes, see Strabo, lib. i., p. 49–56 ; lib. ii., p. 108.

† Strabo, lib. xi., p. 519 ; Agathem, in Hudson, *Geogr. Græc. Min.*, vol. ii., p. 4. On the accuracy of the grand orographic views of Eratosthenes, see my *Asie Centrale*, t. i., p. 104–150, 198, 208–227, 413–415 ; t. ii., p. 367 and 414–435 ; and *Examen Critique de l'Hist. de la Géogr.*, t. i., p. 152–154. I have purposely called the measurement of a degree made by Eratosthenes as the first Hellenic one, since a very ancient Chaldean determination of the magnitude of a degree in camels' paces is not improbable. See Chasles, *Recherches sur l'Astronomie Indienne et Chaldéenne*, in the *Comptes Rendus de l'Acad. des Sciences*, t. xxiii., 1846, p. 851.

vestigation of the construction of the universe, and who was the first to recognize the immeasurable distance of the region of fixed stars from our small planetary system ; nay, he even conjectured the two-fold motion of the earth round its axis and round the sun ; to Seleucus of Erythræa (or of Babylon),* who, a century subsequent to this period, endeavored to establish the hypothesis of the Samian philosopher, which, resembling the views of Copernicus, met with but little attention during that age ; and, lastly, to Hipparchus, the founder of scientific astronomy, and the greatest astronomical observer of antiquity. Hipparchus was the actual originator of astronomical tables among the Greeks,† and was also the discoverer of the precession of the equinoxes. On comparing his own observations of fixed stars (made at Rhodes, and not at Alexandria) with those made by Timochares and Aristyllus, he was led, probably without the apparition of a new star,‡ to this great discovery, to which, indeed, the earlier Egyptians might have attained by a long-continued observation of the heliacal rising of Sirius.§

A peculiar characteristic of the labors of Hipparchus is the use he made of his observations of celestial phenomena for the determination of geographical position. Such a connection between the study of the earth and of the celestial regions, mutually reflected on each other, animated through its uniting influences the great idea of the Cosmos. In the new map of the world constructed by Hipparchus, and founded upon that of Eratosthenes, the geographical degrees of longitude and latitude were based on lunar observations and on the measurements of shadows, wherever such an application of astronom-

---

* The latter appellation appears to me the more correct, since Strabo, lib. xvi., p. 739, quotes, " Seleucus of Seleucia, among several very honorable men, as a Chaldean, skilled in the study of the heavenly bodies." Seleucia, on the Tigris, a flourishing commercial city, is probably the one meant. It is indeed singular that Strabo also speaks of a Seleucus, an exact observer of the tides, and terms him, too, a Babylonian (lib. i., p. 6), and subsequently (lib. iii., p. 174), perhaps from carelessness, an Erythræan. (Compare Stobæus, *Ecl. Phys.*, p. 440.)

† Ideler, *Handbuch der Chronologie*, bd. i., s. 212 und 329.

‡ Delambre, *Histoire de l'Astronomie Ancienne*, t. i., p. 290.

§ Böckh has entered into a discussion, in his *Philolaos*, s. 118, as to whether the Pythagoreans were early acquainted, through Egyptian sources, with the precession, under the name of the motion of the heavens of the fixed stars. Letronne (*Observations sur les Représentations Zodiacales qui nous restent de l'Antiquité*, 1824, p. 62) and Ideler (in his *Handbuch der Chronol.*, bd. i., s. 192) vindicate the exclusive claim of Hipparchus to this discovery.

ical observations was admissible. While the hydraulic clock of Ctesibius, an improvement on the earlier clepsydra, must have yielded more exact measurements of time, determinations in space must likewise have improved in accuracy, in consequence of the better modes of measuring angles, which the Alexandrian astronomers gradually possessed, from the period of the ancient gnomon and the scaphe to the invention of astrolabes, solstitial armils, and linear dioptrics. It was thus that man, and step by step, as it were, by the acquisition of new organs, arrived at a more exact knowledge of the movements of the planetary system. Many centuries, however, elapsed before any advance was made toward a knowledge of the absolute size, form, mass, and physical character of the heavenly bodies.

Many of the astronomers of the Alexandrian Museum were not only distinguished as geometricians, but the age of the Ptolemies was, moreover, a most brilliant epoch in the prosecution of mathematical investigations. In the same century there appeared Euclid, the creator of mathematics as a science, Apollonius of Perga, and Archimedes, who visited Egypt, and was connected through Conon with the school of Alexandria. The long period of time which leads from the so-called geometrical analysis of Plato, and the three conic sections of Menæchmes,* to the age of Kepler and Tycho Brahe, Euler and Clairaut, D'Alembert and Laplace, is marked by a series of mathematical discoveries, without which the laws of the motion of the heavenly bodies and their mutual relations in the regions of space would not have been revealed to mankind. While the telescope serves as a means of penetrating space, and of bringing its remotest regions nearer to us, mathematics, by inductive reasoning, have led us onward to the remotest regions of heaven, and brought a portion of them within the range of our possession ; nay, in our own times—so propitious to extension of knowledge—the application of all the elements yielded by the present condition of astronomy has even revealed to the intellectual eye a heavenly body, and assigned to it its place, orbit, and mass, before a single telescope had been directed toward it.†

* Ideler, on *Eudoxus*, s. 23.
† The planet discovered by Le Verrier.

UNIVERSAL DOMINION OF THE ROMANS.—INFLUENCE OF A VAST PO-
LITICAL UNION ON COSMICAL VIEWS.—ADVANCE OF GEOGRAPHY
BY MEANS OF INLAND TRADE.—STRABO AND PTOLEMY.—THE FIRST
ATTEMPTS TO APPLY MATHEMATICS TO OPTICS AND CHEMISTRY.—
PLINY'S ATTEMPTS TO GIVE A PHYSICAL DESCRIPTION OF THE UNI-
VERSE.—THE RISE OF CHRISTIANITY PRODUCTIVE OF, AND FAVOR-
ABLE TO, THE FEELING OF THE UNITY OF MANKIND.

In tracing the intellectual advance of mankind and the
gradual extension of cosmical views, the period of the universal dominion of the Romans presents itself to our consideration as one of the most important epochs in the history of the
world. We now, for the first time, find all the fruitful districts which surround the basin of the Mediterranean associated together in one great bond of political union, and even
connected with many vast territories in the East.

The present would seem a fitting place again to remind my
readers* that the general picture I have endeavored to draw
of the history of the contemplation of the universe acquires,
from this condition of political association, an objective unity
of presentation. Our civilization, understanding the term as
being synonymous with the intellectual development of all the
nations included in the European Continent, may be regarded
as based on that of the inhabitants of the shores of the Mediterranean, and more directly on that of the Greeks and Romans. That which we, perhaps too exclusively, term classical
literature, received the appellation from the fact of its being
recognized as the source of a great portion of our early knowledge, and as the means by which the first impulse was awakened in the human mind to enter upon a sphere of ideas and
feelings most intimately connected with the social and intellectual elevation of the different races of men.† In these
considerations we do not by any means disregard the importance of those elements which have flowed in a variety of different directions—from the Valley of the Nile, Phœnicia, the
Euphrates, and the Indus, into the great stream of Greek and
Roman civilization; but even for these elements we are originally indebted to the Greeks and to the Romans, who were
surrounded by Etruscans and other nations of Hellenic descent. How recent is the date of any direct investigation,
interpretation, and secular classification of the great monuments of more anciently civilized nations! How short is the
time that has elapsed since hieroglyphics and arrow-headed

* See *ante*, p. 110, 113, 117, and 141.
† Wilhelm von Humboldt, *Ueber die Kawi-Sprache*, bd. i., s. xxxvii.

characters were first deciphered, and how numerous are the armies and the caravans which, for thousands of years, have passed and repassed without ever divining their import !

The basin of the Mediterranean, more especially in its varied northern peninsulas, certainly constituted the starting point of the intellectual and political culture of those nations who now possess what we may hope is destined to prove an imperishable and daily increasing treasure of scientific knowledge and of creative artistic powers, and who have spread civilization, and, with it, servitude at first, but subsequently freedom, over another hemisphere.   Happily, in our hemisphere, under the favor of a propitious destiny, unity and diversity are gracefully blended together.   The elements taken up have been no less heterogeneous in their nature than in the affinities and transformations effected under the influence of the sharply-contrasting peculiarities and individual characteristics of the several races of men by whom Europe has been peopled.   Even beyond the ocean, the reflection of these contrasts may still be traced in the colonies and settlements which have already become powerful free states, or which, it is hoped, may still develop for themselves an equal amount of political freedom.

The Roman dominion, in its monarchical form under the Cæsars, considered according to its area,* was certainly exceeded in absolute magnitude by the Chinese empire under the dynasty of Thsin and the Eastern Han (from thirty years before to one hundred and sixteen years after our era), by the Mongolian empire under Genghis Khan, and by the present area of the Russian empire in Europe and Asia ; but, with the single exception of the Spanish monarchy—as long as it extended over the new world—there has never been combined under one scepter a greater number of countries favored by climate, fertility, and position, than those comprised under the Roman empire from Augustus to Constantine.

This empire, extending from the western extremity of Europe to the Euphrates, from Britain and part of Caledonia to Gætulia and the confines of the Libyan desert, manifested not

---

* The superficial area of the Roman empire under Augustus is calculated by Professor Berghaus, the author of the excellent *Physical Atlas,* at rather more than 400,000 geographical square miles (according to the boundaries assumed by Heeren, in his *Geschichte der Staaten des Alterthums.*, s. 403–470), or about one fourth greater than the extent of 1,600,000 square miles assigned by Gibbon, in his *History of the Decline and Fall of the Roman Empire*, vol. i., chap. i., p. 39, but which he indeed gives as a very uncertain estimate.

only the greatest diversities in the form of the ground, in organic products and physical phenomena, but it also exhibited mankind in all the various gradations from civilization to barbarism, and in the possession of ancient knowledge and long-practiced arts, no less than in the imperfectly-lighted dawn of intellectual awakening. Distant expeditions were prosecuted with various success to the north and south, to the amber lands, and under Ælius Gallus and Balbus, to Arabia, and to the territory of the Garamantes. Measurements of the whole empire were begun even under Augustus, by the Greek geometricians Zenodoxus and Polycletus, while itineraries and special topographies were prepared for the purpose of being distributed among the different governors of the provinces, as had already been done several hundred years before in the Chinese empire.* These were the first statistical labors instituted in Europe. Many of the prefectures were traversed by Roman roads, divided into miles, and Adrian even visited his extensive dominions from the Iberian Peninsula to Judea, Egypt, and Mauritania, in an eleven years' journey, which was not, however, prosecuted without frequent interruptions. Thus the large portion of the earth's surface, which was subject to the dominion of the Romans, was opened and rendered accessible, realizing the idea of the *pervius orbis* with more truth than we can attach to the prophecy in the chorus of the Medea as regards the whole earth.†

The enjoyment of a long peace might certainly have led us to expect that the union under one empire of extensive countries having the most varied climates, and the facility with which the officers of state, often accompanied by a numerous train of learned men, were able to traverse the provinces, would have been attended, to a remarkable extent, by an advance not only in geography, but in all branches of natural science, and by the acquisition of a more correct knowledge of the connection existing among the phenomena of nature : yet such high expectations were not fulfilled. In this long period of undivided Roman empire, embracing a term of almost four centuries, the names of Dioscorides the Cilician and Galen of Pergamus have alone been transmitted to us as those of observers of nature. The first of these, who increased so considerably the number of the described species of plants, is far

* Veget., *De Re Mil.*, iii., 6.
† Act ii., v. 371, in the celebrated prophecy which, from the time of the son of Columbus, was interpreted to relate to the discovery of America.

inferior to the philosophically combining Theophrastes, while the delicacy of his manner of dissecting, and the extent of his physiological discoveries, place Galen, who extended his observations to various genera of animals, " very nearly as high as Aristotle, and, in some respects, even above him." This judgment embodies the views of Cuvier.*

Besides Dioscorides and Galen, our attention is called to a third and great name—that of Ptolemy. I do not mention him here as an astronomical systematizer or as a geographer, but as an experimental physicist, who measured refraction, and who may, therefore, be regarded as the founder of an important branch of optical science, although his incontestable claim to this title has been but recently admitted.† However important were the advances made in the sphere of organic life and in the general views of comparative zootomy, our attention is yet more forcibly arrested by those physical experiments on the passage of a ray of light, which, preceding the period of the Arabs by an interval of five hundred years, mark the first step in a newly-opened course, and the earliest indication of a striving toward the establishment of mathematical physics.

The distinguished men whom we have already named as shedding a scientific luster on the age of the imperial rulers of Rome were all of Greek origin. The profound arithmetician and algebraist Diophantus (who was still unacquainted with the use of symbols) belonged to a later period.‡ Amid the different directions presented by intellectual cultivation in the Roman empire, the palm of superiority remained with the Hellenic races, as the older and more happily-organized peo-

* Cuvier, *Hist. des Sciences Naturelles*, t. i., p. 312–328.

† *Liber Ptholemei de Opticis sive Aspectibus;* a rare manuscript of the Royal Library at Paris (No. 7310), which I examined on the occasion of discovering a remarkable passage on the refraction of rays in Sextus Empiricus (*adversus Astrologos*, lib. v., p. 351, Fabr.). The extracts which I made from the Paris manuscript in 1811 (therefore before Delambre and Venturi) will be found in the introduction to my *Recueil d'Observations Astronomiques*, t. i., p. lxv.–lxx. The Greek original has not been preserved to us, and we have only a Latin translation of two Arabic manuscripts of Ptolemy's *Optics*. The name of the Latin translator was Amiracus Eugenius, Siculus. Compare Venturi, *Comment. sopra la Storia e le Teorie dell' Ottica*, Bologna, 1814, p. 227 ; Delambre, *Hist. de l'Astronomie Ancienne*, 1817, t. i., p. 61, and t. ii., p. 410–432.

‡ Letronne shows, from the occurrence of the fanatical murder of the daughter of Theon of Alexandria, that the much-contested epoch of Diophantus can not be placed later than the year 389 (*Sur l'Origine Grecque des Zodiaques prétendus Egyptiens*, 1837, p. 26).

ple ; but, after the gradual downfall of the Egypto-Alexandrian school, the dimmed sparks of knowledge and of intellectual investigation were scattered abroad, and it was not until a later period that they reappeared in Greece and Asia Minor. As is the case in all unlimited monarchies embracing a vast extent of the most heterogeneous elements, the efforts of the Roman government were mainly directed to avert, by military restraint and by means of the internal rivalry existing in their divided administration, the threatened dismemberment of the political bond ; to conceal, by an alternation of severity and mildness, the domestic feuds in the house of the Cæsars ; and to give to the different dependencies such an amount of peace, under the sway of noble rulers, as an unchecked and patiently-endured despotism is able periodically to afford.

The attainment of universal sway by the Romans certainly emanated from the greatness of the national character, and from the continued maintenance of rigid morals, coupled with a high sense of patriotism. When once universal empire was attained, these noble qualities were gradually weakened and altered under the unavoidable influence of the new relations induced. The characteristic sensitiveness of separate individuals became extinguished with the national spirit, and thus vanished the two main supports of free institutions, publicity and individuality. The eternal city had become the center of too extended a sphere, and the spirit was wanting which ought to have permanently animated so complicated a state. Christianity became the religion of the state when the empire was already profoundly shaken, and the beneficent effects of the mildness of the new doctrine were frustrated by the dogmatic dissensions awakened by party spirit. That dreary contest of knowledge and of faith had already then begun, which continued through so many centuries, and proved, under various forms, so detrimental to intellectual investigation.

If the Roman empire, from its extent and the form of constitution necessitated by its relations of size, was wholly unable to animate and invigorate the intellectual activity of mankind, as had been done by the small Hellenic republics in their partially-developed independence, it enjoyed, on the other hand, peculiar advantages, to which we must here allude. A rich treasure of ideas was accumulated as a consequence of experience and numerous observations. The objective world became considerably enlarged, and was thus prepared for that meditative consideration of natural phenomena which has characterized recent times. National intercourse was animated by the Roman dominion, and the Latin tongue spread over the whole West, and over a portion of Northern Africa. In the East, Hellenism still predominated long after the destruction of the Bactrian empire under Mithridates I., and thirteen years before the irruption of the Sacæ or Scythians.

With respect to geographical extent, the Latin tongue gained upon the Greek, even before the seat of empire had been removed to Byzantium. The reciprocal transfusion of these two highly-organized forms of speech, which were so rich in literary memorials, became a means for the more complete amalgamation and union of different races, while it was likewise conducive to an increase of civilization, and to a greater susceptibility for intellectual cultivation, tending, as Pliny says, " to humanize men and to give them one common country."*

However much the languages of the barbarians, the dumb, ἄγλωσσοι, as Pollux terms them, may have been generally despised, there were some cases in which, according to the examples of the Lagides, the translation of a literary work from the Punic was undertaken in Rome by order of the authorities ; thus, for instance, we find that Mago's treatise on agriculture was translated at the command of the Roman Senate.

While the empire of the Romans extended in the Old Continent as far westward as the northern shores of the Mediterranean—reaching to its extremest confines at the holy promontory—its eastern limit, even under Trajan, who navigated the Tigris, did not advance beyond the meridian of the Persian Gulf. It was in this direction that the progress of the international contact produced by inland trade, whose results were so important with respect to geography, was most strongly manifested during the period under consideration. After the downfall of the Græco-Bactrian empire, the reviving power of the Arsacidæ favored intercourse with the Seres, although only by indirect channels, as the Romans were impeded by the active commercial intervention of the Parthians

---

* This beneficial influence of civilization, exemplified by the extension of a language in exciting feelings of general good will, is finely characterized in Pliny's praise of Italy : " omnium terrarum alumna eadem et parens, numine Deûm electa, quæ sparsa congregaret imperia, ritusque molliret, et tot populorum discordes ferasque linguas sermonis commercio contraheret, colloquia, et humanitatem homini daret, breviterque una cunctarum gentium in toto orbe patria fieret." (Plin., *Hist. Nat.*, iii., 5.)

from entering into relations of direct intercourse with the inhabitants of the interior of Asia. Movements, which emanated from the remotest parts of China, produced the most
rapid, although not long-persisting changes in the political
condition of the vast territories which lie between the volcanic
celestial mountains (Thian-schan) and the chain of the Kuen-
lun in the north of Thibet. A Chinese expedition subdued
the Hiungnu, levied tribute from the small territory of Kho-
tan and Kaschgar, and carried its victorious arms as far as
the eastern shores of the Caspian. This great expedition,
which was made in the time of Vespasian and Domitian, was
headed by the general Pantschab, under the Emperor Mingti,
of the dynasty of Han, and Chinese writers ascribe a grand
plan to the bold and fortunate commander, maintaining that
he designed to attack the Roman empire (Tathsin), but was
deterred by the admonitory counsel of the Persians.* Thus
there arose connections between the shores of the Dead Sea,
the Schensi, and those territories on the Oxus in which an
animated trade had been prosecuted from an early age with
the nations inhabiting the coasts of the Black Sea.

The direction in which the stream of immigration inclined
in Asia was from east to west, while in the New Continent
it was from north to south. A century and a half before our
era, about the time of the destruction of Corinth and Carthage, the first impulse to that "immigration of nations,"
which did not, however, reach the borders of Europe until
five hundred years afterward, was given, by the attack of the
Hiungnu (a Turkish race confounded by De Guignes and Johann Müller with the Finnish Huns) on the fair-haired and
blue-eyed Yueti (Getæ ?), probably of Indo-Germanic descent,†
and on the Usun, who dwelt near the wall of China. In this
manner the stream of population flowed from the upper river

* Klaproth, *Tableaux Historiques de l'Asie,* 1826, p. 65–67.
† To this fair-haired, blue-eyed Indo-Germanic, Gothic, or Arian race
of Eastern Asia, belong the Usun, Tingling, Hutis, and great Yueti. The
last are called by the Chinese writers a Thibetian nomadic race, who,
three hundred years before our era, migrated to the district between
the upper course of the Hoang-ho and the snowy mountains of Nan-
schan. I here recall this descent, as the Seres (Plin., vi., 22) are also
described as "rutilis comis et cæruleis oculis." (Compare Ukert,
Geogr. der Griech. und Römer, th. iii., abth. 2, 1845, s. 275.) We are
indebted for the knowledge of these fair-haired races (who, in the
most eastern part of Asia, gave the first impulse to what has been called
"the great migration of nations") to the researches of Abel Remusat
and Klaproth, which belong to the most brilliant historical discoveries
of our age.

valleys of the Hoang-ho westward to the Don and the Danube, and the opposite tendencies of these currents, which at first brought the different races into antagonist conflict in the northern parts of the Old Continent, ended in establishing friendly relations of peace and commerce. It is when considered from this point of view that great currents of migration, advancing like oceanic currents between masses which are themselves unmoved, become objects of cosmical importance.

In the reign of the Emperor Claudius, the embassy of Rachias of Ceylon came to Rome by way of Egypt. Under Marcus Aurelius Antoninus (named An-tun by the writers of the history of the dynasty of Han), Roman legates visited the Chinese court, having come by sea by the route of Tunkin. We here observe the earliest traces of the extended intercourse of the Romans with China and India, since it is highly probable that the knowledge of the Greek sphere and zodiac, as well as that of the astrological planetary week, was not generally diffused until the first century of our era, and that it was then effected by means of this intercourse between the two countries.* The great Indian mathematicians Warahamihira, Brahmagupta, and perhaps even Aryabhatta, lived at a more recent period than that under consideration ;† but the elements of knowledge, either discovered by Indian nations, frequently in different and wholly independent directions, or existing among these ancient civilized races from primitive ages, may have penetrated into the West even before the time of Diophantus, by means of the extended commercial relations existing between the Ptolemies and the Cæsars. I will not here attempt to determine what is due to each individual race and epoch, my object being merely to indicate the different channels by which an interchange of ideas has been effected.

The strongest evidence of the multiplicity of means, and the extent of the advance that had been made in general intercourse, is testified by the colossal works of Strabo and Ptolemy. The gifted geographer of Amasea does not possess the numerical accuracy of Hipparchus, or the mathematical and

---

* See Letronne, in the *Observations Critiques et Archéologiques sur les Représentations Zodiacales de l'Antiquité*, 1824, p. 99, as well as his later work, *Sur l'Origine Grecque des Zodiaques prétendus Egyptiens*, 1837, p. 27.

† The sound inquirer, Colebrooke, places Warahamihira in the fifth, Brahmagupta at the end of the sixth century, and Aryabhatta rather indefinitely between the years 200 and 400 of our era. (Compare Holtzmann, *Ueber den Griechischen Ursprung des Indischen Thierkreises*, 1841, s. 23.)

geographical information of Ptolemy ; but his work surpasses all other geographical labors of antiquity by the diversity of the subjects and the grandeur of the composition. Strabo, as he takes pleasure in informing us, had seen with his own eyes a considerable portion of the Roman empire, " from Armenia to the Tyrrhenian coasts, and from the Euxine to the borders of Ethiopia." After he had completed the historical work of Polybius by the addition of forty-three books, he had the courage, in his eighty-third year,* to begin his work on geography. He remarks, " that in his time the empire of the Romans and Parthians had extended the sphere of the known world more even than Alexander's campaigns, from which Eratosthenes derived so much aid." The Indian trade was no longer in the hands of the Arabs alone ; and Strabo, when in Egypt, remarked with astonishment the increased number of vessels passing directly from Myos Hormos to India.† In imagination he penetrated beyond India as far as the eastern shores of Asia. At this point, in the parallel of the Pillars of Hercules and the island of Rhodes, where, according to his idea, a connected mountain chain, a prolongation of the Taurus, traversed the Old Continent in its greatest width, he conjectured the existence of *another continent* between the west of Europe and Asia. " It is very possible," he writes,‡ " that

* On the reasons on which we base our assertion of the exceedingly late commencement of Strabo's work, see Groskurd's German translation, th. i., 1831, s. xvii.

† Strabo, lib. i., p. 14 ; lib. ii., p. 118; lib. xvi., p. 781 ; lib. xvii., p. 789 and 815.

‡ Compare the two passages of Strabo, lib. i., p. 65, and lib. ii., p. 118 (Humboldt, *Examen Critique de l'Hist. de la Géographie*, t. i., p. 152–154). In the important new edition of Strabo, published by Gustav Kramer, 1844, th. i., p. 100, " the parallel of Athens is read for the parallel of Thinæ, as if Thinæ had first been named in the Pseudo-Arrian, in the *Periplus Maris Rubri*." Dodwell places the *Periplus* under Marcus Aurelius and Lucius Verus, while, according to Letronne, it was written under Septimius Severus and Caracalla. Although five passages in Strabo, according to all our manuscripts, have *Thinæ*, yet lib. ii., p. 79, 86, 87, and, above all, 82, in which Eratosthenes himself is named, prove decidedly that the reading should be the " parallel of Athens and Rhodes." These two places were confounded, as old geographers made the peninsula of Attica extend too far toward the south. It would also appear surprising, supposing the usual reading Θινῶν κύκλος to be the more correct, that a particular parallel, the Diaphragm of Dicæarchus, should be called after a place so little known as that of the Sines (Tsin). However, Cosmas Indicopleustes also connects his Tzinitza (Thinæ) with the chain of mountains which divides Persia and the Romanic districts no less than the whole habitable world into two parts, subjoining the remarkable observation that this division is accord-

in the same temperate zone, near the parallel of Thinæ or Athens, which passes through the Atlantic Ocean, besides the world we inhabit, there may be one or more other worlds peopled by beings different from ourselves." It is astonishing that this expression did not attract the attention of Spanish writers, who, at the beginning of the sixteenth century, believed that they every where, in classical authors, found the traces of a knowledge of the New World.

" Since," as Strabo well observes, " in all works of art which are designed to represent something great, the object aimed at is not the completeness of the individual parts," his chief desire, in his gigantic work, is pre-eminently to direct attention to the form of the whole. This tendency toward a generalization of ideas did not prevent him, at the same time, from prosecuting researches which led to the establishment of a large number of admirable physical results, referring more especially to geognosy.* He entered, like Posidonius and Polybius, into the consideration of the influence of the longer or shorter interval that occurred between each passage of the sun across the zenith ; of the maximum of atmospheric heat under the tropics and the equator ; of the various causes which give rise to the changes experienced by the earth's surface ;

ing to the " belief of the Indian philosophers and Brahmins." Compare Cosmas, in Montfaucon, *Collect. nova Patrum.*; t. ii., p. 137 ; and my *Asie Centrale*, t. i., p. xxiii., 120–129, and 194–203 ; t. ii., p. 413. Cosmas and the Pseudo-Arrian, Agathemeros, according to the learned investigations of Professor Franz, decidedly ascribe to the metropolis of the Sines a high northern latitude (nearly in the parallel of Rhodes and Athens) ; while Ptolemy, misled by the accounts of mariners, has no knowledge except of a Thinæ three degrees south of the equator (*Geogr.*, i., 17). I conjecture that Thinæ merely meant, generally, a Chinese emporium, a harbor in the land of Tsin, and that, therefore, one Thinæ (Tzinitza) may have been designated north of the equator, and another south of the equator.

* Strabo, lib. i., p. 49–60 ; lib. ii., p. 95 and 97 ; lib. vi., p. 277 ; lib. xvii., p. 830. On the elevation of islands and of continents, see particularly lib. i., p. 51, 54, and 59. The old Eleat Xenophanes was led to conclude, from the numerous fossil marine productions found at a distance from the sea, that " the present dry ground had been raised from the bottom of the sea" (Origen, *Philosophumena*, cap. 4). Apuleius collected fossils at the time of the Antonines from the Gætulian (Mauritanian) Mountains, and attributed them to the Deucalion flood, to which he ascribed the same character of universality as the Hebrews to the Deluge of Noah, and the Mexican Azteks to that of the Coxcox. Professor Franz, by means of very careful investigation, has refuted the belief entertained by Beckmann and Cuvier, that Apuleius possessed a collection of specimens of natural history. (See Beckmann's *History of Inventions*, Bohn's Standard Library (1846), vol. i., p. 285 ; and *Hist. des Sciences Nat.*, t. i., p. 350.)

of the breaking forth of originally closed seas ; of the general
level of the sea, which was already recognized by Archimedes ;
of oceanic currents ; of the eruption of submarine volcanoes ;
of the petrifactions of shells and the impressions of fishes ; and,
lastly, of the periodic oscillations of the earth's crust, a subject
that most especially attracts our attention, since it constitutes
the germ of modern geognosy.   Strabo expressly remarks that
the altered limits of the sea and land are to be ascribed less
to small inundations than to the upheaval and depression of
the bottom, for " not only separate masses of rock and islands
of different dimensions, but entire continents, may be upheaved." Strabo, like Herodotus, was an attentive observer of the
descent of nations, and of the diversities of the different races
of men, whom he singularly enough calls " land and air animals, which require much light."*   We find the ethnological
distinction of races most sharply defined in the Commentaries of Julius Cæsar, and in the noble eulogy on Agricola by
Tacitus.

Unfortunately, Strabo's great work, which was so rich in
facts, and whose cosmical views we have already alluded to,
remained almost wholly unknown in Roman antiquity until
the fifth century, and was not even then made use of by that
universal collector, Pliny.   It was not until the close of the
Middle Ages that Strabo exercised any essential influence on
the direction of ideas, and even then in a less marked degree
than that of the more mathematical and more tabularly concise geography of Claudius Ptolemæus, which was almost
wholly wanting in views of a truly physical character.   This
latter work served as a guide to travelers as late as the sixteenth century, while every new discovery of places was always supposed to be recognized in it under some other appellation.

In the same manner as natural historians long continued
to include all recently-discovered plants and animals under
the classifying definitions of Linnæus, the earliest maps of the
New Continent appeared in the Atlas of Ptolemy, which
Agathodæmon prepared at the same time that, in the remotest part of Asia among the highly-civilized Chinese, the western provinces of the empire were already marked in forty-four
divisions.†   The Universal Geography of Ptolemy has indeed
the advantage of presenting us with a picture of the whole
world represented graphically in outlines, and numerically in
determinations of places, according to their parallels of longi-

---

* Strabo, lib. xvii., p. 810.   † Carl Ritter. *Asien*, th. v., s. 560.

tude and latitude, and to the length of the day ; but, notwith-standing the constant reference to the advantages of astro-nomical results over mere itinerary measurements by land and sea, it is, unfortunately, impossible to ascertain, among these uncertain positions (upward of 2500 of which are given), the nature of the data on which they are based, and the relative probability which may be ascribed to them, from the itinera-ries then in existence.

The entire ignorance of the polarity of the magnetic needle, and, consequently, of the use of the compass (which, twelve centuries and a half before the time of Ptolemy, under the Chinese Emperor Tsing-wang, had been used, together with a *way measurer*, in the construction of the magnetic cars), caused the most perfect of the itineraries of the Greeks and Romans to be extremely uncertain, owing to the deficiency of means for learning with certainty the direction or the line which formed the angle with the meridian.*

In proportion as a better knowledge has been acquired, in modern times, of the Indian and ancient Persian (or Zend) languages, we are more and more astonished to find that a great portion of the geographical nomenclature of Ptolemy may be regarded as an historical monument of the commercial relations existing between the West and the remotest regions of Southern and Central Asia.† We may reckon the knowl-edge of the complete insulation of the Caspian Sea as one of the most important results of these relations, but it was not

* See a collection of the most striking instances of Greek and Roman errors, regarding the directions of different mountain chains, in the in-troduction to my *Asie Centrale*, t. i., p. xxxvii.-xl. Most satisfactory investigations respecting the uncertainty of the numerical bases of Ptol-emy's positions are to be found in a treatise of Ukert, in the *Rheinische Museum für Philologie*, Jahrg., vi., 1838, s. 314-324.

† For examples of Zend and Sanscrit words which have been pre-served to us in Ptolemy's Geography, see Lassen, *Diss. de Taproban. Insula*, p. 6, 9, and 17; Burnouf's *Comment. sur le Yaçna*, t. i., p. xciii.-cxx. and clxxxi.-clxxxv.; and my *Examen Crit. de l'Hist. de la Géogr.*, t. i., p. 45-49. In a few cases Ptolemy gives both the Sanscrit names and their significations, as, for the island of Java, "barley island," Ἰαβαδίου, ὁ σημαινει κριδῆς νῆσος, Ptol., vii., 2 (Wilhelm von Humboldt, *Ueber die Kawi-Sprache*, bd. i., s. 60-63). According to Buschmann, the two-stalked barley, *Hordeum distichon*, is still termed in the princi-pal Indian languages (as in Hindostanee, Bengalee, and Nepaulese, and in the Mahratta, Guzerat, and Cingalese languages), as well as in Per-sian and Malay, *yava, dschav,* or *dschau,* and in the language of Orissa, *yaa.* (Compare the Indian translation of the Bible, in the passage Joh., vi., 9, and 13, and Ainslie, *Materia Medica of Hindostan*, Madras, 1813, p. 217.)

until after a period of five hundred years that the accuracy of the fact was re-established by Ptolemy. Herodotus and Aristotle entertained correct views regarding this subject, and the latter fortunately wrote his *Meteorologica* before the Asiatic campaigns of Alexander. The Olbiopolites, from whose lips the father of history derived his information, were well acquainted with the northern shores of the Caspian Sea, between Cuma, the Volga (Rha), and the Jaik (Ural), but there were no indications that could lead to the supposition of its connection with the Icy Sea. Very different causes led to the deception of Alexander's army, when, passing through Hecatompylos (Damaghan) to the humid forests of Mazanderan, at Zadrakarta, a little to the west of the present Asterabad, they saw the Caspian Sea stretching northward in an apparently boundless expanse of waters. This sight first gave rise, as Plutarch remarks in his Life of Alexander, to the conjecture that the sea they beheld was a bay of the Euxine.* The Macedonian expedition, although, on the whole, extremely favorable to the advance of geographical knowledge, nevertheless gave rise to some errors which long held their ground. The Tanais was confounded with the Jaxartes (the Araxes of Herodotus), and the Caucasus with the Paropanisus (the Hindoo-Coosh). Ptolemy was enabled, during his residence in Alexandria, as well as from the expeditions of the Aorsi, whose camels brought Indian and Babylonian goods to the Don and the Black Sea,† to obtain accurate knowledge of the countries which immediately surrounded the Caspian (as, for instance, Albania, Atropatene, and Hyrcania). If Ptolemy, in contradiction to the more correct knowledge of Herodotus, believed that the greater diameter of the Caspian Sea inclined from west to east, he might, perhaps, have been misled by a vague knowledge of the former great extension of the Scythian gulf (Karabogas), and the existence of Lake Aral, the earliest definite notice of which we find in the work of a Byzantine author, Menander, who wrote a continuation of Agathias.‡

It is to be regretted that Ptolemy, who had arrived at so correct a knowledge of the complete insulation of the Caspian (after it had long been considered to be open, in accordance with the hypothesis of four gulfs, and even according to sup-

* See my *Examen Crit. de l'Hist. de la Géographie*, t. ii., p. 147–188
† Strabo, lib. xi., p. 506.
‡ Menander, *De Legationibus Barbarorum ad Romanos, et Romanorum ad Gentes*, e rec. Bekkeri et Niebuhr, 1829, p. 300, 619, 623, and 628.

posed reflections of similar forms on the moon's disk),* should not have relinquished the myth of the unknown south land connecting Cape Prasum with Cattigara and Thinæ (*Sinarum Metropolis*), joining, therefore, Eastern Africa with the land of Tsin (China). This myth, which supposes the Indian Ocean to be an inland sea, was based upon views which may be traced from Marinus of Tyre to Hipparchus, Seleucus the Babylonian, and even to Aristotle.† We must limit ourselves, in these cosmical descriptions of the progress made in the contemplation of the universe, to a few examples illustrative of the fluctuations of knowledge, by which imperfectly-recognized facts were so often rendered still more obscure. The more the extension of navigation and of inland trade led to a hope that the whole of the earth's surface might become known, the more earnestly did the ever-wakeful imagination of the Greeks, especially in the Alexandrian age under the Ptolemies and under the Roman empire, strive by ingenious combinations to fuse ancient conjectures with newly-acquired knowledge, and thus speedily to complete the scarcely sketched map of the earth. We have already briefly noticed that Claudius Ptolemæus, by his optical inquiries, which have been in part preserved to us by the Arabians, became the founder of one branch of mathematical physics, which, according to Theon of Alexandria, had already been noticed, with reference to the refraction of rays of light, in the *Catoptrica* of Archimedes.‡ We may esteem it as an important advance when physical phenomena, instead of being simply observed and compared together (of which we have memorable examples in Greek antiquity, in the comprehensive pseudo-Aristotelian problems, and in Roman antiquity in the works of Seneca), are intentionally evoked under altered conditions, and are then measured.§

* Plutarch, *De Facie in Orbe Lunæ*, p. 921, 19 (compare my *Examen Crit.*, t. i., p. 145–191). I have myself met, among highly-informed Persians, with a repetition of the hypothesis of Agesianax, according to which, the marks on the moon's disk, in which Plutarch (p. 935, 4) thought he saw "a peculiar kind of shining mountains" (volcanoes ?), were merely the reflected images of terrestrial lands, seas, and isthmuses. These Persians would say, for instance, "What we see through telescopes on the surface of the moon are the reflected images of our own country."

† Ptolem., lib. iv., cap. 9; lib. vii., cap. 3 and 5. Compare Letronne, in the *Journal des Savans*, 1831, p. 476–480, and 545–555; Humboldt, *Examen Crit.*, t. i., p. 144, 161, and 329; t. ii., p. 370–373.

‡ Delambre, *Hist. de l'Astronomie Ancienne*, t. i., p. liv.; t. ii., p. 551. Theon never makes any mention of Ptolemy's *Optics*, although he lived fully two centuries after him.

§ It is often difficult in reading ancient works on physics, to decide

This latter mode of proceeding characterizes the investigations of Ptolemy on the refraction of rays in their passage
through media of unequal density. Ptolemy caused the rays
to pass from air into water and glass, and from water into
glass, under different angles of incidence, and he finally arranged the results of these physical experiments in tables.
This measurement of a physical phenomenon called forth at
will, of a process of nature not dependent upon a movement
of the waves of light (Aristotle, assuming a movement of the
medium between the eye and the object), stands wholly isolated in the period which we are now considering.* This age
presents, with respect to investigation into the elements of nature, only a few chemical experiments by Dioscorides, and, as
I have already elsewhere noticed, the technical art of collecting fluids by the process of distillation.† Chemistry can not
be said to have begun until man learned to obtain mineral
acids, and to employ them for the solution and liberation of
substances, and it is on this account that the distillation of sea
water, described by Alexander of Aphrodisias under Caracalla,
is so worthy of notice.   It designates the path by which man
gradually arrived at a knowledge of the heterogeneous nature
of substances, their chemical composition, and their mutual
affinities.

The only names which we can bring forward in connection
with the study of organic nature are the anatomist Marinus ;
Rufus of Ephesus, who dissected apes, and distinguished between nerves of sensation and of motion ; and Galen of Pergamus, who eclipsed all others.   The natural history of animals by Ælian of Præneste, and the poem on fishes by Op-
pianus of Cilicia, contain scattered notices, but no facts based
on personal examination.   It is impossible to comprehend how
the enormous multitudes of elephants, rhinoceroses, hippopot-
amuses, elks, lions, tigers, panthers, crocodiles, and ostriches,
which for upward of four centuries were slain in the Roman

whether a particular result has sprung from a phenomenon purposely
called forth or accidentally observed.   Where Aristotle (De Cœlo, iv.,
4) treats of the weight of the atmosphere, which, however, Ideler appears to deny (Meteorologia veterum Græcorum et Romanorum, p. 23),
he says distinctly, " an inflated bladder is heavier than an empty one."
The experiment must have been made with condensed air, if actually
tried.

    * Aristot., De Anima., ii., 7 ; Biese, Die Philosophie des Aristot., bd.
ii., s. 147.

    † Joannis (Philoponi) Grammatici, in libr. De Generat., and Alex-
andri Aphrodis., in Meteorol. Comment. (Venet., 1527), p. 97, b.   Compare my Examen Critique, t. ii., p. 306–312.

circus, should have failed to advance the knowledge of comparative anatomy.* I have already noticed the merit of
Dioscorides in regard to the collection and study of plants,
and it only remains, therefore, to observe that his works exercised the greatest influence on the botany and pharmaceutical
chemistry of the Arabs. The botanical garden of the Roman physician Antonius Castor, who lived to be upward of a
hundred years of age, was perhaps laid out in imitation of the
botanical gardens of Theophrastes and Mithridates, but it did
not, in all probability, lead to any further advancement in
science than did the collection of fossil bones formed by the
Emperor Augustus, or the museum of objects and products of
nature which has been ascribed on very slight foundation to
Apuleius of Madaura.†

The representation of the contributions made by the epoch
of the Roman dominion to cosmical knowledge would be incomplete were I to omit mentioning the great attempt made
by Caius Plinius Secundus to comprise a description of the
universe in a work consisting of thirty-seven books. In the
whole of antiquity, nothing similar had been attempted ; and
although the work grew, from the nature of the undertaking,
into a species of encyclopædia of nature and art (the author
himself, in his dedication to Titus, not scrupling to apply to
his work the then more noble Greek expression ἐγκυκλοπαι-
δεία, or conception and popular sphere of universal knowledge),
yet it must be admitted that, notwithstanding the deficiency of
an internal connection among the different parts of which the
whole is composed, it presents the plan of a physical description of the universe.

The *Historia Naturalis* of Pliny, entitled, in the tabular
view which forms what is known as the first book, *Historia
Mundi*, and in a letter of his nephew to his friend Macer still
more aptly, *Naturæ Historia*, embraces both the heavens and
the earth, the position and course of the heavenly bodies, the
meteorological processes of the atmosphere, the form of the

* The Numidian Metellus caused 142 elephants to be killed in the
circus. In the games which Pompey gave, 600 lions and 406 panthers
were assembled. Augustus sacrificed 3500 wild beasts in the national
festivities, and a tender husband laments that he could not celebrate
the day of his wife's death by a sanguinary gladiatorial fight at Verona,
" because contrary winds had detained in port the panthers which had
been bought in Africa !" (Plin., *Epist.*, vi., 34.)

† See *ante*, p. 190. Yet Apuleius, as Cuvier remarks (*Hist. des Sciences Naturelles*, t. i., p. 287), was the first to describe accurately the bony
hook in the second and third stomach of the Aplysiæ.

earth's surface, and all terrestrial objects, from the vegetable mantle with which the land is covered, and the mollusca of the ocean, up to mankind. Man is considered, according to the variety of his mental dispositions and his exaltation of these spiritual gifts, in the development of the noblest creations of art. I have here enumerated the elements of a general knowledge of nature which lie scattered irregularly throughout different parts of the work. "The path on which I am about to enter," says Pliny, with a noble self-confidence, "is untrodden (*non trita auctoribus via*); no one among my own countrymen, or among the Greeks, has as yet attempted to treat of the whole of nature under its character of universality (*nemo apud Græcos qui unus omnia tractaverit*). If my undertaking should not succeed, it is, at any rate, both beautiful and noble (*pulchrum atque magnificum*) to have made the attempt."

A grand and single image floated before the mind of the intellectual author; but, suffering his attention to be distracted by specialities, and wanting the living contemplation of nature, he was unable to hold fast this image. The execution was incomplete, not merely from a superficiality of views, and a want of knowledge of the objects to be treated of (here we, of course, can only judge of the portions that have come down to us), but also from an erroneous mode of arrangement. We discover in the author the busy and occupied man of rank, who prided himself on his wakefulness and nocturnal labors, but who, undoubtedly, too often confided the loose web of an endless compilation to his ill-informed dependents, while he was himself engaged in superintending the management of public affairs, when holding the place of Governor of Spain, or of a superintendent of the fleet in Lower Italy. This taste for compilation, for the laborious collection of the separate observations and facts yielded by science as it then existed, is by no means deserving of censure, but the want of success that has attended Pliny's undertaking is to be ascribed to his incapacity of mastering the materials accumulated, of bringing the descriptions of nature under the control of higher and more general views, or of keeping in sight the point of view presented by a comparative study of nature. The germs of such nobler, not merely orographic, but truly geognostic views, were to be met with in Eratosthenes and Strabo; but Pliny never made use of the works of the latter, and only on one occasion of those of the former; nor did Aristotle's History of Animals teach him their division into large classes based upon internal organization, or lead him to adopt the method of induction, which is the only safe means of generalizing results.

Beginning with pantheistic considerations, Pliny descends from the celestial regions to terrestrial objects. He recognizes the necessity of representing the forces and the glory of nature (*naturæ vis atque majestas*) as a great and comprehensive whole (I would here refer to the motto on the title of my work), and at the beginning of the third book he distinguishes between general and special geography ; but this distinction is again soon neglected when he becomes absorbed in the dry nomenclature of countries, mountains, and rivers. The greater portions of Books VIII.-XXVII., XXXIII. and XXXIV., XXXVI. and XXXVII., consist of categorical enumerations of the three kingdoms of nature. Pliny the Younger, in one of his letters, justly characterizes the work of his uncle as " learned and full of matter, no less various than Nature herself (*opus diffusum, eruditum, nec minus varium quam ipsa natura*)." Many things which have been made subjects of reproach against Pliny as needless and irrelevant admixtures, rather appear to me deserving of praise. It has always afforded me especial gratification to observe that he refers so frequently, and with such evident partiality, to the influence exercised by nature on the civilization and mental development of mankind. It must, however, be admitted, that his points of connection are seldom felicitously chosen (as, for instance, in VII., 24-47 ; XXV., 2 ; XXVI., 1 ; XXXV., 2 ; XXXVI., 2-4 ; XXXVII., 1). Thus the consideration of the nature of mineral and vegetable substances leads to the introduction of a fragment of the history of the plastic arts, but this brief notice has become more important, in the present state of our knowledge, than all that we can gather regarding descriptive natural history from the rest of the work.

The style of Pliny evinces more spirit and animation than true dignity, and it is seldom that his descriptions possess any degree of pictorial distinctness. We feel that the author has drawn his impressions from books and not from nature, however freely it may have been presented to him in the different regions of the earth which he visited. A grave and somber tone of color pervades the whole composition, and this sentimental feeling is tinged with a touch of bitterness whenever he enters upon the consideration of the conditions of man and his destiny. On these occasions, almost as in the writings of Cicero, although with less simplicity of diction,* the aspect of

* " Est enim animorum ingeniorumque naturale quoddam quasi pab-

the grand unity of nature is adduced as productive of encouragement and consolation to man.

The conclusion of the *Historia Naturalis* of Pliny—the greatest Roman memorial transmitted to the literature of the Middle Ages—is composed in a true spirit of cosmical description. It contains, in the condition in which we have possessed it since 1831,* a brief consideration of the comparative natural history of countries in different zones, a eulogium of Southern Europe between the Mediterranean and the chain of the Alps, and a description in praise of the Hesperian sky, "where the temperate and gentle mildness of the climate had," according to a dogma of the older Pythagoreans, "early hastened the liberation of mankind from barbarism."

The influence of the Roman dominion as a constant element of union and fusion required the more urgently and forcibly to be brought forward in a history of the contemplation of the universe, since we are able to recognize the traces of this influence in its remotest consequences even at a period when the bond of political union had become less compact, and was even partially destroyed by the inroads of barbarians. Claudian, who stands forth in the decline of literature during the latter and more disturbed age of Theodosius the Great and his sons, distinguished for the endowment of a revived poetic productiveness, still sings, in too highly laudatory strains, of the dominion of the Romans.†

> *Hæc est, in gremium victos quæ sola recepit,*
> *Humanumque genus communi nomine fovit,*
> *Matris, non dominæ, ritu; civesque vocavit*
> *Quos domuit, nexuque pio longinqua revinxit.*
> *Hujus pacificis debemus moribus omnes*
> *Quod veluti patriis regionibus utitur hospes. . . .*

External means of constraint, artificially-arranged civil institutions, and long-continued servitude, might certainly tend to unite nations by destroying the individual existence of each one ; but the feeling of the unity and common condition of the whole human race, and of the equal rights of all men, has a nobler origin, and is based on the internal promptings of the

ulum consideratio contemplatioque naturæ. Erigimur, elatiores fieri videmur, humana despicimus, cogitantesque supera atque cœlestia hæc nostra, ut exigua et minima, contemnimus." (Cic., *Acad.*, ii., 41.)

* Plin., xxxvii., 13 (ed. Sillig., t. v., 1836, p. 320). All earlier editions closed with the words "Hispaniam quacunque ambitur mari." The conclusion of the work was discovered in 1831, in a Bamberg Codex, by Herr Ludwig v. Jan, professor at Schweinfurt.

† Claudian, in *Secundum Consulatum Stillichonis*, v. 150–155.

spirit and on the force of religious convictions. Christianity has materially contributed to call forth this idea of the unity of the human race, and has thus tended to exercise a favorable influence on the *humanization* of nations in their morals, manners, and institutions. Although closely interwoven with the earliest doctrines of Christianity, this idea of humanity met with only a slow and tardy recognition ; for at the time when the new faith was raised at Byzantium, from political motives, to be the established religion of the state, its adherents were already deeply involved in miserable party dissensions, while intercourse with distant nations was impeded, and the foundations of the empire were shaken in many directions by external assaults. Even the personal freedom of entire races of men long found no protection in Christian states from ecclesiastical land-owners and corporate bodies.

Such unnatural impediments, and many others which stand in the way of the intellectual advance of mankind and the ennoblement of social institutions, will all gradually disappear. The principle of individual and political freedom is implanted in the ineradicable conviction of the equal rights of one sole human race. Thus, as I have already remarked,[*] mankind presents itself to our contemplation as one great fraternity and as one independent unity, striving for the attainment of one aim—the free development of moral vigor. This consideration of humanity, or, rather, of the tendency toward it, which, sometimes checked, and sometimes advancing with a rapid and powerful progressive movement—and by no means a discovery of recent times—belongs, by the generalizing influence of its direction, most specially to that which elevates and animates cosmical life. In delineating the great epoch of the history of the universe, which includes the dominion of the Romans and the laws which they promulgated, together with the beginning of Christianity, it would have been impossible not to direct special attention to the manner in which the religion of Christ enlarged these views of mankind, and to the mild and long-enduring, although slowly-operating, influence which it exercised on general, intellectual, moral, and social development.

[*] See vol. i., p. 358 ; and compare, also, Wilhelm von Humboldt, *Ueber die Kawi-Sprache*, bd. i., s. xxxviii.

INVASION OF THE ARABS.—INTELLECTUAL APTITUDE OF THIS BRANCH
OF THE SEMITIC RACES.—INFLUENCE OF FOREIGN ELEMENTS ON
THE DEVELOPMENT OF EUROPEAN CULTURE.—THE INDIVIDUALITY
OF THE NATIONAL CHARACTER OF THE ARABS.—TENDENCY TO A
COMMUNION WITH NATURE AND PHYSICAL FORCES.—MEDICINE AND
CHEMISTRY.—EXTENSION OF PHYSICAL GEOGRAPHY.—ASTRONOMY
AND MATHEMATICAL SCIENCES IN THE INTERIOR OF CONTINENTS.

IN the preceding sketch of the history of the physical contemplation of the universe we have already considered four
principal momenta in the gradual development of the recognition of the unity of nature, viz. :

1. The attempts made to penetrate from the basin of the
Mediterranean eastward to the Euxine and Phasis ; southward to Ophir and the tropical gold lands ; and westward,
through the Pillars of Hercules, into the " all-encircling
ocean."

2. The Macedonian campaign under Alexander the Great.

3. The age of the Ptolemies.

4. The universal dominion of the Romans.

We now, therefore, proceed to consider the important influence exercised on the general advancement of the physical and
mathematical sciences, first, by the admixture of the foreign
elements of Arabian culture with European civilization, and,
six or seven centuries later, by the maritime discoveries of the
Portuguese and Spaniards ; and likewise their influence on
the knowledge of the earth and the regions of space, with respect to form and measurement, and to the heterogeneous
nature of matter, and the forces inherent in it.   The discovery and exploration of the New Continent, through the
range of its volcanic Cordilleras and its elevated plateaux,
where climates are ranged in strata, as it were, above one
another, and the development of vegetation within 120 degrees of latitude, undoubtedly indicates the period which has
presented, in the shortest period of time, the greatest abundance of new physical observations to the human mind.

From this period, the extension of cosmical knowledge
ceased to be associated with separate and locally-defined political occurrences.   Great inventions now first emanated from
spontaneous intellectual power, and were no longer solely
excited by the influence of separate external causes.   The
human mind, acting simultaneously in several directions,
created, by new combinations of thought, new organs, by
which the human eye could alike scrutinize the remote re-

gions of space, and the delicate tissues of animal and vegetable structures, which serve as the very substratum of life.
Thus the whole of the seventeenth century, whose commencement was brilliantly signalized by the great discovery of the
telescope, together with the immediate results by which it was
attended—from Galileo's observation of Jupiter's satellites,
of the crescentic form of the disk of Venus, and the spots on
the sun, to the theory of gravitation discovered by Newton—
ranks as the most important epoch of a newly-created physical
astronomy. This period constitutes, therefore, from the unity
of the efforts made toward the observation of the heavenly
bodies, and in mathematical investigations, a sharply-defin-
ed section in the great process of intellectual development,
which, since then, has been characterized by an uninterrupted progress.

In more recent times, the difficulty of signalizing separate
momenta increases in proportion as human activity becomes
more variously directed, and as the new order of social and
political relations binds all the various branches of science in
one closer bond of union. In some few sciences, whose development has been considered in the history of the physical contemplation of the universe, as, for instance, in chemistry and
descriptive botany, individual periods may be instanced, even
in the most recent time, in which great advancement has been
rapidly made, or new views suddenly opened ; but, in the history of the contemplation of the universe, which, from its very
nature, must be limited to the consideration of those facts regarding separate branches of science which most directly relate
to the extension of the idea of the Cosmos considered as one
natural whole, the connection of definite epochs becomes impracticable, since that which we have named the process of
intellectual development presupposes an uninterrupted simultaneous advance in all spheres of cosmical knowledge. At
this important point of separation between the downfall of the
universal dominion of the Romans and the introduction of a
new and foreign element of civilization by means of the first
direct contact of our continent with the land of the tropics, it
appears desirable that we should throw a general glance over
the path on which we are about to enter.

The Arabs, a people of Semitic origin, partially dispelled
the barbarism which had shrouded Europe for upward of two
hundred years after the storms by which it had been shaken,
from the aggressions of hostile nations. The Arabs lead us
back to the imperishable sources of Greek philosophy ; and,

besides the influence thus exercised on scientific cultivation, they have also extended and opened new paths in the domain of natural investigation. In our continent these disturbing storms began under Valentinian I., when the Huns (of Finnish, not Mongolian origin) penetrated beyond the Don in the closing part of the fourth century, and subdued, first the Alani, and subsequently, with their aid, the*Ostrogoths. In the remote parts of Eastern Asia, the stream of migratory nations had already been moved in its onward course for several centuries before our era. The first impulse was given, as we have already remarked, by the attack of the Hiungnu, a Turkish race, on the fair-haired and blue-eyed Usuni, probably of Indo-Germanic origin, who bordered on the Yueti (Geti), and dwelt in the upper river valley of the Hoang-ho, in the northwest of China. The devastating stream of migration directed from the great wall of China, which was erected as a protection against the inroads of the Hiungnu (214 B.C.), flowed on through Central Asia, north of the chain of the Celestial Mountains. These Asiatic hordes were uninfluenced by any religious zeal before they entered Europe, and some writers have even attempted to show that the Moguls were not as yet Buddhists when they advanced victoriously to Poland and Silesia.* Wholly different relations imparted a peculiar character to the warlike aggressions of a more southern race—the Arabs.

Remarkable for its form, and distinguished as a detached branch of the slightly-articulated continent of Asia, is situated the peninsula of Arabia, between the Red Sea and the Persian Gulf, the Euphrates and the Syro-Mediterranean Sea.† It is the most western of the three peninsulas of Southern Asia,

* If, as has often been asserted, Charles Martel, by his victory at Tours, protected Central Europe against the Mussulman invasion, it can not be maintained, with equal justice, that the retreat of the Moguls after the battle of Liegnitz prevented Buddhism from penetrating to the shores of the Elbe and the Rhine. The Mongolian battle, which was fought in the plain of Wahlstatt, near Liegnitz, and in which Duke Henry the Pious fell fighting bravely, took place on the 9th of April, 1241, four years after Kaptschak (Kamtschatka) and Russia became subject to the Asiatic horde, under Batu, the grandson of Genghis Khan. But the earliest introduction of Buddhism among the Mongolians took place in the year 1247, when, in the east at Leang-tscheu, in the Chinese province of Schensi, the sick Mongolian prince Godan caused the Sakya Pandita, a Thibetian archbishop, to be sent for, in order to cure and convert him. (Klaproth, in a MS. fragment, "*Ueber die Verbreitung des Buddhismus im östlichen und nördlichen Asien.*") The Mongolians have never occupied themselves with the conversion of conquered nations. † See vol. i., p. 291.

and its vicinity to Egypt, and to a European sea-basin, gives it signal advantages in a political no less than a commercial point of view. In the central parts of the Arabian Peninsula lived the tribe of the Hedschaz, a noble and valiant race, unlearned, but not wholly rude, imaginative, and, at the same time, devoted to the careful observation of all the processes of free nature manifested in the ever-serene vault of heaven and on the surface of the earth. This people, after having continued for thousands of years almost without contact with the rest of the world, and advancing chiefly in nomadic hordes, suddenly burst forth from their former mode of life, and, acquiring cultivation from the mental contact of the inhabitants of more ancient seats of civilization, converted and subjected to their dominion the nations dwelling between the Pillars of Hercules and the Indus, to the point where the Bolor chain intersects the Hindoo-Coosh. They maintained relations of commerce as early as the middle of the ninth century simultaneously with the northern countries of Europe, with Madagascar, Eastern Africa, India, and China; diffused languages, money, and Indian numerals, and founded a powerful and long-enduring communion of lands united together by one common religion. In these migratory advances great provinces were often only temporarily occupied. The swarming hordes, threatened by the natives, only rested for a while, according to the poetical diction of their own historians, "like groups of clouds which the winds ere long will scatter abroad." No other migratory movement has presented a more striking and instructive character; and it would appear as if the depressive influence manifested in circumscribing mental vigor, and which was apparently inherent in Islamism, acted less powerfully on the nations under the dominion of the Arabs than on Turkish races. Persecution for the sake of religion was here, as every where, even among Christians, more the result of an unbounded, dogmatizing despotism than the consequence of any original form of belief or any religious contemplation existing among the people. The anathemas of the Koran are especially directed against superstition and the worship of idols among races of Aramæic descent.*

* Hence the contrast between the tyrannical measures of Motewekkil, the tenth calif of the house of the Abbassides, against Jews and Christians (Joseph von Hammer, *Ueber die Länderverwaltung unter dem Chalifate*, 1835, s. 27, 85, und 117), and the mild tolerance of wiser rulers in Spain (Conde, *Hist. de la Dominacion de los Arabes en España*, t. i., 1820, p. 67). It should also be remembered that Omar, after the taking of Jerusalem, tolerated every rite of Christian worship, and con-

As the life of nations is, independently of mental culture, determined by many external conditions of soil, climate, and vicinity to the sea, we must here remember the great varieties presented by the Arabian peninsula. Although the first impulse toward the changes effected by the Arabs in the three continents emanated from the Ismaelitish Hedschaz, and owed its principal force to one sole race of herdsmen, the littoral portions of the peninsula had continued for thousands of years open to intercourse with the rest of the world. In order to understand the connection and existence of great and singular occurrences, it is necessary to ascend to the primitive causes by which they have been gradually prepared.

Toward the southwest, on the Erythrean Sea, lies Yemen, the ancient seat of civilization (of Saba), the beautiful, fruitful, and richly-cultivated land of the Joctanidæ.* It produced incense (the *lebonah* of the Hebrews, perhaps the Boswellia thurifera of Colebrooke),† myrrh (a species of Amyris, first ac-

cluded a treaty with the patriarch favorable to the Christians. (*Fundgruben des Orients*, bd. v., s. 68.)

* It would appear from tradition that a branch of the Hebrews migrated to Southern Arabia, under the name of Jokthan (Qachthan), before the time of Abraham, and there founded flourishing kingdoms. (Ewald, *Geschichte des Volkes Israel*, bd. i., s. 337 und 450.)

† The tree which furnishes the Arabian incense of Hadramaut, celebrated from the earliest times, and which is never to be found in the island of Socotora, has not yet been discovered and determined by any botanist, not even by the laborious investigator Ehrenberg. An article similar to this incense is found in Eastern India, and particularly in Bundelcund, and is exported in considerable quantities from Bombay to China. This Indian incense is obtained, according to Colebrooke (*Asiatic Researches*, vol. ix., p. 377), from a plant made known by Roxburgh, *Boswellia thurifera* or *serrata* (included in Kunth's family of *Burseraceæ*). As, from the very ancient commercial connections between the coasts of Southern Arabia and Western India (Gildemeister, *Scriptorum Arabum Loci de Rebus Indicis*, p. 35), doubts might be entertained as to whether the λίβανος of Theophrastus (the *thus* of the Romans) belonged originally to the Arabian peninsula, Lassen's remark (*Indische Alterthumskunde*, bd. i., s. 286), that incense is called "*yàwana*, Javanese, i. e., Arabian," in Amara-Koscha, itself becomes very important, apparently implying that this product is brought to India from Arabia. It is called *Turuschka' pindaka' sihlô* (three names signifying incense) "*yàwanô*" in Amara-Koscha. (*Amarakocha*, publ. par A. Loiseleur Deslongchamps, Part i., 1839, p. 156.) Dioscorides also distinguishes Arabian from Indian incense. Carl Ritter, in his excellent monograph on the kinds of incense (*Asien*, bd. viii., abth. i., s. 356–372,) remarks very justly, that, from the similarity of climate, this species of plant (*Boswellia thurifera*) might be diffused from India through the south of Persia to Arabia. The American incense (*Olibanum Americanum* of our Pharmacopœias) is obtained from *Icica gujanensis*, Aubl., and *Icica tacamahaca*, which Bonpland and myself fre-

curately described by Ehrenberg), and the so-called balsam of
Mecca (the Balsamodendron Gileadense of Kunth). These
products constituted an important branch of commerce between the contiguous tribes and the Egyptians, Persians, and
Indians, as well as the Greeks and Romans ; and it was
owing to their abundance and luxuriance that the country
acquired the designation of "Arabia Felix," which occurs as
early as in the writings of Diodorus and Strabo. In the
southeast of the peninsula, on the Persian Gulf, and opposite
the Phœnician settlements of Aradus and Tylus, lay Gerrha,
an important emporium for Indian articles of commerce.

Although the greater part of the interior of Arabia may be
termed a barren, treeless, and sandy waste, we yet meet in
Oman, between Jailan and Basna, with a whole range of
well-cultivated oases, irrigated by subterranean canals ; and
we are indebted to the meritorious activity of the traveler
Wellsted for the knowledge of three mountain chains, of which
the highest and wood-crowned summit, named Dschebel-Akhdar, rises six thousand feet above the level of the sea near
Maskat.* In the hilly country of Yemen, east of Loheia, and
in the littoral range of Hedschaz, in Asyr, and also to the east
of Mecca, at Tayef, there are elevated plateaux, whose perpetually low temperature was known to the geographer Edrisi.†

The same diversity of mountain landscape characterizes the
peninsula of Sinai, the Copper-land of the Egyptians of the
old kingdom (before the time of the Hyksos), and the stony
valleys of Petra. I have already elsewhere spoken of the
Phœnician commercial settlements on the most northern portion of the Red Sea, and of the expeditions to Ophir under
Hiram and Solomon, which started from Ezion-Geber.‡ Arabia, and the neighboring island of Socotora (the island of Dioscorides), inhabited by Indian colonists, participated in the

quently found growing on the vast grassy plains (llanos) of Calaboso, in
South America. *Icica*, like *Boswellia*, belongs to the family of *Burseraceæ*.

The red pine (*Pinus abies*, Linn.) produces the common incense of
our churches. The plant which bears myrrh, and which Bruce thought
he had seen (Ainslie, *Materia Medica of Hindostan*, Madras, 1813, p.
29), has been discovered by Ehrenberg near el-Gisan in Arabia, and
has been described by Nees von Esenbeck, from the specimens collected by him, under the name of *Balsamodendron myrrha*. The *Balsamodendron Kotaf* of Kunth, an Amyris of Forskaal, was long erroneously regarded as the true myrrh-tree.

* Wellsted, *Travels in Arabia*, 1838, vol. i., p. 272–289.

† Jomard, *Etudes Géogr. et Hist. sur l'Arabie*, 1839, p. 14 and 32.

‡ See *ante*, p. 136.

universal traffic with India and the eastern coasts of Africa.
The natural products of these countries were interchanged for
those of Hadramaut and Yemen. "All they from Sheba
shall come," sings the Prophet Isaiah of the dromedaries of
Midian ; "they shall bring gold and incense."* Petra was
the emporium for the costly wares destined for Tyre and Sidon,
and the principal settlement of the Nabatæi, a people once
mighty in commerce, whose primitive seat is supposed by the
philologist Quatremère to have been situated among the Ger-
rhœan Mountains, on the Lower Euphrates. This northern
portion of Arabia maintained an active connection with other
civilized states, from its vicinity to Egypt, the diffusion of
Arabian tribes over the Syro-Palestinian boundaries and the
districts around the Euphrates, as well as by means of the
celebrated caravan track from Damascus through Emesa and
Tadmor (Palmyra) to Babylon. Mohammed himself, who
had sprung from a noble but impoverished family of the Ko-
reischite tribe, in his mercantile occupation, visited, before he
appeared as an inspired prophet and reformer, the fair at Bosra
on the Syrian frontier, that at Hadramaut, the land of incense,
and more particularly that held at Okadh, near Mecca, which
continued during twenty days, and whither poets, mostly Bed-
ouins, assembled annually, to take part in the lyric competi-
tions. I mention these individual facts referring to international relations of commerce, and the causes from which they
emanated, in order to give a more animated picture of the
circumstances which conduced to prepare the way for a universal change.

The spread of Arabian population toward the north reminds
us most especially of two events, which, notwithstanding the
obscurity in which their more immediate relations are shrouded, testify that even thousands of years before Mohammed,
the inhabitants of the peninsula had occasionally taken part
in the great universal traffic, both toward the West and East,
in the direction of Egypt and of the Euphrates. The *Semitic*
or Aramæic origin of the Hyksos, who put an end to the old
kingdom under the twelfth dynasty, two thousand two hundred years before our era, is now almost universally admitted
by all historians. Even Manetho says, " Some maintain that
these herdsmen were Arabians." Other authorities call them
Phœnicians, à term which was extended in antiquity to the
inhabitants of the Valley of the Jordan, and to all Arabian
races. The acute Ewald refers especially to the Amalekites,

* Isaiah, ch. lx., v. 6.

who originally lived in Yemen, and then spread themselves beyond Mecca and Medina to Canaan and Syria, appearing in the Arabian annals as rulers over Egypt in the time of Joseph.* It seems extraordinary that the nomadic races of the Hyksos should have been able to subdue the ancient powerful and well-organized kingdom of the Egyptians. Here the more freely-constituted nation entered into a successful contest with another long habituated to servitude, but yet the victorious Arabian immigrants were not then, as in more modern times, inspired by religious enthusiasm. The Hyksos, actuated by fear of the Assyrians (races of Arpaschschad), established their festivals and place of arms at Avaris, on the eastern arm of the Nile. This circumstance seems to indicate attempted advances on the part of hostile warlike bodies, and a great migration westward. A second event, which occurred probably a thousand years later, is mentioned by Diodorus on the authority of Ctesias.† Ariæus, a powerful prince of the Himyarites, entered into an alliance with Ninus, on the Tigris, and after they had conjointly defeated the Babylonians, he returned laden with rich spoils to his home in Southern Arabia.‡

Although a free pastoral mode of life may be regarded as predominating in the Hedschaz, and as constituting that of a great and powerful majority, the cities of Medina and of Mecca, with its ancient and mysterious temple holiness, the Kaaba, are mentioned as important places, much frequented by foreigners. It is probable that the complete and savage wildness generated by isolation was unknown in those districts which we term river valleys, and which were contiguous to coasts or to caravansery tracks. Gibbon, who knew so well how to consider the conditions of human life, draws attention to the essential differences existing between a nomadic life in the Arabian peninsula and that described by Herodotus and Hippocrates, in the so-called land of the Scythians, since in the latter region no portion of the pastoral people ever settled

---

* Ewald, *Gesch. des Volkes Israel*, bd. i., s. 300 und 450; Bunsen, *Ægypten*, buch iii., s. 10 und 32. The traditions of Medes and Persians in Northern Africa indicate very ancient migrations toward the West. They have been connected with the various versions of the myth of Hercules, and with the Phœnician Melkarth. (Compare Sallust, *Bellum Jugurth.*, cap. 18, drawn from Punic writings by Hiempsal; and Pliny, v. 8.) Strabo even terms the Maurusians (inhabitants of Mauritania) "Indians who had come with Hercules."

† Diod. Sic., lib. ii., cap. 2 and 3.

‡ *Ctesiæ Cnidii Operum Reliquiæ*, ed. Baehr, *Fragmenta Assyriaca*, p. 421; and Carl Müller, in Dindorf's edition of Herodotus (Par., 1844), p. 13–15.

in cities, while in the great Arabian peninsula the country people still hold communion with the inhabitants of the towns, whom they regard as of the same origin as themselves.* In the Kirghis steppe, a portion of the plain inhabited by the ancient Scythians (the Scoloti and Sacæ), and which exceeds in extent the area of Germany, there has never been a city for thousands of years, and yet, at the time of my journey in Siberia, the number of the tents (Yurti or Kibitkes) occupied by the three nomadic hordes exceeded 400,000, which would give a population of 2,000,000.† It is hardly necessary to enter more circumstantially into the consideration of the effect produced on mental culture by such great contrasts in the greater or less isolation of a nomadic life, even where equal mental qualifications are presupposed.

In the more highly-gifted race of the Arabs, natural adapt-ability for mental cultivation, the geographical relations we have already indicated, and the ancient commercial intercourse of the littoral districts with the highly-civilized neighboring states, all combine to explain how the irruption into Syria and Persia, and the subsequent possession of Egypt, were so speedily able to awaken in the conquerors a love for science and a tendency to the pursuit of independent observation. It was ordained in the wonderful decrees by which the course of events is regulated, that the Christian sects of Nestorians, which exercised a very marked influence on the geographical diffusion of knowledge, should prove of use to the Arabs even before they advanced to the erudite and contentious city of Alexandria, and that, protected by the armed followers of the creed of Islam, these Nestorian doctrines of Christianity were enabled to penetrate far into Eastern Asia. The Arabs were first made acquainted with Greek literature through the Syrians, a kindred Semitic race, who had themselves acquired a knowledge of it only about a hundred and fifty years earlier through the heretical Nestorians.‡ Physicians, who had been educated in the scholastic establishments of the Greeks, and in the celebrated school of medicine founded by the Nestorian Christians at Edessa in Mesopotamia, were settled at Mecca as early as Mohammed's time, and there lived on a footing of friendly intercourse with the Prophet and Abu-Bekr.

* Gibbon, *Hist. of the Decline and Fall of the Roman Empire*, vol. ix., chap. 50, p. 200 (Leips., 1829).

† Humboldt, *Asie Centr.*, t. ii., p. 128.

‡ Jourdain, *Recherches Critiques sur l'Age des Traductions d'Aristote.* 1819, p. 81 and 87.

The school of Edessa, a prototype of the Benedictine schools of Monte Cassino and Salerno, gave the first impulse to a scientific investigation of remedial agents yielded from the mineral and vegetable kingdoms. When these establishments were dissolved by Christian fanaticism, under Zeno the Isaurian, the Nestorians were scattered over Persia, where they soon attained to political importance, and founded at Dschondisapur, in Khusistan, a medical school, which was afterward much frequented. They succeeded, toward the middle of the seventh century, in extending their knowledge and their doctrines as far as China, under the Thang dynasty, 572 years after Buddhism had penetrated thither from India.

The seeds of Western civilization, which had been scattered over Persia by learned monks and by the philosophers of the last Platonic school at Athens, persecuted by Justinian, had exercised a beneficial influence on the Arabs during their first Asiatic campaigns. However faint the sparks of knowledge diffused by the Nestorian priesthood might have been, their peculiar tendency to the investigation of medical pharmacy could not fail to influence a race which had so long lived in the enjoyment of a free communion with nature, and which preserved a more vivid feeling for every kind of natural investigation than the Greek and Italian inhabitants of cities. The cosmical importance attached to the age of the Arabs depends in a great measure on the national characteristics which we are here considering. The Arabs, I would again remark, are to be regarded as the actual founders of physical science, considered in the sense which we now apply to the words.

It is undoubtedly extremely difficult to associate any absolute beginning with any definite epoch of time in the history of the mental world and of the intimately-connected elements of thought. Individual luminous points of knowledge, and the processes by which knowledge was gradually attained, may be traced scattered through very early periods of time. How great is the difference that separates Dioscorides, who distilled mercury from cinnabar, from the Arabian chemist Dscheber; how widely is Ptolemy, as an optician, removed from Alhazen; but we must, nevertheless, date the foundation of the physical sciences, and even of natural science, from the point where new paths were first trodden by many different investigators, although with unequal success. To the mere contemplation of nature, to the observation of the phenomena accidentally presented to the eye in the terrestrial and celestial regions of space, succeeds investigation into the actual, an estimate by the measurement of magnitudes and the duration of motion. The earliest epoch of such a species of natural observation, although principally limited to organic substances, was the age of Aristotle. There remains a third and higher stage in the progressive advancement of the knowledge of physical phenomena, which embraces an investigation into natural forces, and the powers by which these forces are enabled to act, in order to be able to bring the substances liberated into new combinations. The means by which this liberation is effected are experiments, by which phenomena may be called forth at will.

The last-named stage of the process of knowledge, which was almost wholly disregarded in antiquity, was raised by the Arabs to a high degree of development. This people belonged to a country which enjoyed, throughout its whole extent, the climate of the region of palms, and in its greater part that of tropical lands (the tropic of Cancer intersecting the peninsula in the direction of a line running from Maskat to Mecca), and this portion of the world was therefore characterized by the highly-developed vital force pervading vegetation, by which an abundance of aromatic and balsamic juices was yielded to man from various beneficial and deleterious vegetable substances. The attention of the people must early have been directed to the natural products of their native soil, and those brought as articles of commerce from the accessible coasts of Malabar, Ceylon, and Eastern Africa. In these regions of the torrid zone, organic forms become individualized within very limited portions of space, each one being characterized by individual products, and thus increasing the communion of men with nature by a constant excitement toward natural observation. Hence arose the wish to distinguish carefully from one another these precious articles of commerce, which were so important to medicine, to manufactures, and to the pomp of temples and palaces, and to discover the native region of each, which was often artfully concealed from motives of avarice. Starting from the staple emporium of Gerrha, on the Persian Gulf, and from Yemen, the native district of incense, numerous caravan tracks intersected the whole interior of the Arabian peninsula to Phœnicia and Syria, and thus every where diffused a taste for and a knowledge of the names of these powerful natural products.

The science of medicine, which was founded by Dioscorides in the school of Alexandria, when considered with reference to its scientific development, is essentially a creation of the Arabs,

to whom the oldest, and, at the same time, one of the richest
sources of knowledge, that of the Indian physicians, had been
early opened.* Chemical pharmacy was created by the Arabs,
while to them are likewise due the first official prescriptions
regarding the preparation and admixture of different remedial
agents—the dispensing recipes of the present day. These
were subsequently diffused over the south of Europe by the
school of Salerno. Pharmacy and *Materia Medica*, the first
requirements of practical medicine, led simultaneously, in two
directions, to the study of botany and to that of chemistry.
From its narrow sphere of utility and its limited application,
botany gradually opened a wider and freer field, comprehending investigations into the structure of organic tissues and
their connection with vital forces, and into the laws by which
vegetable forms are associated in families, and may be distinguished geographically according to diversities of climate and
differences of elevation above the earth's surface.

From the time of the Asiatic conquests, for the maintenance of which Bagdad subsequently constituted a central point
of power and civilization, the Arabs spread themselves, in the
short space of seventy years, over Egypt, Cyrene, and Carthage, through the whole of Northern Asia to the far remote
western peninsula of Iberia. The inconsiderable degree of
cultivation possessed by the people and their leaders might
certainly incline us to expect every demonstration of rude barbarism ; but the mythical account of the burning of the Alexandrian Library by Amru, including the account of its application, during six months, as fuel to heat 4000 bathing rooms,
rests on the sole testimony of two writers who lived 580 years
after the alleged occurrence took place.† We need not here
describe how, in more peaceful times, during the brilliant
epoch of Al-Mansur, Haroun Al-Raschid, Mamun, and Mota-
sem, the courts of princes, and public scientific institutions,
were enabled to draw together large numbers of the most distinguished men, although without imparting a freer develop-

* On the knowledge which the Arabs derived from the Hindoos regarding the Materia Medica, see Wilson's important investigations in the
Oriental Magazine of Calcutta, 1823, February and March ; and those
of Royle, in his *Essay on the Antiquity of Hindoo Medicine*, 1837, p. 56–
59, 64–66, 73, and 92. Compare an account of Arabic pharmaceutical
writings, translated from Hindostanee, in Ainslie (Madras edition), p.
289.

† Gibbon, vol. ix., chap. li., p. 392 ; Heeren, *Gesch. des Studiums der
Classischen Litteratur*, bd. i., 1797, s. 44 und 72; Sacy, *Abd-Allatif*, p.
240 ; Parthey, *Das Alexandrinische Museum*, 1838, s. 106.

ment to the mental culture of the mass of the people. It is
not my object in the present work to give a characteristic
sketch of the far-extended and variously-developed literature
of the Arabs, or to distinguish the elements that spring from
the hidden depths of the organization of races, and the natural unfolding of their character, from those which are owing to
external inducements and accidental controlling causes. The
solution of this important problem belongs to another sphere of
ideas, while our historical considerations are limited to a fragmentary enumeration of the various elements which have contributed, in mathematical, astronomical, and physical science,
toward the diffusion of a more general contemplation of the
universe among the Arabs.

Alchemy, magic, and mystic fancies, deprived by scholastic
phraseology of all poetic charm, corrupted here, as elsewhere,
in the Middle Ages, the true results of inquiry ; but still the
Arabs have enlarged the views of nature, and given origin to
many new elements of knowledge, by their indefatigable and
independent labors, while, by means of careful translations into
their own tongue, they have appropriated to themselves the
fruits of the labors of earlier cultivated generations.   Attention has been justly drawn to the great difference existing in
the relations of civilization between immigrating Germanic
and Arabian races.* The former became cultivated after their
immigration ; the latter brought with them from their native country not only their religion, but a highly-polished language, and the graceful blossoms of a poetry which has not
been wholly devoid of influence on the Provençals and Minnesingers.

The Arabs possessed remarkable qualifications alike for ap-
propriating to themselves, and again diffusing abroad, the seeds
of knowledge and general intercourse, from the Euphrates to
the Guadalquivir, and to the south of Central Africa.   They
exhibited an unparalleled mobility of character, and a tendency to amalgamate with the nations whom they conquered,
wholly at variance with the repelling spirit of the Israelitish
castes, while, at the same time, they adhered to their national
character, and the traditional recollections of their original
home, notwithstanding their constant change of abode.   No
other race presents us with more striking examples of extensive land journeys, undertaken by private individuals, not only
for purposes of trade, but also with the view of collecting in-

* Heinrich Ritter, *Gesch. der Christlichen Philosophie*, th. iii., 1844,
s. 669–676.

formation, surpassing in these respects the travels of the Buddhist priests of Thibet and China, Marco Polo, and the Christian missionaries, who were sent on an embassy to the Mongolian princes. Important elements of Asiatic knowledge reached Europe through the intimate relations existing between the Arabs and the natives of India and China (for at the close of the seventh century, under the califate of the Ommajades, the Arabs had already extended their conquests to Kaschgar, Kabul, and the Punjaub).* The acute investigations of Reinaud have taught us the amount of knowledge regarding India that may be derived from Arabian sources. The incursion of the Moguls into China certainly disturbed the intercourse with the nations beyond the Oxus, but the Moguls soon served to extend the international relations of the Arabs, from the light thrown on geography by their observations and careful investigations, from the coasts of the Dead Sea to those of Western Africa, and from the Pyrenees to Scherif Edrisi's marsh lands of Wangarah, in the interior of Africa.† According to the testimony of Frähn, Ptolemy's geography was translated into Arabic by order of the Calif Mamun, between the years 813 and 833 ; and it is not improbable that several fragments of Marinus Tyrius, which have not come down to us, were employed in this translation.‡

Of the long series of remarkable geographers presented to us in the literature of the Arabs, it will be sufficient to name the first and last, El-Istâchri and Alhassan (Johannes Leo Africanus).§ Geography never acquired a greater acquisition

* Reinaud, in three late writings, which show how much may still be derived from Arabic and Persian, as well as Chinese sources ; *Fragments Arabes et Persans inédits relatifs à l'Inde antérieurement au XIe Siècle de l'ère Chrétienne*, 1845, p. xx.–xxxiii. ; *Relation des Voyages faits par les Arabes et les Persans dans l'Inde et à la Chine dans le IXe Siecle de notre ère*, 1845, t. i., p. xlvi. ; *Mémoire Géog. et Hist. sur l'Inde d'après les écrivains Arabes, Persans, et Chinois, anterieurement au milieu du onzième Siècle de l'ère Chretienne*, 1846, p. 6. The second of these memoirs of the learned Oriental scholar is based on the incomplete treatise of the Abbé Renaudot, *Anciennes Relations des Indes, et de la Chine, de deux Voyageurs Mahométans*, 1718. The Arabic manuscript contains only one notice of a voyage, that of the merchant Soleiman, who embarked on the Persian Gulf in the year 851. To this notice is added what Abu-Zeyd-Hassan, of Syraf, in Farsistan, who had never traveled to India or China, had learned from other well-informed merchants. † Reinaud et Favé, *Du Feu Grégeois*, 1845, p. 200.

‡ Ukert, *Ueber Marinus Tyrius und Ptolemaus die Geographen*, in the *Rheinische Museum für Philologie*, 1839, s. 329–332 ; Gildemeister, *De Rebus Indicis*, pars 1, 1838, p. 120 ; *Asie Centrale*, t. ii., p. 191.

§ The "*Oriental Geography* of Ebn-Haukal," which Sir William

of facts, even from the discoveries of the Portuguese and Spaniards. Within fifty years after the death of the Prophet, the Arabs had already reached the extremest western coasts of Africa and the port of Asfi. Whether the islands of the Guansches were visited by Arabian vessels subsequently, as I was long disposed to conjecture, to the expedition of the so-called Almagrurin adventurers to the *Mare tenebrosum*, is a question that has again been lately regarded as doubtful.* The presence of a great quantity of Arabian coins, found buried in the lands of the Baltic, and in the extreme northern parts of Scandinavia, is not to be ascribed to direct intercourse with Arabian vessels in those regions, but to the widely-diffused inland trade of the Arabs.†

Geography was no longer limited to a representation of the relations of space, and the determinations of latitude and longitude, which had been multiplied by Abul-Hassan, or to a description of river districts and mountain chains ; but it rather led the people, already familiar with nature, to an acquaintance with the organic products of the soil, especially those of the vegetable world.‡ The repugnance entertained by all the

Ouseley published in London in 1800, is that of Abu-Ishak el-Istâchri, and, as Frähn has shown (Ibn Fozlan, p. ix., xxii., and 256–263), is half a century older than Ebn-Haukal. The maps which accompany the " *Book of Climates*" of the year 920, and of which there is a fine manuscript copy in the library of Gotha, have afforded me much aid in my observations on the Caspian Sea and the Sea of Aral (*Asie Centrale*, t. ii., p. 192–196). We have lately been put in possession of an edition of Istâchri, and a German translation (*Liber Climatum, ad similitudinem Codicis Gothani delineandum*, cur. J. H. Moeller, Goth., 1839 ; *Das Buch der Länder*, translated from the Arabic by A. D. Mordtmann, Hamb., 1845).

* Compare Joaquim Jose da Costa de Macedo, *Memoria em que se pretende provar que os Arabes não conhecerão as Canarias antes dos Portuguezes* (Lisboa, 1844, p. 86–99, 205–227, with Humboldt, *Examen Crit. de l'Hist. de la Géographie*, t. ii., p. 137–141.

† Leopold von Ledebur, *Ueber die in den Baltischen Ländern gefundenen Zeugnisse eines Handels-Verkehrs mit dem Orient zur Zeit der Arabischen Weltherrschaft*, 1840, s. 8 und 75.

‡ The determinations of longitude which Abul-Hassan Ali of Morocco, an astronomer of the thirteenth century, has embodied in his work on the astronomical instruments of the Arabs, are all calculated from the first meridian of Arin. M. Sédillot the younger first directed the attention of geographers to this meridian. I have also made it an object of careful inquiry, because Columbus, who was always guided by Cardinal d'Ailly's *Imago Mundi*, in his fantasies regarding the difference of form between the eastern and western hemispheres, makes mention of an Isla de Arin: "centro de el hemispherio del qual habla Toloméo y quès debaxo la linea equinoxial entre el Sino Arabico y aquel de Persia." (Compare J. J. Sédillot, *Traité des Instrumens Astronomiques*

adherents of Islamism toward anatomical investigations impeded their advance in zoology. They remained contented with that which they were able to appropriate to themselves from translations of the works of Aristotle and Galen ;* but,

des Arabes, publ. par L. Am. Sédillot, t. i., 1834, p. 312–318; t. ii., 1835, preface, with Humboldt's *Examen Crit. de l'Hist. de la Géogr.*, t. iii., p. 64, and *Asie Centrale*, t. iii., p. 593–596, in which the data occur which I derived from the *Mappa Mundi* of Alliacus of 1410, in the "*Alphonsine Tables*," 1483, and in Madrignano's *Itinerarium Portugallensium*, 1508. It is singular that Edrisi appears to know nothing of Khobbet Arin (Cancadora, more properly Kankder). Sédillot the younger (in the *Mémoire sur les Systèmes Géographiques des Grecs et des Arabes*, 1842, p. 20–25) places the meridian of Arin in the group of the Azores, while the learned commentator of Abulfeda, Reinaud (*Mémoire sur l'Inde anterieurement au XIe siècle de l'ère Chrétienne d'après les écrivains Arabes et Persans*, p. 20–24), assumes that "the word Arin has originated by confusion from *Azyn*, *Ozein*, and *Odjein*, an old seat of cultivation (according to Burnouf, Udjijayani in Malwa), the 'Οζήνη of Ptolemy. This Ozene was supposed to be in the meridian of Lanka, and in later times Arin was conjectured to be an island on the coast of Zanguebar. perhaps the Εσσυνον of Ptolemy." Compare, also, Am. Sédillot, *Mém. sur les Instr. Astron. des Arabes*, 1841, p. 75.

* The Calif Al-Mamun caused many valuable Greek manuscripts to be purchased in Constantinople, Armenia, Syria, and Egypt, and to be translated direct from Greek into Arabic, in consequence of the earlier Arabic versions having long been founded on Syrian translations (Jourdain, *Recherches Crit. sur l'Age et sur l'Origine des Traductions Latines d'Aristote*, 1819, p. 85, 88, and 226). Much has thus been rescued by the exertions of Al-Mamun, which, without the Arabs, would have been wholly lost to us. A similar service has been rendered by Armenian translations, as Neumann of Munich was the first to show. Unhappily, a notice by the historian Guezi of Bagdad, which has been preserved by the celebrated geographer Leo Africanus, in a memoir entitled *De Viris inter Arabes illustribus*, leads to the conjecture that at Bagdad itself many Greek originals, which were believed to be useless, were burned; but this passage may not, perhaps, refer to important manuscripts already translated. It is capable of several interpretations, as has been shown by Bernhardy (*Grundriss der Griech. Litteratur*, th. i., s. 489), in opposition to Heeren's *Geschichte der Classischen Litteratur*, bd. i., s. 135. The Arabic translations of Aristotle have often been found serviceable in executing Latin versions of the original, as, for instance, the eight books of Physics, and the History of Animals; but the larger and better part of the Latin translations have been made direct from the Greek (Jourdain, *Rech. Crit. sur l'Age des Traductions d'Aristote*, p. 230–236). An allusion to the same two-fold source may be recognized in the memorable letter of the Emperor Frederic II. of Hohenstaufen, in which he recommends the translations of Aristotle which he presents, in 1232, to his universities, and especially to that of Bologna. This letter expresses noble sentiments, and shows that it was not only the love of natural history which taught Frederic II. to appreciate the philosophical value of the "Compilationes varias quæ ab Aristotele aliisque philosophis sub Græcis Arabicisque vocabulis antiquitus editæ sunt." He writes as follows : "We have from our earliest

nevertheless, the zoological history of Avicenna, in the possession of the Royal Library at Paris, differs from Aristotle's
work on the same subject.* As a botanist, we must name
Ibn-Baithar of Malaga, whose travels in Greece, Persia, India, and Egypt entitle him to be regarded with admiration
for the tendency he evinced to compare together, by independent observations, the productions of different zones in the East
and West.† The point from whence all these efforts emanated was the study of medicine, by which the Arabs long
ruled the Christian schools, and for the more perfect development of which Ibn-Sina (Avicenna), a native of Aschena, near
Bokhara, Ibn-Roschd (Averroes) of Cordova, the younger Sera-
pion of Syria, and Mesue of Maridin on the Euphrates, availed themselves of all the means yielded by the Arabian caravan and sea trade. I have purposely enumerated the widely-
removed birth-places of celebrated Arabian literati, since they
are calculated to remind us of the great area over which the
peculiar mental direction and the simultaneous activity of the
Arabian race extended the sphere of ideas.

The scientific knowledge of a more anciently-civilized race
—the Indians—was also drawn within this circle, when, un-

youth striven to attain to a more intimate acquaintance with science,
although the cares of government have withdrawn us from it ; we have
delighted in spending our time in the careful reading of excellent works,
in order that our soul might be enlightened and strengthened by exercise, without which the life of man is wanting both in rule and in freedom (ut animæ clarius vigeat instrumentum in acquisitione scientiæ,
sine qua mortalium vita non regitur liberaliter). Libros ipsos tamquam
præmium amici Cæsaris gratulantur accipite, et ipsos antiquis philosophorum operibus, qui vocis vestræ ministerio reviviscunt, aggregantes
in auditorio vestro." (Compare Jourdain, p. 169–178, and Friedrich
von Raumer's excellent work *Geschichte der Hohenstaufen*, bd. iii.,
1841, s. 413.) The Arabs have served as a uniting link between ancient and modern science. If it had not been for them and their love
of translation, a great portion of that which the Greeks had either
formed themselves, or derived from other nations, would have been
lost to succeeding ages. It is when considered from this point of view
that the subjects which have been touched upon, though apparently
merely linguistic, acquire general cosmical interest.

* Jourdain, in his *Traductions d'Aristote*, p. 135–138, and Schneider,
Adnot. ad Aristotelis de Animalibus Hist., lib. ix., cap. 15, speak of Michael Scot's translation of Aristotle's *Historia Animalium*, and of a similar work by Avicenna (Manuscript No. 6493, in the Paris Library).

† On Ibn-Baithar, see Sprengel, *Gesch. der Arzneykunde*, th. ii., 1823,
s. 468; and Royle, *On the Antiquity of Hindoo Medicine*, p. 28. We
have possessed, since 1840, a German translation of Ibn-Baithar, under
the title *Grosse Zusammenstellung über die Kräfte der bekannten ein-
fachen Heil- und Nahrungs-mittel.*, translated from the Arabic by J. v.
Sontheimer, 2 bändes.

der the Califate of Haroun Al-Raschid, several important
works, probably those known under the half-fabulous name of
Tscharaka and Susruta,* were translated from the Sanscrit
into Arabic. Avicenna, who possessed a powerful grasp of
mind, and who has often been compared to Albertus Magnus,
affords, in his work on Materia Medica, a striking proof of the
influence thus exercised by Indian literature. He is acquainted, as the learned Royle observes, with the true Sanscrit name
of the Deodwar of the snow-crowned Himalayan Alps, which
had certainly not been visited by any Arab in the eleventh
century, and he regards this tree as an alder, a species of ju-
niper, from which oil of turpentine was extracted.† The sons
of Averroes lived at the court of the great Hohenstaufen, Frederic II., who owed a portion of his knowledge of the natural
history of Indian animals and plants to his intercourse with
Arabian literati and Spanish Jews, versed in many languages.‡
The Calif Abdurrahman I. himself laid out a botanical garden at Cordova,§ and caused rare seeds to be collected by his
own travelers in Syria and other countries of Asia. He planted, near the palace of Rissafah, the first date-tree known in
Spain, and sang its praises in a poem expressive of plaintive
longing for his native Damascus.

The most powerful influence exercised by the Arabs on
general natural physics was that directed to the advances of

* Royle, p. 35–65. Susruta, the son of Visvamitra, is considered by
Wilson to have been a cotemporary of Rama. We have a Sanscrit edition of his work (*The Sus'ruta, or System of Medicine taught by Dhan-
wantari, and composed by his disciple Sus'ruta*, ed. by Sri Madhusûdana
Gupta, vol. i., ii., Calcutta, 1835, 1836), and a Latin translation, *Sus'ru-
tas. áyurvedas. Id est Medicinæ Systema a venerabili D'havantare demon-
stratum, a Susruta discipulo compositum*. Nunc pr. ex Sanskrïta in Lat-
inum sermonem vertit Franc. Hessler, Erlangæ, 1844, 1847, 2 vols.
† Avicenna speaks of the Deiudur (*Deodar*), of the genus *'abhel* (Ju-
niperus); and also of an Indian pine, which gives a peculiar milk, *syr
deiudar* (fluid turpentine).
‡ Spanish Jews from Cordova transmitted the opinions of Avicenna
to Montpellier, and principally contributed to the establishment of its
celebrated medical school, which was framed according to Arabian
models, and belongs to the twelfth century. (Cuvier, *Hist. des Sciences
Naturelles*, t. i., p. 387.)
§ Respecting the gardens of the palace of Rissafah, which was built
by Abdurrahman Ibn-Moawijeh, see *History of the Mohammedan Dynasties in Spain extracted from Ahmed Ibn-Mohammed Al-Makkari*, by
Pascual de Gayangos, vol. i., 1840, p. 209–211. " En su Huerta planto
el Rey Abdurrahman una palma que era entonces (756) unica, y de
ella procediéron todas las que huy en España. La vista del arbol acren-
taba mas que templaba su melancolia." (Antonio Conde, *Hist. de la
Dominacion de los Arabes en España*, t. i. p. 169.)

chemistry, a science for which this race created a new era. It must be admitted that alchemistic and new Platonic fancies were as much blended with chemistry as astrology with astronomy. The requirements of pharmacy, and the equally urgent demands of the technical arts, led to discoveries which were promoted, sometimes designedly, and sometimes by a happy accident depending upon alchemistical investigation into the study of metallurgy. The labors of Geber, or rather Djaber (Abu-Mussah-Dschafar-al-Kufi), and the much more recent ones of Razes (Abu-Bekr Arrasi), have been attended by the most important results. This period is characterized by the preparation of sulphuric and nitric acids,* aqua regia, preparations of mercury, and of the oxyds of other metals, and by the knowledge of the alcoholic process of fermentation.† The first scientific foundation, and the subsequent advances of chemistry, are so much the more important, as they imparted a knowledge of the heterogeneous character of matter, and the nature of forces not made manifest by motion, but which now led to the recognition of the importance of *composition*, no less than to that of the perfectibility of form assumed in accordance with the doctrines of Pythagoras and Plato. Differences of form and of composition are, however, the elements of all our knowledge of matter—the abstractions which we believe capable, by means of measurement and analysis, of enabling us to comprehend the whole universe.

It is difficult, at present, to decide what the Arabian chemists may have acquired through their acquaintance with Indian literature (the writings on the *Rasayana*) ;‡ from the

---

* The preparation of nitric acid and aqua regia by Djaber (more properly Abu-Mussah-Dschafar) dates back more than five hundred years before Albertus Magnus and Raymond Lully, and almost seven hundred years before the Erfurt monk, Basilius Valentinus. The discovery of these decomposing (dissolving) acids, which constitutes an epoch in the history of science, was, however, long ascribed to the three last-named experimentalists.

† For the rules given by Razes for the vinous fermentation of amylum and sugar, and for the distillation of alcohol, see Höfer, *Hist. de la Chimie*, t. i., p. 325. Although Alexander of Aphrodisias (*Joannis Philoponi Grammatici*, in libr. *de Generatione et Interitu Comm.*, Venet., 1527, p. 97), properly speaking, only gives a circumstantial description of distillation from sea water, he also draws attention to the fact that wine may likewise be distilled. This statement is the more remarkable, because Aristotle (*Meteorol.*, ii., 3, p. 358, Bekker) had advanced the erroneous opinion that in natural evaporation fresh water only rose from wine, as from the salt water of the sea.

‡ The chemistry of the Indians, embracing alchemistic arts, is called *rasáyana* (*rasa*, juice or fluid, also quicksilver; and *áyana*, course or

ancient technical arts of the Egyptians ; the new alchemistic precepts of the pseudo-Democritus and the Sophist Synesius ; or even from Chinese sources, through the agency of the Moguls. According to the recent and very careful investigations of a celebrated Oriental scholar, M. Reinaud, the invention of gunpowder,* and its application to the discharge of hollow projectiles, must not be ascribed to the Arabs. Hassan Al-Rammah, who wrote between 1285 and 1295, was not acquainted with this application ; while, even in the twelfth century, and, therefore, nearly two hundred years before Berthold Schwarz, a species of gunpowder was used to blast the rock in the Rammelsberg, in the Harz Mountains. The invention of an air thermometer is also ascribed to Avicenna from a notice by Sanctorius, but this notice is very obscure, and six centuries passed before Galileo, Cornelius Drebbel, and the *Academia del Cimento*, by the establishment of an exact measurer of heat, created an important means for penetrating into a world of unknown phenomena, and comprehending the cosmical connection of effects in the atmosphere, the superimposed strata of the ocean, and the interior of the earth, thus revealing phenomena whose regularity and periodicity excite our astonishment. Among the advances which science owes to the Arabs, it will be sufficient to mention Alhazen's work on Refraction, partly borrowed, perhaps, from Ptolemy's Optics, and the knowledge and first application of the pendulum as a means of measuring time, due to the great astronomer Ebn-Junis.†

<hr>

process), and forms, according to Wilson, the seventh division of the *àyur-Veda*, the " science of life, or of the prolongation of life." (Royle, *Hindoo Medicine*, p. 39–48.) The Indians have been acquainted from the earliest times (Royle, p. 131) with the application of mordants in calico or cotton printing, an Egyptian art, which is most clearly described in Pliny, lib. xxxv., cap. 11, No. 150. The word *" chemistry"* indicates literally " Egyptian art," the art of the black land ; for Plutarch (*De Iside et Osir.*, cap. 33) knew that the Egyptians called their country Χημία, from the black earth. The inscription on the Rosetta stone has *Chmi*. I find this word, as applied to the analytic art, first in the decrees of Diocletian against " the old writings of the Egyptians which treat of the ' χημία ' of gold and silver" (περι χημίας ἀργύρου καὶ χρυσοῦ). Compare my *Examen Crit. de l' Hist. de la Géographie et de l'Astronomie Nautique,* t. ii., p. 314.

* Reinaud et Favé, *Du Feu Grégeois, des Feux de Guerre et des Origines de la Poudre à Canon*, t. i., 1845, p. 89, 97, 201, and 211 ; Piobert, *Traité d'Artillerie*, 1836, p. 25 ; Beckmann, *Technologie*, s. 342.

† Laplace, *Précis de l'Hist. de l'Astronomie*, 1821, p. 60 ; and Am. Sédillot, *Mémoire sur les Instrumens Astr. des Arabes*, 1841, p. 44. Thomas Young (*Lectures on Natural Philosophy and the Mechanical*

Although the purity and rarely-disturbed transparency of
the sky of Arabia must have especially directed the attention

Arts, 1807, vol. i., p. 191) does not either doubt that Ebn-Junis, at the
end of the tenth century, applied the pendulum to the measurement of
time, but he ascribes the first combination of the pendulum with wheel-
work to Sanctorius, in 1612, therefore forty-four years before Huygens.
With reference to the very elaborately constructed clock included in
the presents which Haroun Al-Raschid, or, rather, the Calif Abdallah,
sent, two hundred years earlier, from Persia to Charlemagne at Aix-la-
Chapelle, Eginhard distinctly says that it was moved by water (Horologium ex aurichalco arte mechanica mirifice compositum, in quo duo-
decim horarum cursus ad clepsidram vertebatur); *Einhardi Annales*,
in Pertz's *Monumenta Germaniæ Historica, Scriptorum*, t. i., 1826, p.
195.   Compare H. Mutius, *De Germanorum Origine, Gestis*, &c.
Chronic., lib. viii., p. 57, in *Pistorii Germanicorum Scriptorum*, t. ii.,
Francof., 1584 ; Bouquet, *Recueil des Historiens des Gaules*, t. v., p.
333 and 354.   The hours were indicated by the sound of the fall of
small balls, and by the coming forth of small horsemen from as many
opening doors.   The manner in which the water acted in such clocks
may indeed have been very different among the Chaldeans, who
" weighed time " (determining it by the weight of fluids), and in the
clepsydras of the Greeks and the Indians ; for the hydraulic clock-work
of Ctesibius, under Ptolemy Euergetes II., which marked the (civil)
hours throughout the year at Alexandria, was never known, according
to Ideler, under the common denomination of κλεψύδρα.   (Ideler's
Handbuch der Chronologie, 1825, bd. i., s. 231.)   According to the description of Vitruvius (lib. ix., cap. 4), it was an actual astronomical
clock, a " horologium ex aqua," a very complicated " machina hydrau-
lica," working by toothed wheels (versatilis tympani denticuli æquales
alius alium impellentes).   It is therefore not improbable that the Arabs,
who were acquainted with the improved mechanical constructions in
use under the Roman empire, may have succeeded in constructing an
hydraulic clock with wheel-work (tympana quæ nonnulli rotas appel-
lant, Græci autam περίτοχα.   Vitruvius, x., 4).   Leibnitz (*Annales Im-
perii Occidentis Brunsvicenses*, ed. Pertz, t. i., 1843, p. 247) expresses
his admiration of the construction of the clock of Haroun Al-Raschid
(*Abd-Allatif*, trad. par Silvestre de Sacy, p. 578).   The piece of mechanism which the sultan sent from Egypt, in 1232, to the Emperor Frederic II., seems, however, to have been much more remarkable.   It was
a large tent, in which the sun and moon were moved by mechanism,
and made to rise and set, and show the hours of the day and night at
correct intervals of time.   In the *Annales Godefridi Monachi S. Panta-
leonis apud Coloniam Agrippinam*, it is said to have been a " tentorium,
in quo imagines solis et lunæ artificialiter motæ cursum suum certis et
debitis spaciis peragrant, et horas diei et noctis infallibiliter indicant."
(*Freheri Rerum Germanicarum Scriptores*, t. i., Argentor., 1717, p. 398.)
The monk Godefridus, or whoever else may have written the annals
of those years in the chronicle composed for the convent of St. Panta-
leon at Cologne, which was probably the work of many different authors
(see Böhmer, *Fontes Rerum Germanicarum*, bd. ii., 1845, s. xxxiv.-
xxxvii.), lived in the time of the great Emperor Frederic II. himself.
The emperor caused this curious work, the value of which was estimated at 20,000 marks, to be preserved at Venusium, with other treasures.   (Fried. von Raumer, *Gesch. der Hohenstaufen*, bd. iii., s. 430.)

of the people, in their early uncultivated condition, to the motions of the stars, as we learn from the fact that the stellar worship of Jupiter, practiced under the Lachmites by the race of the Asedites, included Mercury, which, from its proximity to the sun, is less frequently visible, it would nevertheless appear that the remarkable scientific activity manifested by the Arabs in all branches of practical astronomy is to be ascribed less to native than to Chaldean and Indian influences. Atmospheric conditions merely favored that which had been called forth by mental qualifications, and by the contact of highly-gifted races with more civilized neighboring nations. How many rainless portions of tropical America, as Cumana, Coro, and Payta, enjoy a still more transparent atmosphere than Egypt, Arabia, and Bokhara ! A tropical sky, and the eternal clearness of the heavens, radiant in stars and nebulous spots, undoubtedly every where exercise an influence on the mind, but they can only lead to thought, and to the solution of mathematical propositions, where other internal and external incitements, independent of climatic relations, affect the national character, and where the requirements of religious and agricultural pursuits make the exact division of time a necessity prompted by social conditions. Among calculating commercial nations (as the Phœnicians) ; among constructive nations, partial to architecture and the measurement of land (as the Chaldæans and Egyptians), empirical rules of arithmetic and geometry were early discovered ; but these are merely capable of preparing the way for the establishment of mathematical and astronomical science. It is only in the later phases of civilization that the established regularity of the changes in the heavens is known to be reflected, as it were, in terrestrial phenomena, and that, in accordance with the words of our great poet, we seek the " fixed pole." The conviction entertained in all climates of the regularity of the planetary movements has contributed more than any thing else to lead man to seek similar laws of order in the moving atmosphere, in the oscillations of the ocean, in the

That a movement like that of the vault of heaven should have been given to the whole tent, as has often been asserted, appears to me very improbable. In the *Chronica Monasterii Hirsaugiensis*, edited by Trithemius, we find scarcely any thing beyond a mere repetition of the passage in the *Annales Godefridi*, without any information regarding the mechanical construction. (*Joh. Trithemii Opera Historica*, Part ii., Francof., 1601, p. 180.) Reinaud says that the movement was imparted " par des ressorts cachés." (*Extraits des Historiens Arabes relatifs aux Guerres des Croisades*, 1829, p. 435.)

periodic course of the magnetic needle, and in the distribution of organisms over the earth's surface.

The Arabs were in possession of planetary tables* as early as the close of the eighth century. We have already observed that the *Susruta*, the ancient incorporation of all the medical knowledge of the Indians, was translated by learned men belonging to the court of the Calif Haroun Al-Raschid—a proof of the early introduction of Sanscrit literature. The Arabian mathematician Albiruni even went to India for the purpose of studying astronomy. His writings, which have only recently been made accessible to us, prove how intimately he had made himself acquainted with the country, traditions, and comprehensive knowledge of the Indians.†

However much the Arabian astronomers may have owed to the earlier civilized nations, and especially to the Indian and Alexandrian schools, they have, nevertheless, considerably extended the domain of astronomy by their own practical endowments of mind ; by the number and direction of their observations ; the improvement of their instruments for angular measurement ; and their zealous efforts to rectify the older tables by a comparison with the heavens. In the seventh book of the *Almagest* of Abul-Wefa, Sédillot found a notice of the important inequality in the moon's longitude, which disappears at the syzygies and quadratures, attains its maximum at the octants, and has long been regarded, under the name of *variation*, as the discovery of Tycho Brahe.‡ The observations of

---

* On the Indian tables which Alphazari and Alkoresmi translated into Arabic, see Chasles, *Recherches sur l'Astronomie Indienne*, in the *Comptes Rendus des Séances de l'Acad. des Sciences*, t. xxiii., 1846. p. 846–850. The substitution of the sine for the arc, which is usually ascribed to Albategnius, in the beginning of the tenth century, also belongs originally to the Indians; tables of sines are to be found in the *Surya-Siddhanta*.

† Reinaud, *Fragments Arabes relatifs à l'Inde*, p. xii.–xvii., 96–126, and especially 135–160. Albiruni's proper name was Abul-Ryhan. He was a native of Byrun, in the Valley of the Indus, and a friend of Avicenna, with whom he lived at the Arabian academy which had been formed in Charezm. His stay in India, and the composition of his history of that country (*Tarikhi-Hind*), of which Reinaud has made known the most remarkable fragments, belong to the years 1030–1032.

‡ Sédillot, *Matériaux pour servir à l'Histoire comparée des Sciences Mathématiques chez les Grecs et les Orientaux*, t. i., p. 50–89; also in the *Comptes Rendus de l'Acad. des Sciences*, t. ii., 1836, p. 202 ; t. xvii., 1843, p. 163–173 ; t. xx., 1845, p. 1308. In opposition to this opinion, Biot maintains that the fine discovery of Tycho Brahe by no means belongs to Abul-Wefa, and that the latter was acquainted, not with the " variation," but only with the second part of the " evection." (*Journal*

Ebn-Junis in Cairo have become extremely important with reference to the perturbations and secular changes of the orbits of the two largest planets, Jupiter and Saturn.* The measurement of a degree, which the Calif Al-Mamun caused to be made in the great plain of Sindschar, between Tadmor and Rakka, by observers whose names have been transmitted to us by Ebn-Junis, has proved less important in its results than by the evidence which it affords of the scientific culture of the Arabian race.

We must regard among the results yielded by the reflection of this culture, in the West, the astronomical congress held at Toledo, in Christian Spain, under Alfonso of Castile, in which the Rabbin Isaac Ebn Sid Hazan played an important part; and in the far East, the observatory founded by Ilschan Holagu, the grandson of the great conqueror Genghis Khan, on a hill near Meraghar, and supplied with many instruments. It was here that Nassir Eddin, of Tus, in Khorassan, made his observations. These individual facts deserve to be noticed in a history of the contemplation of the universe, since they tend vividly to remind us of how much the Arabs have effected in diffusing knowledge over vast tracts of territory, and in accumulating those numerical data which contributed, in a great degree, during the important period of Kepler and Tycho, to lay the foundation of theoretical astronomy, and of correct views of the movements of the heavenly bodies. The spark kindled in those parts of Asia which were peopled by Tartars spread, in the fifteenth century, westward to Samarcand, where Ulugh Beig, of the race of Timour, established, besides an observatory, a gymnasium after the manner of the Alexandrian Museum, and caused a catalogue of stars to be drawn up, which was based on wholly new and independent observations.†

Besides making laudatory mention of that which we owe to the natural science of the Arabs in both the terrestrial and celestial spheres, we must likewise allude to their contributions in separate paths of intellectual development to the gen-

---

des Savans, 1843, p. 513–532, 609–626, 719–737; 1845, p. 146–166; and Comptes Rendus, t. xx., 1845, p. 1319–1323.)

* Laplace, Expos. du Système du Monde, note 5, p. 407.

† On the observatory of Meragha, see Delambre, Histoire de l'Astronomie du Moyen Age, p. 198–203; and Am. Sédillot, Mém. sur les Instrumens Arabes, 1841, p. 201–205, where the gnomon is described with a circular opening. On the peculiarities of the star catalogue of Ulugh Beig, see J.-J. Sédillot, Traité des Instrumens Astronomiques des Arabes, 1834, p. 4.

eral mass of mathematical science. According to the most recent works which have appeared in England, France, and Germany[*] on the history of mathematics, we learn that " the algebra of the Arabs originated from an Indian and a Greek source, which long flowed independently of one another." The Compendium of Algebra which the Arabian mathematician, Mohammed Ben-Musa (the Chorowazneir), framed by command of the Calif Al-Mamun, was not based on Diophantus, but on Indian science, as has been shown by my lamented and too-early deceased friend, the learned Friedrich Rosen ;[†] and it would even appear that Indian astronomers had been called to the brilliant court of the Abbassides as early as the close of the eighth century, under Almansur. Diophantus was, according to Castri and Colebrooke, first translated into Arabic, by Abul-Wefa Buzjani, toward the close of the tenth century. The process of establishing a conclusion by a progressive advance from one proposition to another, which seems to have been unknown to the ancient Indian algebraists, was acquired by the Arabs from the Alexandrian school. This noble inheritance, enriched by their additions, passed in the twelfth century, through Johannes Hispalensis and Gerhard of Cremona, into the European literature of the Middle Ages.[‡] " In the algebraic works of the Indians, we find the general solution of indeterminate equations of the first degree, and a far more elaborate mode of treating those of the second, than has been transmitted to us in the writings of the Alexandrian philosophers ; there is, therefore, no doubt, that if the works of the Indians had reached us two hundred years earlier, and were not now first made known to Europeans, they might have acted very beneficially in favoring the development of modern analysis."

The same channels and the same relations which led the

---

* Colebrooke, *Algebra with Arithmetic and Mensuration*, from the Sanscrit of Brahmagupta and Bhascara, Lond., 1817. Chasles, *Aperçu Historique sur l'Origine et le Développement des Méthodes en Géométrie*, 1837, p. 416–502 ; Nesselmann, *Versuch einer kritischen Geschichte der Algebra*, th. i., s. 30–61, 273–276, 302–306.

† *Algebra of Mohammed Ben-Musa*, edited and translated by F. Rosen, 1831, p. viii., 72, and 196–199. The mathematical knowledge of India was extended to China about the year 720 ; but this was at a period when many Arabians were already settled in Canton and other Chinese cities. Reinaud, *Relation des Voyages faits par les Arabes dans l'Inde et à la Chine*, t. i., p. cix. ; t. ii., p. 36.

‡ Chasles, *Histoire de l'Algèbre*, in the *Comptes Rendus*, t. xiii., 1841, p. 497–524, 601–626. Compare, also, Libri, in the same volume, p. 559–563.

Arabs to a knowledge of Indian algebra, enabled them also to obtain, in the ninth century, Indian numerals from Persia and the shores of the Euphrates. Persians were established at that period as revenue collectors on the Indus, and the use of Indian numerals was gradually transmitted to the revenue officers of the Arabs in Northern Africa, opposite the shores of Sicily. Nevertheless, the important historical investigations of the distinguished mathematician Chasles* have rendered it more than probable, according to his correct interpretation of the so-called Pythagorean table in the Geometry of Boëthius, that the Christians in the West were familiar with Indian numerals even earlier than the Arabs, and that they were acquainted with the use of nine figures or characters, according to their position value, under the name of the system of the abacus.

The present is not a fitting place to enter more fully into the consideration of this subject, which I have already treated of in two papers (written in 1819 and 1829), and presented to the *Academie des Inscriptions* at Paris, and the Academy of Sciences at Berlin ;† but, in our attempts to solve a historical

* Chasles, *Aperçu Historique des Methodes en Géométrie*, 1837, p. 464–472 ; also in the *Comptes Rendus de l'Acad. des Sciences*, t. viii., 1839, p. 78; t. ix., 1839, p. 449 ; t. xvi., 1843, p. 156–173, and 218–246 ; t. xvii., 1843, p. 143–154.

† Humboldt, *Ueber die bei verschiedenen Völkern üblichen Systeme von Zahlezeichen und über den Ursprung des Stellenwerthes in den Indischen Zahlen, in Crell's Journal für die reine und angewandte Mathematik*, bd. iv. (1829), s. 205–231. Compare, also, my *Examen Crit. de l'Hist. de la Géographie*, t. iv., p. 275. The simple enumeration of the different methods which nations, to whom the Indian arithmetic by position was unknown, employed for expressing the multiplier of the fundamental groups, furnishes, in my opinion, an explanation of the gradual rise or origin of the Indian system. If we express the number 3568, either perpendicularly or horizontally, by means of " indicators," corresponding to the different divisions of the abacus (thus, $M^3C^5X^6I^8$), we shall easily perceive that the group-signs (MCXI) might be omitted. But our Indian numbers are, however, nothing more than these indicators —the multipliers of the different groups. We are also reminded of this designation by indicators by the ancient Asiatic Suanpan (the reckoning machine which the Moguls introduced into Russia), which has successive rows of strings, to represent thousands, hundreds, tens, and units. These strings would bear in the numerical example just cited, 3, 5, 6, and 8 balls. In the Suanpan there is no apparent group-sign ; the group-signs are the positions themselves; and these positions (strings) are occupied by units (3, 5, 6, and 8) as multipliers or indicators. In both ways, whether by the figurative (the written) or by the palpable arithmetic, we arrive at the value of position and at the simple use of nine numbers. If a string be without any ball, the place will be left blank in writing. If a group (a member of the progression) be want-

problem, concerning which much yet remains to be elucidated, the question arises, whether position-value—the ingenious

ing, the vacuum is graphically filled by the symbol of a vacuum (*sûnya, sifron, tzúphra*). In the "*Method of Eutocius*," I find in the group of the myriads the first trace of the exponential or indicational system of the Greeks, which was so influential in the East: M^α, M^β, M^γ, designate 10,000, 20,000, 30,000. That which is here alone applied to the myriads, passes among the Chinese and the Japanese, who derived their knowledge from the Chinese two hundred years before the Christian era, through all the multiples of the groups. In the Gobar, the Arabian "dust-writing" (discovered by my deceased friend and teacher Silvestre de Sacy, in a manuscript in the library of the old Abbey of St. Germain des Près), the group-signs are points—therefore zeros or ciphers; for in India, Thibet, and Persia, zeros and points are identical. In the Gobar, 3 · is written for 30; 4 · · for 400; and 6 ·· for 6000. The Indian numbers, and the knowledge of the value of position, must be more modern than the separation of the Indians and the Arians; for the Zend nation only used the far less convenient Pehlwi numbers. The conjecture of the successive improvements that have been made in the Indian notation appears to me to be supported by the Tamul system, which expresses units by nine characters, and all other values by group-signs for 10, 100, and 1000, with multipliers added to the left. The singular ἀριθμοί Ἰνδικοί, in a scholium of the monk Neophytos, discovered by Prof. Brandis in the library of Paris, and kindly communicated to me for publication, appear to corroborate the opinion of such a gradual process of improvement. The nine characters of Neophytos are, with the exception of the 4, quite similar to the present Persian; but the value of these nine units is raised to 10, 100, 1000 fold by writing one, two, or three ciphers or zero-signs above them; as $\overset{o}{2}$ for 20, $\overset{o}{2}$ 4 for 24, $\overset{o}{5}$ for 500, and $\overset{oo}{3}$ 6 for 306. If we suppose points to be used instead of zeros, we have the Arabic dust-writing, Gobar. As my brother, Wilhelm von Humboldt, has often remarked of the Sanscrit, that it is very inappropriately designated by the terms "Indian" and "ancient Indian" language, since there are in the Indian peninsula several very ancient languages not at all derived *from* the Sanscrit, so the expression Indian or ancient Indian arithmetical characters is also very vague, and this vagueness applies both to the form of the characters and to the spirit of the methods, which sometimes consist in mere juxtaposition, sometimes in the employment of coefficients and indicators, and sometimes in the actual value of position. Even the existence of the cipher or zero is, as the scholium of Neophytos shows, not a necessary condition of the simple position-value in Indian numerical characters. The Indians who speak the Tamul language have arithmetical symbols which differ from their alphabetical characters, and of which the 2 and the 8 have a faint resemblance to the 2 and the 5 of the Devanagari figures (Rob. Anderson, *Rudiments of Tamul Grammar*, 1821, p. 135); and yet an accurate comparison proves that the Tamul arithmetical characters are derived from the Tamul alphabetical writing. According to Carey, the Cingalese are still more different from the Devanagari characters. In the Cingalese and in the Tamul, there is no position-value or zero-sign, but symbols for the groups of tens, hundreds, and thousands. The Cingalese work, like the Romans, by juxtaposition, the Tamuls by coefficients. Ptolemy uses the present zero-

application of position—which occurs in the Tuscan abacus, and in the Suampan of Inner Asia, has been twice independently invented, in the East under the Ptolemies, and in the West? or whether the system of position-value may not have been transferred by the direction of universal traffic from the Indian western peninsula to Alexandria, and subsequently have been given out amid the renewed dreams of the Pythagoreans as an invention of the founder of their sect? The bare possibility of ancient and wholly unknown combinations anterior to the sixtieth Olympiad is scarcely worthy of notice. Wherefore should a feeling of similar requirements not have severally given rise, among highly-gifted nations of different origin, to combinations of the same ideas?

While the algebra of the Arabs, by means of that which they had acquired from the Greeks and Indians, combined with the portions due to their own invention, acted so beneficially on the brilliant epoch of the Italian mathematicians of the Middle Ages, notwithstanding a great deficiency in symbolical designations, we likewise owe to the same people the merit of having furthered the use of the Indian numerical system from Bagdad to Cordova by their writings and their extended commercial relations. Both these effects—the simultaneous diffusion of the knowledge of the science of numbers and of numerical symbols with value by position—have variously, but powerfully, favored the advance of the mathematical portion of natural science, and facilitated access to the more abstruse departments of astronomy, optics, physical geography, and the theories of heat and magnetism, which, without such aids, would have remained unopened.

The question has often been asked, in the history of nations, what would have been the course of events if Carthage had conquered Rome and subdued the West? "We may ask with equal justice," as Wilhelm von Humboldt[*] observes, "what would be the condition of our civilization at the present day if the Arabs had remained, as they long did, the sole possessors of scientific knowledge, and had spread themselves permanently over the West? A less favorable result would

sign to represent the descending negative scale for degrees and minutes both in his Almagest and in his Geography. The zero-sign was consequently in use in the West much earlier than the epoch of the invasion of the Arabs. (See my work above cited, and the memoir printed in Crell's *Mathematical Journal*, p. 215, 219, 223, and 227.)

[*] Wilhelm von Humboldt, *Ueber die Kawi-Sprache*, bd. i., s. cclxii. Compare, also, the excellent description of the Arabs in Herder's *Ideen zur Gesch. der Menscheit*, book xix., 4 and 5.

probably have supervened in both cases. It is to the same causes which procured for the Romans a dominion over the world—the Roman spirit and character—and not to external and merely adventitious chances, that we owe the influence exercised by the Romans on our civil institutions, our laws, languages, and culture. It was owing to this beneficial influence, and to the intimate alliance of races, that we were rendered susceptible to the influence of the Greek mind and language, while the Arabs directed their consideration principally only to those scientific results of Greek investigation which referred to the description of nature, and to physical, astronomical, and purely mathematical science." The Arabs, by carefully preserving the purity of their native tongue, and the delicacy of their figurative modes of expression, were enabled to impart the charm of poetic coloring to the expression of feeling and of the noble axioms of wisdom ; but, to judge from what they were under the Abbassides, had they built on the same foundation with which we find them familiar, it is scarcely probable that they could have produced those works of exalted poetic and creative art, which, fused together in one harmonious accord, are the glorious fruits of the mature season of our European culture.

PERIOD OF OCEANIC DISCOVERIES.—OPENING OF THE WESTERN HEM-
ISPHERE.—EXTENSION OF SCIENTIFIC KNOWLEDGE, AND THOSE
EVENTS WHICH LED TO OCEANIC DISCOVERIES.—COLUMBUS, SE-
BASTIAN CABOT, AND GAMA.—AMERICA AND THE PACIFIC.—CABRIL-
LO, SEBASTIAN VIZCAINO, MENDAÑA, AND QUIROS.—THE RICHEST
ABUNDANCE OF MATERIALS FOR THE FOUNDATION OF PHYSICAL GE-
OGRAPHY IS PRESENTED TO THE NATIONS OF WESTERN EUROPE.

THE fifteenth century belongs to those remarkable epochs in which all the efforts of the mind indicate one determined and general character, and one unchanging striving toward the same goal. The unity of this tendency, and the results by which it was crowned, combined with the activity of whole races, give to the age of Columbus, Sebastian Cabot, and Gama, a character both of grandeur and enduring splendor. In the midst of two different stages of human culture, the fifteenth century may be regarded as a period of transition, which belongs both to the Middle Ages and to the beginning of more recent times. It is the age of the greatest discoveries in space, embracing almost all degrees of latitude and all elevations of the earth's surface. While this period doubled the number of the works of creation known to the inhabitants of Europe, it likewise offered to the intellect new and powerful incitements toward the improvement of natural sciences, in the departments of physics and mathematics.*

The world of objects now, as in Alexander's campaigns, although with still more overwhelming power, manifested itself to the combining mind in individual forms of nature, and in the concurrent action of vital forces. The scattered images of sensuous perception were gradually fused together into one concrete whole, notwithstanding their abundance and diversity, and terrestrial nature was conceived in its general character, and made an object of direct observation, and not of vague presentiments, floating in varying forms before the imagination. The vault of heaven revealed to the eye, which was as yet unaided by telescopic powers, new regions, unknown constellations, and separate revolving nebulous masses. At no other period, as we have already remarked, were a greater abundance of facts, and a richer mass of materials for the establishment of comparative physical geography, presented to any one portion of the human race. At no other period have discoveries in the material world of space called forth more extraordinary changes in the manners and well-being of men, and in the long-enduring condition of slavery of a portion of the human race, and their late awakening to political freedom ; nor has any other age afforded so large an extension to the field of view by the multiplication of products and objects of barter, and by the establishment of colonies of a magnitude hitherto unknown.

On investigating the course of the history of the universe, we shall discover that the germ of those events which have imparted any strongly-marked progressive movement to the human mind may be traced deeply rooted in the track of preceding ages. It does not lie in the destinies of mankind that all should equally experience mental obscuration. A principle of preservation fosters the eternal vital process of advancing reason. The age of Columbus attained the object of its destination so rapidly because a track of fruitful germs had already been cast abroad by a number of highly-gifted men, who formed, as it were, a lengthened beam of light amid the darkness of the Middle Ages. One single century—the thirteenth—shows us Roger Bacon, Nicolaus Scotus, Albertus Magnus, and Vincentius of Beauvais. The mental activity,

* Compare Humboldt, *Examen Crit. de l'Hist. de la Géographie*, t. i., p. viii. and xix.

once awakened, was soon followed by an extension of geographical knowledge. When Diego Ribero returned, in the year 1525, from the geographical and astronomical congress which had been held at the Puente de Caya, near Yelves, for the purpose of settling the contentions that had arisen regarding the boundaries of the two empires of the Portuguese and the Spaniards, the outlines of the new continent had been already laid down from Terra del Fuego to the coasts of Labrador. On the western side of America, opposite to Asia, the advance was, of course, less rapid, although Rodriguez Cabrillo had penetrated further northward than Monterey as early as 1543 ; and notwithstanding that this great and daring mariner met his death in the Canal of Santa Barbara, in New California, the pilot, Bartholomeus Ferreto, conducted the expedition to the 43d degree of latitude, where Vancouver's Cape Oxford is situated. The emulous enterprise of the Spaniards, English, and Portuguese, directed to one and the same object, was then so great, that fifty years sufficed to determine the external configuration or the general direction of the coasts of the countries in the Western hemisphere.

Although the acquaintance of the nations of Europe with the western part of the earth is the main subject of our consideration in this section, and that around which the numerous relations of a more correct and a grander view of the universe are grouped, we must yet draw a strong line of separation between the undoubted first discovery of America, in its northern portions, by the Northmen, and its subsequent rediscovery in its tropical regions. While the Califate still flourished under the Abbassides at Bagdad, and Persia was under the dominion of the Samanides, whose age was so favorable to poetry, America was discovered in the year 1000 by Leif, the son of Eric the Red, by the northern route, and as far as 41° 30′ north latitude.[*] The first, although accidental, incitement toward this event emanated from Norway. Toward the close of the ninth century, Naddod was driven by

* Parts of America were seen, although no landing was made on them, fourteen years before Leif Ericksson, in the voyage which Bjarne Herjulfsson undertook from Greenland to the southward in 986. Leif first saw the land at the island of Nantucket, 1° south of Boston ; then in Nova Scotia ; and, lastly, in Newfoundland, which was subsequently called " Litla Helluland," but never " Vinland." The gulf, which divides Newfoundland from the mouth of the great river St. Lawrence, was called by the Northmen, who had settled in Iceland and Greenland, Markland's Gulf. See *Caroli Christiani Rafn Antiquitates Americanæ*, 1845, p. 4, 421, 423, and 463.

storms to Iceland while attempting to reach the Färoë Islands, which had already been visited by the Irish. The first settlement of the Northmen was made in 875 by Ingolf. Greenland, the eastern peninsula of a land which appears to be every where separated by the sea from America proper, was early seen,* although it was first peopled from Iceland a hundred years later (983). The colonization of Iceland, which Naddod first called Snow-land, *Snjoland*, was carried through Greenland in a southwestern direction to the New Continent.

The Färoë Islands and Iceland must be considered as intermediate stations and starting points for attempts made to reach Scandinavian America. In a similar manner, the settlement at Carthage served the Tyrians in their efforts to reach the Straits of Gadeira and the port of Tartessus ; and thus, too, Tartessus, in its turn, led this enterprising people from station to station on to Cerne, the Gauleon (Ship Island) of the Carthaginians.†

Notwithstanding the proximity of the opposite shores of Labrador (*Helluland it mikla*), one hundred and twenty-five years elapsed from the first settlement of the Northmen in Iceland to Leif's great discovery of America. So small were the means possessed by a noble, enterprising, but not wealthy race for furthering navigation in these remote and dreary regions of the earth. The littoral tracts of Vinland, so called by the German Tyrker from the wild grapes which were found there, delighted its discoverers by the fruitfulness of the soil and the mildness of its climate when compared with Iceland and Greenland. This tract, which was named by Leif the "Good Vinland" (*Vinland it goda*), comprised the coast line between Boston and New York, and consequently parts of the present states of Massachusetts, Rhode Island, and Connecticut, between the parallels of latitude of Civita Vecchia and Terracina, which, however, correspond there only to mean annual temperatures of 47°·8 and 52°·1.‡ This was the prin-

* Gunnbjörn was wrecked, in 876 or 877, on the rocks subsequently called by his name, which were lately rediscovered by Captain Graah. Gunnbjörn saw the east coast of Greenland, but did not land upon it. (Rafn, *Antiquit. Amer.*, p. 11, 93, and 304.)

† See *ante*, p. 132.

‡ These mean annual temperatures of the eastern coast of America, under the parallels of 42° 25′ and 41° 15′, correspond in Europe to the latitudes of Berlin and Paris, places which are situated 8° or 10° more to the north. Besides, the decrease of mean annual temperature from lower to higher latitudes is here so rapid, that, in the interval of latitude between Boston and Philadelphia, which is 2° 41′, an increase of one

cipal settlement of the Northmen.   The colonists had often
to contend with a very warlike race of Esquimaux, who then
extended further to the south under the name of the Skralinger.   The first Bishop of Greenland, Eric Upsi, an Icelander,
undertook, in 1121, a Christian mission to Vinland ; and the
name of the colonized country has even been discovered in old
national songs of the inhabitants of the Färoë Islands.*

The activity and bold spirit of enterprise manifested by the
Greenland and Icelandic adventurers are proved by the circumstance that, after they had established settlements south
of 41° 30' north latitude, they erected three boundary pillars
on the eastern shores of Baffin's Bay, at the latitude of 72°
55', on one of the Woman's Islands,† northwest of the present
most northern Danish colony of Upernavick.   The Runic inscriptions, which were discovered in the autumn of the year
1824, contain, according to Rask and Finn Magnusen, the
date 1135.   From this eastern coast of Baffin's Bay, more
than six hundred years before the bold expeditions of Parry
and Ross, the colonists very regularly visited Lancaster Sound
and a part of Barrow's Straits for the purpose of fishing.   The
locality of the fishing ground is very definitely described, and
Greenland priests, from the Bishopric of Gardar, conducted
the first voyage of discovery (1266).   This northwestern summer station was called the Kroksfjardar Heath.   Mention is
even made of the drift-wood (undoubtedly from Siberia) collected there, and of the abundance of whales, seals, walruses,
and sea bears.‡

degree of latitude corresponds to a decrease in the mean annual temperature of almost 3°.6, while, according to my researches, on the system of isothermal lines in Europe, the same decrease of temperature
scarcely amounts to half a degree for the same interval.   (*Asie Centrale*,
t. iii., p. 227.)

* See *Carmen Færöicum in quo Vinlandiæ mentio fit.*   (Rafn, *Antiquit. Amer.*, p. 320–332.)

† The Runic stone was placed on the highest point of the island of
Kingiktorsoak "on the Saturday before the day of victory," *i. e.*, before
the 21st of April, a great heathen festival of the ancient Scandinavians,
which, at their conversion to Christianity, was changed into a Christian
festival.   (Rafn, *Antiquit. Amer.*, p. 347–355.)   On the doubts which
Brynjulfsen, Mohnike, and Klaproth express respecting the Runic numbers, see my *Examen Crit.*, t. ii., p. 97–101 ; yet, from other indications,
Brynjulfsen and Graah are led to regard the important monument on
the Woman's Islands (as well as the Runic inscriptions found at Igalik-
ko and Egegeit, lat. 60° 51' and 60° 0', and the ruins of buildings near
Upernavik, lat. 72° 50') as belonging undoubtedly to the eleventh and
twelfth centuries.

‡ Rafn, *Antiquit. Amer.*, p. 20, 274, and 415–418 (Wilhelmi, *Ueber Isl-*

Certain accounts of the intercourse maintained between the extreme north of Europe, or between Greenland and Iceland with the American Continent, properly so called, do not extend beyond the fourteenth century. In the year 1347, a ship was sent from Greenland to Markland (Nova Scotia) to collect building timber and other necessary articles. On the return voyage the ship encountered heavy storms, and was obliged to take refuge at Straumfjord in the west of Iceland. These are the latest accounts preserved to us by ancient Scandinavian authorities of the visits of Northmen to America.*

We have hitherto kept strictly on historical ground. By means of the critical and highly praiseworthy efforts of Christian Rafn, and of the Royal Society of Northern Antiquities at Copenhagen, the sagas and narratives of the voyages of the Northmen to Helluland (Newfoundland), to Markland (the mouth of the St. Lawrence and Nova Scotia), and to Vinland (Massachusetts), have been separately printed, accompanied

and, Hvitramannaland, Greenland, und Vinland, s. 117–121). According to a very ancient saga, the most northern part of the east coast of Greenland was also visited in 1194, under the name of Svalbard, at a part which corresponds to Scoresby's Land, near the point 73° 16', where my friend Col., then Capt. Sabine, made his pendulum observations, and where there is a very dreary cape bearing my name. (*Rafn, Antiquit. Amer.*, p. 303, and *Aperçu de l'Ancienne Géographie des Régions Arctiques de l'Amerique*, 1847, p. 6.)

* Wilhelmi, op. cit., s. 226; Rafn, *Antiquit. Amer.*, p. 264 and 453. The settlements on the west coast of Greenland, which, until the middle of the fourteenth century, were in a very flourishing condition, fell gradually to decay, from the ruinous operation of commercial monopolies, from the attacks of Esquimaux (Skralinger), the "black death," which, according to Hecker, depopulated the north during the years 1347 to 1351, and from the invasion of a hostile fleet, regarding whose course nothing is known. At the present day no faith is any longer attached to the meteorological myth of a sudden alteration of climate, and of the formation of a barrier of ice, which was immediately followed by the entire separation from their mother country of the colonies established in Greenland. As these colonies were only on the more temperate district of the west coast of Greenland, it can not be possible that a bishop of Skalholt, in 1540, should have seen "shepherds feeding their flocks" on the east coast of Greenland, beyond the icy wall. The accumulation of masses of ice on the east coast opposite to Iceland depends on the configuration of the land, the neighborhood of a chain of mountains having glaciers and running parallel to the coast line, and on the direction of the oceanic current. This state of things can not be solely referred to the close of the fourteenth or the beginning of the fifteenth century. As Sir John Barrow has very justly shown, it has been subject to many accidental alterations, particularly in the years 1815–1817. (See Barrow, *Voyages of Discovery within the Arctic Regions*, 1846, p. 2–6.) Pope Nicholas V. appointed a bishop for Greenland as late as 1448.

by able commentaries.* The length of the voyage, the direction of its course, and the times of the rising and setting of the sun, are all minutely detailed.

Less certainty appertains to the traces which have been supposed to be found of a discovery of America before the year 1000 by the Irish. The Skralinger related to the Northmen settled in Vinland, that further southward, beyond the Chesapeake Bay, there dwelt " white men, who clothed themselves in long white garments, carried before them poles to which cloths were attached, and called with a loud voice." This account was interpreted by the Christian Northmen to indicate processions, in which banners were borne accompanied by singing. In the oldest sagas, the historical narrations of Thorfinn Karlsefne, and the Icelandic Landnama book, these southern coasts, lying between Virginia and Florida, are designated under the name of the *Land of the White Men*. They are expressly called Great Ireland (*Irland it mikla*), and it is maintained that they were peopled by the Irish. According to testimonies which extend to 1064, before Leif discovered Vinland, and probably about the year 982, Ari Marsson, of the powerful Icelandic race of Ulf the squint-eyed, was driven in a voyage from Iceland to the south by storms on the coasts of the Land of the White Men, and there baptized in the Christian faith ; and, not being allowed to depart, was recognized by men from the Orkney Islands and Iceland.†

An opinion has been advanced by some northern antiquarians that, as in the oldest Icelandic documents the first inhabitants of the island are called " West Men, who had come across the sea" (emigrants settled in Papyli on the southeast coast, and on the neighboring small island of Papar), Iceland was not at first peopled directly from Europe, but from Virginia and Carolina (Great Ireland, the American White Men's Land), by Irishmen who had earlier emigrated to America.

* The main sources of information are the historic narrations of Eric the Red, Thorfinn Karlsefne, and Snorre Thorbrandsson, probably written in Greenland itself as early as the twelfth century, and partly by descendants of settlers born in Vinland (Rafn, *Antiquit. Amer.*, p. vii., xiv., and xvi.). The care with which genealogical tables were kept was so great, that that of Thorfinn Karlsefne, whose son, Snorre Thorbrandsson, was born in America, has been brought down from 1007 to 1811.

† *Hvitramannaland*, the Land of the White Men. Compare the original sources of information, in Rafn, *Antiquit. Amer.*, p. 203–206, 211, 446–451; and Wilhelmi, *Ueber Island, Hvitramannaland*, &c., s. 75–81.

The important work, *De Mensura Orbis Terræ*, composed by the Irish monk Dicuil about the year 825, and, therefore, thirty-eight years before the Northmen acquired their knowledge of Iceland from Naddod, does not, however, confirm this opinion.

Christian anchorites in the north of Europe, and pious Buddhist monks in the interior of Asia, explored and opened to civilization regions that had previously been inaccessible. The eager striving to diffuse religious opinions has sometimes paved the way for warlike expeditions, and sometimes for the introduction of peaceful ideas and the establishment of relations of commerce. Religious zeal, which so strongly characterizes the doctrines promulgated in the systems of India, Palestine, and Arabia, and which is so widely opposed to the indifference of the ancient polytheistic Greeks and Romans, was the means of furthering the advance of geographical knowledge in the earlier portions of the Middle Ages. Letronne, the commentator on Dicuil, has shown much ingenuity in his attempts to prove that after the Irish missionaries had been driven from the Färoë Islands by the Northmen, they began, about the year 795, to visit Iceland. The Northmen, when they first reached Iceland, found Irish books, mass bells, and other objects, which had been left by the earlier settlers, called Papar. These *Papæ*, fathers, are the *Clerici* of Dicuil.* If, as his testimony would lead us to conclude, these objects had belonged to Irish monks, who had come from the Färoë Islands, the question naturally arises, why these monks (*Papar*) should be termed in the native sagas Westmen (Vestmenn), who had "come from the West across the sea? (*Kommir til vestan um haf*)." The deepest obscurity still shrouds every thing connected with the voyage of the Gaelic chief Madoc, son of Owen Guineth, to a great western land in the year 1170, and the connection of this event with the Great Ireland of the Icelandic Saga. In like manner, the race of Celto-Americans, whom credulous travelers have professed to discover in many parts of the United States, have also disappeared since the establishment of an earnest and scientific ethnology, based, not on accidental similarities of sounds, but on grammatical forms and organic structure.†

* Letronne, *Recherches Géogr. et Crit. sur le Livre "de Mensura Orbis Terræ*," composé en Irlande, par Dicuil, 1814, p. 129–146. Compare my *Examen Crit. de l'Hist. de la Géogr.*, t. ii., p. 87–91.
† The statements which have been advanced from the time of Raleigh, of natives of Virginia speaking pure Celtic; of the supposition of the

That this first discovery of America in or before the eleventh century should not have produced the important and

Gaelic salutation, *hao, hui, iach,* having been heard there; of Owen
Chapelain, in 1669, saving himself from the hands of the Tuscaroras,
who were about to scalp him, " because he addressed them in his native Gaelic," have all been appended to the ninth book of my travels
(*Relation Historique,* t. iii., 1825, p. 159).   These Tuscaroras of North
Carolina are now, however, distinctly recognized by linguistic investi
gations as an Iroquois tribe.   See Albert Gallatin on *Indian Tribes,* in
the *Archæologia Americana,* vol. ii. (1836), p. 23 and 57.   An extensive
catalogue of Tuscarora words is given by Catlin, one of the most admirable observers of manners who ever lived among the aborigines of
America.   He, however, is inclined to regard the rather fair, and often
blue-eyed nation of the Tuscaroras as a mixed people, descended from
the ancient Welsh, and from the original inhabitants of the American
continent.   See his *Letters and Notes on the Manners, Customs, and Conditions of the North American Indians,* 1841, vol. i., p. 207 ; vol. ii., p.
259 and 262-265.   Another catalogue of Tuscarora words is to be found
in my brother's manuscript notes respecting languages, in the Royal
Library at Berlin.   " As the structure of American idioms appears remarkably strange to nations speaking the modern languages of Western
Europe, and who readily suffer themselves to be led away by some
accidental analogies of sound, theologians have generally believed that
they could trace an affinity with *Hebrew,* Spanish colonists with the
Basque, and the English or French settlers with *Gaelic, Erse,* or the
Bas Breton.   I one day met on the coast of Peru a Spanish naval officer
and an English whaling captain, the former of whom declared that he
had heard Basque spoken at Tahiti, and the other Gaelic, or Erse, at
the Sandwich Islands.".—Humboldt, *Voyage aux Régions Equinoctiales,
Relat. Hist.,* t. iii., 1825, p. 160.
    Although no connection of language has yet been proved, I by no
means wish to deny that the Basques and the people of Celtic origin
inhabiting Ireland and Wales, who were early engaged in fisheries on
the most remote coasts, may have been the constant rivals of the Scandinavians in the northern parts of the Atlantic, and even that the Irish
preceded the Scandinavians in the Färoë Islands and in Iceland.   It is
much to be desired that, in our days, when a sound and severe spirit
of criticism, devoid of a character of contempt, prevails, the old investigations of Powel and Richard Hakluyt (*Voyoges and Navigations,* vol.
iii., p. 4) might be resumed in England and in Ireland.   Is the statement based on fact, that the wanderings of Madoc were celebrated in
the poems of the Welsh bard Meredith, fifteen years before Columbus's
discovery ?   I do not participate in the rejecting spirit which has, but
too often, thrown popular traditions into obscurity, but I am, on the
contrary, firmly persuaded that, by greater diligence and perseverance,
many of the historical problems which relate to the maritime expeditions of the early part of the Middle Ages; to the striking identity in
religious traditions, manner of dividing time, and works of art in America and Eastern Asia; to the migrations of the Mexican nations; to the
ancient centers of dawning civilization in Aztlan, Quivira, and Upper
Louisiana, as well as in the elevated plateaux of Cundinamarca and
Peru, will one day be cleared up by discoveries of facts with which
we have hitherto been entirely unacquainted.   See my *Examen Crit.
de l'Hist. de la Géogr. du Nouveau Continent,* t. ii., p. 142-149.

permanent results yielded to the physical contemplation of the universe by the rediscovery of the same continent by Columbus at the close of the fifteenth century, was the necessary consequence of the uncivilized condition of the people, and the nature of the countries to which the early discoveries were limited.   The Scandinavians were wholly unprepared, by previous scientific knowledge, for exploring the countries in which they settled, beyond what was absolutely necessary for the satisfaction of their immediate wants.   Greenland and Iceland, which must be regarded as the actual mother countries of the new colonies, were regions in which man had to contend with all the hardships of an inhospitable climate.   The wonderfully organized free state of Iceland, nevertheless, maintained its independence for three centuries and a half, until civil freedom was annihilated, and the country became subject to Hako VI., king of Norway.   The flower of Icelandic literature, its historical records, and the collection of the Sagas and Eddas, appertain to the twelfth and thirteenth centuries.

It is a remarkable phenomenon in the history of the cultivation of nations, that when the safety of the national treasures of the most ancient records of Northern Europe was endangered at home by domestic disturbances, they should have been transported to Iceland, and have been there carefully preserved, and thus rescued for posterity.   This rescue, the remote consequence of Ingolf's first colonization in Iceland, in the year 875, has proved, amid the vague and misty forms of Scandinavian myths and symbolical cosmogonies, an event of great importance in its influence on the poetic fancy of mankind.   It was natural knowledge alone that acquired no enlargement.   Icelandic travelers certainly occasionally visited the universities of Germany and Italy, but the discoveries of the Greenlanders in the south, and the inconsiderable intercourse maintained with Vinland, whose vegetation presented no remarkable physiognomical character, withdrew colonists and mariners so little from their European interests, that no knowledge of these newly-colonized countries seems to have been diffused among the cultivated nations of Southern Europe.   It would even appear that no tidings of these regions reached the great Genoese navigator in Iceland.   Iceland and Greenland had then been separated upward of two hundred years, since 1261, when the latter country had lost its republican form of government, and when, on its becoming a fief of the crown of Norway, all intercourse with foreigners and even with Iceland was interdicted to it.   Christopher Colum-

bus, in a work "On the five habitable zones of the earth,"
which has now become extremely rare, says that in the month
of February, 1477, he visited Iceland, "where the sea was
not at that time covered with ice, and which had been resorted to by many traders from Bristol."* If he had there heard
tidings of the earlier colonization of an extended and continuous tract of land, situated on the opposite coast, *Helluland
it mikla, Markland*, and the *good Vinland*, and if he connected this knowledge of a neighboring continent with those projects which had already engaged his attention since 1470 and
1473, his voyage to Thule (Iceland) would have been made
so much the more a subject of consideration during the celebrated lawsuit regarding the merit of an earlier discovery,
which did not end till 1517, since the suspicious fiscal officer
mentions a map of the world (*mappa mundo*) which had been
seen at Rome by Martin Alonzo Pinzon, and on which the
New Continent was supposed to be marked. If Columbus
had desired to seek a continent of which he had obtained information in Iceland, he would assuredly not have directed
his course southwest from the Canary Islands. Commercial
relations were maintained between Bergen and Greenland until 1484, and, therefore, until seven years after Columbus's
voyage to Iceland.

Wholly different from the first discovery of the New Continent in the eleventh century, its rediscovery by Christopher
Columbus and his explorations of the tropical regions of America have been attended by events of cosmical importance, and
by a marked influence on the extension of physical views.
Although the mariners who conducted this great expedition
at the end of the fifteenth century were not actuated by the

---

* While this circumstance of the absence of ice in February, 1477,
has been brought forward as a proof that Columbus's Island of Thule
could not be Iceland, Finn Magnusen found in ancient historical sources
that until March, 1477, there was no snow in the northern part of Iceland, and that in February of the same year the southern coast was
free from ice. *Examen Crit.*, t. i., p. 105; t. v., p. 213. It is very remarkable, that Columbus, in the same " *Tratado de las cinco zonas hab-
itables*," mentions a more southern island, Frislanda ; a name which
is not in the maps of Andrea Bianco (1436), or in that of Fra Mauro
(1457–1470), but which plays a great part in the travels, mostly regarded as fabulous, of the brothers Zeni (1388–1404). (Compare *Examen Crit.*, t. ii., p. 114–126.) Columbus can not have been acquainted
with the travels of the Fratelli Zeni, as they even remained unknown
to the Venetian family until the year 1558, in which Marcolini first
published them, fifty-two years after the death of the great admiral.
When came the admiral's acquaintance with the name Frislanda ?

design of attempting to discover a new quarter of the world, and although it would appear to be proved that Columbus and Amerigo Vespucci died in the firm conviction that they had merely touched on portions of Eastern Asia,* yet the expedition manifested the perfect character of being the fulfillment of a plan sketched in accordance with scientific combinations. The expedition was safely conducted westward, through the gate opened by the Tyrians and Colæus of Samos, across the immeasurable dark sea, *mare tenebrosum*, of the Arabian geographers. They strove to reach a goal, with the limits of which they believed themselves acquainted. They were not driven accidentally thither by storms, as Naddod and Gardar had been borne to Iceland, and Gunlijörn, the son of Ulf Kraka, to Greenland. Nor were the discoverers guided on their course by intermediate stations. The great cosmographer, Martin Behaim, of Nürnberg, who accompanied the Portuguese Diego Cam on his expedition to the western coasts of Africa, lived four years, from 1486 to 1490, in the Azores ;

---

* See the proofs, which I have collected from trustworthy documents, for Columbus, in the *Examen Crit.*, t. iv., p. 233, 250, and 261, and for Vespucci, t. v., p. 182–185. Columbus was so fully convinced that Cuba was part of the continent of Asia, and even the south part of Khatai (the province of Mango), that on the 12th of June, 1494, he caused all the crews of his squadron (about 80 sailors) to swear that they were convinced he might go from Cuba to Spain by land, " que esta tierra de Cuba fuese la tierra firme al comienzo de las Indias y fin á quien en estas partes quisiere venir de España por tierra ;" and that " if any who now swore it should at any future day maintain the contrary, they would have to expiate their perjury by receiving one hundred stripes, and having the tongue torn out." (See *Informacion del Escribano publico, Fernando Perez de Luna*, in Navarrete, *Viages y Descubrimientos de los Españoles*, t. ii., p. 143, 149.) When Columbus was approaching the island of Cuba on his first expedition, he believed himself to be opposite the Chinese commercial cities of Zaitun and Quinsay (*y es cierto, dice el Almirante questa es la tierra firme y que estoy, dice el, ante Zayto y Guinsay*). "He intends to present the letters of the Catholic monarchs to the great Mogul Khan (Gran Can) in Khatai, and to return immediately to Spain (but by sea) as soon as he shall have thus discharged the mission intrusted to him. He subsequently sends on shore a baptized Jew, Luis de Torres, because he understands Hebrew, Chaldee, and some Arabic," which are languages in use in Asiatic trading cities. (See Columbus's Journal of his Voyages, 1492, in Navarrete, *Viages y Descubrim.*, t. i., p. 37, 44, and 46.) Even in 1533, the astronomer Schöner maintained that the whole of the so-called New World was a part of Asia (superioris Indiæ), and that the city of Mexico (Temistitan), conquered by Cortes, was no other than the Chinese commercial city of Quinsay, so excessively extolled by Marco Polo. (See *Joannis Schoneri Carlostadii Opusculum Geographicum*, Norimb., 1533, pars ii., cap. 1–20.)

but it was not from these islands, which lie between the coasts
of Spain and Maryland, and only at ⅔ths the distance from
the latter, that America was discovered. The preconception
of this event is celebrated with rich poetical fancy in those
stanzas of Tasso, in which he sings of the deeds which Hercules ventured not to attempt.

> Non osò di tentar l'alto oceano:
> Segnò le mete, en troppo breve chiostri,
> L'ardir ristrinse dell'ingegno umano,
> Tempo verrà che fian d'Ecole i segni
> Favola vile ai naviganti industri
> *Un uom della Liguria* avrà ardimento
> All' incognito corso esporsi in prima.

Tasso, xv. st., 25, 30, et 31.

And yet it was of this "*uom della Liguria*" that the great
Portuguese historical writer, Johannes Barros,* whose first de-
cade appeared in 1552, simply remarked that he was a vain
and fanciful babbler (*homem fallador e glorioso em mostrar
suas habilidades, e mais fantastico, e de imaginaçōes com sua
Ilha Cypango*). Thus, through all ages and through all
stages of civilization, national hatred has striven to obscure
the glory of honorable names.

The discovery of the tropical regions of America by Christopher Columbus, Alonso de Hojeda, and Alvarez Cabral, can
not be regarded in the history of the contemplation of the universe as one isolated event. Its influence on the extension of
physical science, and on the increase of materials yielded to
the ideal world generally, can not be correctly understood
without entering into a brief consideration of the period which
separates the epoch of the great maritime expeditions from
that of the maturity of scientific culture among the Arabs.
That which imparted to the age of Columbus its peculiar
character of uninterrupted and successful efforts toward the
attainment of new discoveries and extended geographical
knowledge, was prepared slowly and in various ways. The
means which contributed most strongly to favor these efforts
were a small number of enterprising men, who early excited
a simultaneous and general freedom of thought, and an independence of investigation into the separate phenomena of nature ; the influence exercised on the deepest sources of mental
vigor by the renewed acquaintance formed in Italy with the
works of ancient Greek literature ; the discovery of an art
which lent to thought at once wings of speed and powers of

* *Da Asia* de Joao de Barros e de Diego de Couto, dec. i., liv. iii.,
cap. 11 (Parte i., Lisboa, 1778, p. 250).

perpetuity; and the more extended knowledge of Eastern
Asia acquired by traveling merchants, and by monks who
had been sent on embassies to the Mogul rulers, and which
was diffused by them among those nations of the southwest
of Europe who maintained extensive commercial relations
with other countries, and who were therefore most anxious
to discover a nearer route to the Spice Islands. To these
means, which most powerfully facilitated the accomplishment
of the wishes so generally entertained at the close of the fifteenth century, we must add the advance in the art of navigation, the gradual perfection of nautical instruments, both
magnetic and astronomical, and, finally, the application of
certain methods for the determination of the ship's place, and
the more general use of the solar and lunar ephemerides of
Regiomontanus.

Without entering into the details of the history of science,
which would be foreign to the present work, I would enumerate, among those who prepared the way for the epoch of
Columbus and Gama, three great names—Albertus Magnus,
Roger Bacon, and Vincenzius of Beauvais. I have named
them according to time, but the most celebrated, influential,
and intellectual was Roger Bacon, a Franciscan monk of
Ilchester, who devoted himself to the study of science at Oxford and Paris. All three were in advance of their age, and
acted influentially upon it. In the long and generally unfruitful contests of the dialectic speculations and logical dog-
matism of a philosophy which has been designated by the indefinite and equivocal name of scholastic, we can not fail to
recognize the beneficial influence exercised by what may be
termed the reflex action of the Arabs. The peculiarity of
their national character, already described in a former section,
and their predilection for communion with nature, procured
for the newly-translated works of Aristotle an extended diffusion which was most instrumental in furthering the establishment of the experimental sciences. Until the close of the
twelfth and the beginning of the thirteenth century, miscon-
ceived dogmas of the Platonic philosophy prevailed in the
schools. Even the fathers of the Church believed that they
could trace in them the prototypes of their own religious
views.* Many of the symbolizing physical fancies of Timæ-

* Jourdain, *Recherch. Crit. sur les Traductions d' Aristote*, p. 230–234,
and 421–423; Letronne, *Des Opinions Cosmographiques des Pères de
l'Eglise, rapprochées des Doctrines philosophiques de la Grèce*, in the
Revue des deux Mondes, 1834, t. i., p. 632.

us were eagerly taken up, and erroneous cosmical views, whose groundlessness had long been shown by the mathematical school of Alexandria, were revived under the sanction of Christian authority. Thus the dominion of Platonism, or, more correctly speaking, the new adaptations of Platonic views, were propagated far into the Middle Ages, under varying forms, from Augustine to Alcuin, Johannes Scotus, and Bernhard of Chartres.*

When the Aristotelian philosophy gained the ascendency by its controlling influence over the direction of the human mind, its effect was manifested in the two-fold channel of investigation into speculative philosophy and a philosophical elaboration of empirical natural science. Although the former of these directions may appear foreign to the object I have had in view in the present work, it must not be passed without notice, since, in the midst of the age of dialectic scholastics, it incited some few noble and highly-gifted men to the exercise of free and independent thought in the most various departments of science. An extended physical contemplation of the universe not only requires a rich abundance of observation as the substratum for a generalization of ideas, but also a preparatory and invigorating training of the human mind, by which it may be enabled, unappalled amid the eternal contest between knowledge and faith, to meet the threatening impediments which, even in modern times, present themselves at the entrance of certain departments of the experimental sciences, and would seem to render them inaccessible. There are two points in the history of the development of man which must not be separated—the consciousness of man's just claims to intellectual freedom, and his long unsatisfied desire of prosecuting discoveries in remote regions of the earth. These free and independent thinkers form a series, which begins in the Middle Ages with Duns Scotus, Wilhelm of Occam, and Nicolas of Cusa, and leads from Ramus, Campanella, and Giordano Bruno to Descartes.†

The seemingly impassable gulf between thought and act-

* Friedrich von Raumer, *Ueber die Philosophie des dreizehnten Jahrhunderts*, in his *Hist. Taschenbuch*, 1840, s. 468. On the tendency toward Platonism in the Middle Ages, and on the contests of the schools, see Heinrich Ritter, *Gesch. der Christl. Philosophie*, th. ii., s. 159; th. iii., s. 131–160, and 381–417.

† Cousin, *Cours de l'Hist. de la Philosophie*, t. i., 1829, p. 360 and 389–436; *Fragmens de Philosophie Cartésienne*, p. 8–12 and 403. Compare also, the recent ingenious work of Christian Bartholomès, entitled *Jordano Bruno*, 1847, t. i., p. 308; t. ii., p. 409–416.

ual being—the relations between the mind that recognizes and the object that is recognized—separated the dialectics into the two celebrated schools of *Realists* and *Nominalists*. The almost forgotten contests of these schools of the Middle Ages deserve a notice here, because they exercised a special influence on the final establishment of the experimental sciences. The Nominalists, who ascribed to general ideas of objects only a subjective existence in the human mind, finally remained the dominant party in the fourteenth and fifteenth centuries, after having undergone various fluctuations of success. From their greater aversion to mere empty abstractions, they urged before all the necessity of experiment, and of the increase of the materials for establishing a sensuous basis of knowledge. This direction was at least influential in favoring the cultivation of empirical science; but even among those with whom the Realistic views were maintained, an acquaintance with the literature of the Arabs had successfully opposed a taste for natural investigation against the all-absorbing sway of theology. Thus we see that in the different periods of the Middle Ages, to which we have perhaps been accustomed to ascribe too strong a character of unity, the great work of discoveries in remote parts of the earth, and their happy adaptation to the extension of the cosmical sphere of ideas, were gradually being prepared on wholly different paths and in purely ideal and empirical directions.

Natural science was intimately associated with medicine and philosophy among the learned Arabs, and in the Christian Middle Ages with theological polemics. The latter, from their tendency to assert an exclusive influence, repressed empirical inquiry in the departments of physics, organic morphology, and astronomy, which was for the most part closely allied to astrology. The study of the comprehensive works of Aristotle, which had been introduced by Arabs and Jewish rabbis, had tended to lead to a philosophical fusion of all branches of study ;* and hence Ibn-Sina (Avicenna) and Ibn-Roschd (Averroes), Albertus Magnus, and Roger Bacon, passed for the representatives of all the knowledge of their time. The fame which in the Middle Ages surrounded the names of these great men, was proportionate to the general diffusion of this opinion of their endowments.

Albertus Magnus, of the family of the Counts of Bollstädt, must also be mentioned as an independent observer in the do-

* Jourdain, *Sur les Trad. d'Aristote*, p. 236 ; and Michael Sachs, *Die religiöse Poesie der Juden in Spanien*, 1845, s. 180–200.

main of analytic chemistry. It is true that his hopes were directed to the transmutation of the metals, but in his attempts to fulfill this object he not only improved the practical manipulation of ores, but he also enlarged the insight of men into the general mode of action of the chemical forces of nature. His works contain some extremely acute observations on the organic structure and physiology of plants. He was acquainted with the sleep of plants, the periodical opening and closing of flowers, the diminution of the sap during evaporation from the surfaces of leaves, and with the influence of the distribution of the vascular bundles on the indentations of the leaves. He wrote commentaries on all the physical works of the Stagirite, although in that on the history of animals he followed the Latin translation of Michael Scotus from the Arabic.* The work of Albertus Magnus, entitled *Liber Cosmographicus de Natura Locorum*, is a kind of physical geography. I have found in it observations, which greatly excited my surprise, regarding the simultaneous dependence of climate on latitude and elevation, and the effect of different angles of incidence of the sun's rays in heating the earth's surface. Albertus probably owes the praise conferred on him by Dante less to himself than to his beloved pupil St. Thomas Aquinas, who accompanied him from Cologne to Paris in 1245, and returned with him to Germany in 1248.

> Questi, che m'è a destra più vicino,
> Frate e maestro fummi; ed esso Alberto
> E' di Cologna, ed io Thomas d'Aquino.
> *Il Paradiso*, x., 97–99.

In all that has directly operated on the extension of the natural sciences, and on their establishment on a mathemat-

* The greater share of merit in regard to the history of animals belongs to the Emperor Frederic II. We are indebted to him for important independent observations on the internal structure of birds. (See Schneider, in *Reliqua Librorum Frederici II., imperatoris de arte venandi cum avibus*, t. i., 1788, in the Preface.) Cuvier also calls this prince of the Hohenstaufen line the "first independent and original zoologist of the scholastic Middle Ages." On the correct view of Albert Magnus, on the distribution of heat over the earth's surface under different latitudes and at different seasons, see his *Liber Cosmographicus de Natura Locorum*, Argent., 1515, fol. 14 b. and 23 a. (*Examen Crit.*, t. i., p. 54–58.) In his own observations, we, however, unhappily too often find that Albertus Magnus shared in the uncritical spirit of his age. He thinks he knows "that rye changes on a good soil into wheat; that from a beech wood which has been hewn down, a birch wood will spring up from the decayed matter; and that from oak branches stuck into the earth vines arise." (Compare, also, Ernst Meyer, *Ueber die Botanik des 13ten Jahrhunderts*, in the *Linnæa*, bd. x., 1836, s. 719.)

ical basis, and by the calling forth of phenomena by the process of experiment, Roger Bacon, the cotemporary of Albertus of Bollstädt, may be regarded as the most important and
influential man of the Middle Ages. These two men occupy
almost the whole of the thirteenth century ; but to Roger Bacon belongs the merit that the influence which he exercised
on the form of the mode of treating the study of nature. has
been more beneficial and lasting than the various discoveries
which, with more or less justice, have been ascribed to him.
Stimulating the mind to independence of thought, he severely condemned the blind faith attached to the authority of the
schools, yet, far from neglecting the investigations of the ancient Greeks, he directed his attention simultaneously to philological researches,* and the application of mathematics and
of the *Scientia experimentalis,* to which last he devoted a
special section of the *Opus Majus.*† Protected and favored
by one pope (Clement IV.), and accused of magic and impris-
oned by two others (Nicholas III. and IV.), he experienced
the changes of fortune common to great minds in all ages.
He was acquainted with the *Optics* of Ptolemy,‡ and with

* So many passages of the *Opus Majus* show the respect which Roger
Bacon entertained for Grecian antiquity, that, as Jourdain has already
remarked (p. 429), we can only interpret the wish expressed by him in
a letter to Pope Clement IV., " to burn the works of Aristotle, in order
to stop the diffusion of error among the scholars," as referring to the
bad Latin translations from the Arabic.

† " Scientia experimentalis a vulgo studentium penitus ignorata; duo
tamen sunt modi cognoscendi, scilicet per argumentum et experientiam
(the ideal path, and the path of experiment). Sine experientia nihil
sufficienter sciri potest. Argumentum concludit, sed non certificat,
neque removet duditationem ; et quiescat animus in intuita veritatis,
nisi eam inveniat via experientiæ." (*Opus Majus,* pars vi., cap. 1.) I
have collected all the passages relating to Roger Bacon's physical
knowledge, and to his proposals for various inventions, in the *Examen
Crit. de l'Hist. de la Geogr.,* t. ii., p. 295–299. Compare, also, Whewell, *Philosophy of the Inductive Sciences,* vol. ii., p. 323–337.

‡ See *ante,* p. 194. I find Ptolemy's *Optics* cited in the *Opus Majus* (ed. Jebb, Lond., 1733), p. 79, 288, and 404. It has been justly
denied (Wilde, *Geschichte der Optik,* th. i., s. 92–96) that the knowledge
derived from Alhazen, of the magnifying power of segments of spheres,
was actually the means of leading Bacon to construct spectacles. This
invention would appear to have been known as early as 1299, or to
belong to the Florentine Salvino degli Armati, who was buried in 1317
in the Church of Santa Maria Maggiore at Florence. If Roger Bacon,
who completed his *Opus Majus* in 1267, speaks of instruments by means
of which small letters appear large, " utiles senibus habentibus oculos
debiles," his words prove, as do also the practically erroneous considerations which he subjoins, that he can not himself have executed that
which obscurely floated before his mind as possible.

the Almagest.   As he, like the Arabs, always calls Hipparchus Abraxis, we may conclude that he also made use of only
a Latin translation from the Arabic.   Next to Bacon's chemical experiments on combustible explosive mixtures, his theoretical optical works on perspective, and the position of the
focus in concave mirrors, are the most important.   His profound *Opus Majus* contains proposals and schemes of practicable execution, but no clear traces of successful optical discoveries.   Profoundness of mathematical knowledge can not be
ascribed to him.   That which characterizes him is rather a
certain liveliness of fancy, which, owing to the impression excited by so many unexplained great natural phenomena, and
the long and anxious search for the solution of mysterious
problems, was often excited to a degree of morbid excess in
those monks of the Middle Ages who devoted themselves to
the study of natural philosophy.

Before the invention of printing, the expense of copyists
rendered it difficult, in the Middle Ages, to collect any large
number of separate manuscripts, and thus tended to produce
a great predilection for encyclopedic works after the extension of ideas in the thirteenth century.   These merit special
consideration, because they led to a generalization of ideas.
There appeared the twenty books *De Rerum Natura* of Thomas Cantipratensis, Professor at Louvain (1230) ; The Mirror of Nature (*Speculum Naturale*), written by Vincenzius of
Beauvais (Bellovacensis) for St. Louis and his consort Mar-
garet of Provence (1250) ; The Book of Nature, by Conrad
von Meygenberg, a priest at Ratisbon (1349) ; and the Picture of the World (*Imago Mundi*) of Cardinal Petrus de Al-
liaco, bishop of Cambray (1410), each work being in a great
measure based upon the preceding ones.   These encyclopedic
compilations were the forerunners of the great work of Father
Reisch, the *Margarita Philosophica*, the first edition of which
appeared in 1486, and which for half a century operated in a
remarkable manner on the diffusion of knowledge.   I must
here pause for a moment to consider the " Picture of the
World" of Cardinal Alliacus (Pierre d'Ailly).   I have elsewhere shown that the work entitled "Imago Mundi" exercised a greater influence on the discovery of America than
did the correspondence with the learned Florentine Toscanelli.*   All that Columbus knew of Greek and Roman writers,

* See my *Examen Crit.*, t. i., p. 61, 64–70, 96–108 ; t. ii., p. 349.
" There are five memoirs *De Concordantia Astronomia cum Theologia*,
by Pierre d'Ailly, whom Don Fernando Colon always calls Pedro de

all those passages of Aristotle, Strabo, and Seneca, on the proximity of Eastern Asia to the Pillars of Hercules, which, as his
son Fernando says, were the means of inciting him to discover
the Indian lands (*autoridad de los escritores para mover al
Almirante á descubrir las Indias*), were gathered by the admiral from the writings of the cardinal. He must have carried these works with him on his voyages ; for, in a letter
which he addressed to the Spanish monarchs from the island
of Haiti, in the month of October, 1498, he translated word
for word a passage from Alliacus's treatise, *De Quantitate
Terræ habitabilis*, which appears to have made a deep impression on his mind. Columbus probably did not know that
Alliacus had also transcribed *verbatim*, from an earlier work,
the *Opus Majus* of Roger Bacon.* Singular age, when the
combined testimony of Aristotle and Averroes (Avenryz), of
Esdras and of Seneca, regarding the small extent of the ocean
in comparison with continental masses, could serve to convince
monarchs of the expediency of a costly enterprise !

I have already drawn attention to the marked predilection
manifested at the close of the thirteenth century for the study
of natural forces, and the progressive and philosophical direction assumed by this study in its scientific establishment on
the basis of experiment. It still remains briefly to consider
the influence exercised by the revival of classical literature, at
the close of the fourteenth century, on the deepest sources of
the mental life of nations, and, therefore, on the general contemplation of the universe. The individuality of certain
highly-gifted men had contributed to increase the rich mass of
facts possessed by the world of ideas. The susceptibility of a
freer intellectual development already existed when Greek
literature, driven from its ancient seats, acquired a firm footing
in Western lands, under the favoring action of apparently accidental relations.

The Arabs, in their classical studies, had remained strangers
to all that appertains to the inspiration of language, their
studies being limited to a very small number of the writers
of antiquity, and, in accordance with their strong national predilection for natural investigation, principally to the physical
books of Aristotle, to the Almagest of Ptolemy, the botanical

Helico. These essays remind us of some very recent ones on the Mosaic Geology, published four hundred years after the cardinal's."
  * Compare Columbus's letter, Navarrete, *Viages y Descubrimientos*,
t. i., p. 244, with the *Imago Mundi* of Cardinal d'Ailly, cap. 8, and
Roger Bacon's *Opus Majus*, p. 183.

and chemical treatises of Dioscorides, and the cosmological fancies of Plato. The dialectics of Aristotle were blended by the Arabs with the study of Physics, as in earlier times, in the Christian mediæval age, they were with that of theology. Men borrowed from the ancients what they judged susceptible of special application, but they were far removed from apprehending the spirit of Hellenism in its general character, from penetrating to the depths of the organic structure of the language, from deriving enjoyment from the poetic creations of the Greek imagination, or of seeking to trace the marvelous luxuriance displayed in the fields of oratory and historical composition.

Almost two hundred years before Petrarch and Boccacio, John of Salisbury and the Platonic Abelard had already exercised a favorable influence with reference to an acquaintance with certain works of classical antiquity. Both possessed the power of appreciating the charm of writings in which freedom and order, nature and mind, were constantly associated together ; but the influence of the æsthetic feeling awakened by them vanished without leaving a trace, and the actual merit of having prepared in Italy a permanent resting-place for the muses exiled from Greece, and of having contributed most powerfully to re-establish classical literature, belongs of right to two poets, linked together by the closest ties of friendship, Petrarch and Boccacio. A monk of Calabria, Barlaam, who had long resided in Greece under the patronage of the Emperor Andronicus, was the instructor of both.* They were the first to begin to make a careful collection of Roman and Greek manuscripts ; and a taste for a comparison of languages had even been awakened in Petrarch,† whose philological acumen seemed to strive toward the attainment of a more general contemplation of the universe. Emanuel Chrysoloras, who was sent as Greek embassador to Italy and England (1391), Cardinal Bessarion of Trebisonde, Gemistus Pletho, and the Athenian Demetrius Chalcondylas, to whom we owe the first printed edition of Homer, were all valuable promoters of the study of the Greek writers.‡ All these came from Greece before the eventful taking of Constantinople (29th May, 1453) ; Constantine Lascaris alone, whose forefathers had once sat on the Byzantine throne, came later to Italy. He brought with

* Heeren, *Gesch. der Classischen Litteratur*, bd. i., s. 284–290.

† Klaproth, *Mémoires relatives à l'Asie*, t. iii., p. 113.

‡ The Florentine edition of Homer of 1488 ; but the first printed Greek book was the grammar of Constantine Lascaris, in 1476.

him a precious collection of Greek manuscripts, now buried in the rarely-used library of the Escurial.* The first Greek book was printed only fourteen years before the discovery of America, although the invention of printing was probably made simultaneously and wholly independently by Guttenberg in Strasburg and Mayence, and by Lorenz Yansson Koster at Haarlem, between 1436 and 1439, and, therefore, in the fortunate period of the first immigration of the learned Greeks into Italy.†

Two centuries before the sources of Greek literature were opened to the nations of the West, and twenty-five years before the birth of Dante—one of the greatest epochs in the history of the civilization of Southern Europe—events occurred in the interior of Asia, as well as in the east of Africa, which, by extending commercial intercourse, accelerated the period of the circumnavigation of Africa and the expedition of Columbus. The advance of the Moguls in twenty-six years from Pekin and the Chinese Wall to Cracow and Liegnitz, terrified Christendom. A number of able monks were sent forth as missionaries and embassadors: John de Plano Carpini and Nicholas Ascelin to Batu Khan, and Ruisbrock (Rubruquis) to Mangu Khan at Karakorum. The last-named of these traveling missionaries has left us many clear and important observations on the distribution of languages and races of men in the middle of the thirteenth century. He was the first who recognized that the Huns, the Baschkirs (inhabitants of Paskatir, the Baschgird of Ibn-Fozlan), and the Hungarians were of Finnish (Uralian) race ; and he even found Gothic tribes who still retained their language in the strong-holds of the Crimea.‡ Rubruquis excited the eager cupidity of the

* Villemain, *Mélanges Historiques et Littéraires*, t. ii., p. 135.

† The result of the investigations of the librarian Ludwig Wachler, at Breslau (see his *Geschichte der Litteratur*, 1833, th. i., s. 12–23). Printing without movable types does not go back, even in China, beyond the beginning of the tenth century of our era. The first four books of Confucius were printed, according to Klaproth, in the province of Szüt-schun, between 890 and 925 ; and the description of the technical manipulation of the Chinese printing-press might have been read in Western countries even as early as 1310, in Raschid-eddin's Persian history of the rulers of Khatai. According to the most recent results of the important researches of Stanislas Julien, however, an iron-smith in China itself, between the years 1041 and 1048 A.D., or almost 400 years before Guttenberg, would seem to have used movable types, made of burned clay. This is the invention of Pi-sching, but it was not brought into application.

‡ See the proofs in my *Examen Crit.*, t. ii., p. 316–320. Josafat Barbaro (1436), and Ghislin von Busbech (1555), still found, between

great maritime nations of Italy—the Venetians and Genoese—
by his descriptions of the inexhaustible treasures of Eastern
Asia. He is acquainted with the " silver walls and golden
towers" of Quinsay, the present Hangtscheufu, although he
does not mention the name of this great commercial mart,
which twenty-five years later acquired such celebrity from
Marco Polo, the greatest traveler of any age.* Truth and
naïve error are singularly intermixed in the Journal of Rubruquis, which has been preserved to us by Roger Bacon. Near
Khatai, which is bounded by the Eastern Sea, he describes a
happy land, "where, on their arrival from other countries, all
men and women cease to grow old."†

More credulous than the monk of Brabant, and therefore,
perhaps, far more generally read, was the English knight Sir

Tana (Asof), Caffa, and the Erdil (the Volga), Alani and Gothic tribes
speaking German. (Ramusio, *Delle Navigationi et Viaggi*, vol. ii., p.
92 b. and 98 a.) Roger Bacon merely terms Rubruquis frater Williel-
mus, quem dominus Rex Franciæ misit ad Tartaros.

* The great and admirable work of Marco Polo (*Il Milione di Messer
Marco Polo*), as we possess it in the correct edition of Count Baldelli,
is inappropriately termed the narrative of " *Travels*." It is, for the
most part, a descriptive, one might say, a statistical work, in which it is
difficult to distinguish what the traveler had seen himself, and what he
had learned from others, and what he derived from topographical descriptions, in which the Chinese literature is so rich, and which might
be accessible to him through his Persian interpreter. The striking
similarity presented by the narratives of the travels of Hiuan-thsung,
the Buddhistic pilgrim of the seventh century, to that which Marco
Polo found in 1277 (respecting the Pamir-Highland), early attracted my
whole attention. Jacquet, who was unhappily too early removed by
a premature death from the investigation of Asiatic languages, and who,
like Klaproth and myself, was long occupied with the work of the great
Venetian traveler, wrote to me as follows shortly before his decease :
" I am as much struck as yourself by the composition of the *Milione*.
It is undoubtedly founded on the direct and personal observation of the
traveler, but he probably also made use of documents either officially
or privately communicated to him. Many things appear to have been
borrowed from Chinese and Mongolian works, although it is difficult
to determine their precise influence on the composition of the *Milione*,
owing to the successive translations from which Polo took his extracts.
While our modern travelers are only too well pleased to occupy their
readers with their personal adventures, Marco Polo takes pains to blend
his own observations with the official data communicated to him, of
which, as governor of the city of Yangui, he was able to have a large
number." (See my *Asie Centrale*, t. ii., p. 395.) The compiling
method of the celebrated traveler likewise explains the possibility of
his being able to dictate his book at Genoa in 1295 to his fellow-prison-
er and friend, Messer Rustigielo of Pisa, as if the documents had been
lying before him. (Compare Marsden, *Travels of Marco Polo*, p.
xxxiii.)

† Purchas, *Pilgrims*, Part iii., ch. 28 and 56 (p. 23 and 34).

John Mandeville. He describes India and China, Ceylon and Sumatra. The comprehensive scope and the individuality of his narratives (like the itineraries of Balducci Pigoletti and the travels of Roy Gonzalez de Clavijo) have contributed considerably to increase a disposition toward a great and general intercourse among different nations.

It has often, and with singular pertinacity, been maintained, that the admirable work of the truthful Marco Polo, and more particularly the knowledge which it diffused regarding the Chinese ports and the Indian Archipelago, exercised great influence on Columbus, who is even asserted to have had a copy of Marco Polo's narratives in his possession during his first voyage of discovery.* I have already shown that Christopher Columbus and his son Fernando make mention of the Geography of Asia by Æneas Sylvius (Pope Pius II.), but never of Marco Polo or Mandeville. What they know of Quinsay, Zaitun, Mango, and Zipangu, may have been learned from the celebrated letter of Toscanelli in 1474 on the facility of reaching Eastern Asia from Spain, and from the relations of Nicolo de Conti, who was engaged during twenty-five years in traveling over India and the southern parts of China, and not through any direct acquaintance with the 68th and 77th chapters of the second book of Marco Polo. The first printed edition of these travels was no doubt the German translation of 1477, which must have been alike unintelligible to Columbus and to Toscanelli. The possibility of a manuscript copy of the narrative of the Venetian traveler being seen by Columbus between the years 1471 and 1492, when he was occupied by his project of "seeking the east by the west" (buscar el levante por el poniente, pasar á donde nacen las especerias, navegando al occidente), can not certainly be denied ;† but wherefore, in a letter written to Ferdinand and Isabella from Jamaica, on the 7th of June, 1503, in which he describes the coast of Veragua as a part of the Asiatic Ciguare near the Ganges, and expresses his hope of seeing horses with golden harness, should he not rath-

---

* Navarrete, *Coleccion de los Viages y Descubrimientos que Hiciéron por mar los Españoles,* t. i., p. 261; Washington Irving, *History of the Life and Voyages of Christopher Columbus,* 1828, vol. iv., p. 297.

† *Examen Crit. de l'Hist. de la Géog.,* t. i., p. 63 and 215; t. ii., p. 350. Marsden, *Travels of Marco Polo,* p. lvii., lxx., and lxxv. The first German Nuremberg version of 1477 (*Das buch des edeln Ritters un landtfarers Marcho Polo*) appeared in print in the life-time of Columbus, the first Latin translation in 1490, and the first Italian and Portuguese translations in 1496 and 1502.

er refer to the Zipangu of Marco Polo than to that of Pope Pius ?

While the diplomatic missions of Christian monks, and the mercantile expeditions by land, which were prosecuted at a period when the universal dominion of the Moguls had made the interior of Asia accessible from the Dead Sea to the Wolga, were the means of diffusing a knowledge of Khatai and Zipangu (China and Japan) among the great sea-faring nations of Europe ; the mission of Pedro de Covilham and Alonzo de Payva (in 1487), which was sent by King John II. to seek for the African Prester John, prepared the way, if not for Bartholomew Diaz, at all events for Vasco de Gama.* Trusting to the reports brought by Indian and Arabian pilots to Calicut, Goa, and Aden, as well as to Sofala, on the eastern shores of Africa, Covilham sent word to King John II., by two Jews from Cairo, that if the Portuguese would prosecute their voyages of discovery southward, along the west coast, they would reach the termination of Africa, from whence the navigation to the *Moon Island*, the Magastar of Polo, to Zanzibar and to Sofala, "rich in gold," would be extremely easy. But, before this news reached Lisbon, it had been already long known there that Bartholomew Diaz had not only made the discovery of the Cape of Good Hope (Cabo tormentoso), but that he had also sailed round it, although only for a short distance.†

* Barros, Dec. i., liv. iii., cap. 4, p. 190, says expressly that Bartholomew Diaz, "e os de sua companhia per causa dos perigos e tormentas, que em o dobrar delle passáram, lhe pazeram nome Tormentoso." The merit of first doubling the Cape does not, therefore, belong, as usually stated, to Vasco de Gama. Diaz was at the Cape in May, 1487, nearly, therefore, at the same time that Pedro de Covilham and Alonzo de Payva set forth from Barcelona on their expedition. In December of the same year (1487), Diaz brought the news of this important discovery to Portugal.

† The planispherium of Sanuto, who speaks of himself as "Marinus Sanuto, dictus Torxellus de Veneicis," appertain to the work entitled *Secreta fidelium Crucis.* "Marinus ingeniously preached a crusade in the interest of commerce, with a desire of destroying the prosperity of Egypt, and directing the course of trade in such a manner as to carry the products of India through Bagdad, Bassora, and Tauris (Tebriz), to Kaffa, Tana (Azow), and the Asiatic coasts of the Mediterranean. Sanuto, who was the cotemporary and compatriot of Polo, with whose *Milione* he was, however, unacquainted, was characterized by grand views regarding commercial policy. He may be regarded as the Raynal of the Middle Ages, without the incredulity of the philosophical abbé of the eighteenth century." (*Examen Critique*, t. i., p. 231, 333–348.) The Cape of Good Hope is set down as Capo di Diab on the map of Fra Mauro, compiled between the years 1457 and 1459. Consult the learned treatise of Cardinal Zurla, entitled *Il Mappamundo di Fra Mauro Camaldolese*, 1806, § 54.

Accounts of the Indian and Arabian trading places on the eastern shores of Africa, and of the configuration of the southern extremity of the continent, may, indeed, early in the Middle Ages, have been transmitted to Venice through Egypt, Abyssinia, and Arabia. The triangular form of Africa is indeed distinctly delineated as early as 1306, on the planispherium of Sanuto, in the Genoese *Portulano della Mediceo-Laurenziana* of 1351, discovered by Count Baldelli, and on the map of the world by Fra Mauro.- I have briefly alluded to these facts, since the history of the contemplation of the universe should indicate the epochs at which the principal details of the configuration of great continental masses were first recognized.

While the gradually developed knowledge of relations in space incited men to think of shorter sea routes, the means for perfecting practical navigation were likewise gradually increased by the application of mathematics and astronomy, the invention of new instruments of measurement, and by a more skillful employment of magnetic forces. It is extremely probable that Europe owes the knowledge of the northern and southern directing powers of the magnetic needle—the use of the mariner's compass—to the Arabs, and that these people were in turn indebted for it to the Chinese. In a Chinese work (the historical Szuki of Szumathsian, a writer who lived in the earlier half of the second century before our era) we meet with an allusion to the "magnetic cars," which the Emperor Tsing-wàng, of the ancient dynasty of the Tscheu, had given more than nine hundred years earlier to the embassadors from Tunkin and Cochin China, that they might not miss their way on their return home. In the third century of our era, under the dynasty of Han, there is a description given in Hiutschin's dictionary Schuewen of the manner in which the property of pointing with one end toward the south may be imparted to an iron rod by a series of methodical blows. Owing to the ordinary southern direction of navigation at that period, the south pointing of the magnet is always the one especially mentioned. A century later, under the dynasty of Tsin, Chinese ships employed the magnet to guide their course safely across the open sea ; and it was by means of these vessels that the knowledge of the compass was carried to India, and from thence to the eastern coasts of Africa. The Arabic designations *Zohron* and *Aphron* (south and north),* which Vincen-

* *Avron*, or *avr* (*aur*), is a more rarely employed term for north, used instead of the ordinary " *schemál;*" the Arabic *Zohron*, or *Zohr*, from

zius of Beauvais gives in his "Mirror of Nature" to the two
ends of the magnetic needle, indicate, like many Arabic names
of stars which we still employ, the channel, and the people
from whom Western countries received the elements of their
knowledge.   In Christian Europe the first mention of the use
of the magnetic needle occurs in the politico-satirical poem
called *La Bible*, by Guyot of Provence, in 1190, and in the
description of Palestine by Jacobus of Vitry, bishop of Ptole-
mais, between 1204 and 1215.   Dante (in his *Parad.*, xii.,
29) refers, in a simile, to the needle (*ago*), " which points to
the star."

The discovery of the mariner's compass was long ascribed
to Flavio Gioja of Positano, not far from the lovely town of
Amalfi, which was rendered so celebrated by its widely-extended maritime laws ; and he may, perhaps, have made some
improvement in its construction (1302).   Evidence of the earlier use of the compass in European seas than at the beginning
of the fourteenth century, is furnished by a nautical treatise of
Raymond Lully of Majorca, the singularly ingenious and eccentric man whose doctrines excited the enthusiasm of Giordano Bruno when a boy,[*] and who was at once a philosophical systematizer and an analytic chemist, a skillful mariner and
a successful propagator of Christianity.   In his book entitled
Fenix de las Maravillas del Orbe, and published in 1286,
Lully remarks, that the seamen of his time employed " instruments of measurement, sea charts, and the magnetic needle."[†]

which Klaproth erroneously endeavors to derive the Spanish *sur* and
the Portuguese *sul*, which, without doubt, like the German *süd*, are true
German words, does not properly refer to the particular designation of
the quarter indicated ; it signifies only the time of high noon; south is
dschenub.   On the early knowledge possessed by the Chinese of the
south pointing of the magnetic needle, see Klaproth's important investigations in his *Lettre à M. A. de Humboldt, sur l'Invention de la Boussole*, 1834, p. 41, 45, 50, 66, 79, and 90; and the treatise of Azuni of
Nice, which appeared in 1805, under the name of *Dissertation sur l'Origine de la Boussole*, p. 35, and 65–68.   Navarrete, in his *Discurso
Historico sobre los Progresos del Arte de Navegar en España*, 1802, p.
28, recalls a remarkable passage in the Spanish *Leyes de las Partidas*
(II., tit. ix., ley 28), of the middle of the thirteenth century : " The
needle, which guides the seaman in the dark night, and shows him,
both in good and in bad weather, how to direct his course, is the inter-
mediary agent (*medianera*) between the loadstone (*la piedra*) and the
north star . . . . ."   See the passage in *Làs siete Partidas del sabio
Rey Don Alonso el IX.* (according to the usually adopted chronological order Alonso the Xth), Madrid, 1829, t. i., p. 473.

    [*] *Jordano Bruno*, par Christian Bartholomès, s. 1847, t. ii., p. 181–
187.
    [†] " Tenian los mareantes instrumento, carta, compas y aguja."—Sal-

The early voyages of the Catalans to the north coast of Scotland and the western shores of tropical Africa (Don Jayme Ferrer reaching the mouth of the Rio de Ouro, in the month of August, 1367), and the discovery of the Azores (the Bracir Islands, on the Atlas of Picigano, 1367) by the Northmen, remind us that the open Western Ocean was navigated long before the time of Columbus. The voyages prosecuted under the Roman dominion in the Indian Ocean, between Ocelis and the coasts of Malabar, in reliance on the regularity of the direction of the winds,* were now conducted by the guidance of the magnetic needle.

The application of astronomy to navigation was prepared by the influence exercised in Italy, from the thirteenth to the fifteenth centuries, by Andalone del Nero and John Bianchini, the corrector of the Alphonsine tables, and in Germany by Nicolaus de Cusa,† George von Peuerbach, and Regiomontanus. Astrolabes designed for the determination of time and of geographical latitudes by meridian altitudes, and capable of being employed at sea, underwent gradual improvement from the time that the astrolabium of the Majorcan pilots was in use, which is described by Raymond Lully,‡ in 1295, in his *Arte de Navegar*, till the invention of the instrument made by Martin Behaim in 1484 at Lisbon, and which was, perhaps, only a simplification of the meteoroscope of his friend Regiomontanus. When the Infante Henry, duke of Viseo, who was himself a navigator, established an academy for pilots at Sagres, Maestro Jayme, of Majorca, was named its director. Martin Behaim received a charge from King John II. of Portugal to compute tables for the sun's declination, and to teach pilots to "navigate by the altitudes of the sun

azar, *Discurso sobre los Progresos de la Hydrografia en España*, 1809, p. 7.   * See *ante, p.* 172.

  † Regarding Cusa (Nicolaus of Cuss, properly of Cues, on the Moselle), see *ante,* p. 109, and also Clemens's treatise, *Ueber Giordano Bruno und Nicolaus de Cusa,* s. 97, where there is given an important fragment, written by Cusa's own hand, and discovered only three years since, respecting a three-fold movement of the earth. (Compare, also, Chasles, *Aperçu sur l'Origine des Méthodes en Géométrie,* 1807, p. 529.)

  ‡ Navarrete, *Dissertacion Historica sobre la parte que tuvieron los Españoles en las Guerras de Ultramar ó de las Cruzadas,* 1816, p. 100 ; and *Examen Crit.,* t. i., p. 274–277. An important improvement in observation, by the use of the plummet, has been ascribed to George von Peuerbach, the instructor of Regiomontanus. The plummet had, however, long been employed by the Arabs, as we learn from Abul-Hassan-Ali's description of astronomical instruments written in the thirteenth century. Sédillot, *Traité des Instrumens Astronomiques des Arabes,* 1835, p. 379 ; 1841, p. 205.

and stars." It can not at present be decided whether, at the close of the fifteenth century, the use of the log was known as a means of estimating the distance traversed while the direction is indicated by the compass ; but it is certain that Piga-fetta, the companion of Magellan, speaks of the log (*la catena a poppa*) as of a well-known means of measuring the course passed over.*

* In all the writings on the art of navigation which I have examined, I have found the erroneous opinion that the log for the measurement of the distance traversed was not used before the end of the sixteenth or the beginning of the seventeenth century. In the *Encyclopædia Britannica* (seventh edition, 1842), vol. xiii., p. 416, it is further stated, " The author of the device for measuring the ship's way is not known, and no mention of it occurs till the year 1607, in an East Indian voyage published by Purchas." This year is also named in all earlier and later dictionaries as the extreme limit (Gehler, bd. vi., 1831, s. 450). Nav-arrete alone, in the *Dissertacion sobre los Progresos del Arte de Navegar*, 1802, places the use of the log-line in English ships in the year 1577. (Duflot de Mofras, *Notice Biographique sur Mendoza et Navarrete*, 1845, p. 64.) Subsequently, in another place (*Coleccion de los Viages de los Españoles*, t. iv., 1837, p. 97), he asserts that, "in Magellan's time, the speed of the ship was only estimated by the eye (*à ojo*), until, in the sixteenth century, the corredera (the log) was devised." The meas-urement of the distance sailed over by means of throwing the log, al-though this means must, in itself, be termed imperfect, has become of such great importance toward a knowledge of the velocity and direc-tion of oceanic currents, that I have been led to make it an object of careful investigation. I here give the principal results which are con-tained in the sixth (still unpublished) volume of my *Examen Critique de l'Histoire de la Géographie et des Progrès de l'Astronomie Nautique*. The Romans, in the time of the republic, had in their ships *way-meas-urers*, which consisted of wheels four feet high, provided with paddles attached to the outside of the ship, exactly as in our steam-boats, and as in the apparatus for propelling vessels, which Blasco de Garay had pro-posed, in 1543, at Barcelona to the Emperor Charles V. (Arago, *An-nuaire du Bur. des Long.*, 1829, p. 152.) The ancient Roman way-measurer (ratio a majoribus tradita, qua in via rheda sedentes vel mari navigantes scire possumus quot millia numero itineris fecerimus) is de-scribed in detail by Vitruvius (lib. x., cap. 14), the credit of whose Au-gustan antiquity has indeed been recently much shaken by C. Schultz and Osann. By means of three-toothed wheels acting on each other, and by the falling of small round stones from a wheel-case (loculamen-tum) having only a single opening, the number of revolutions of the outside wheels which dipped in the sea, and the number of miles pass-ed over in the day's voyage, were given. Vitruvius does not say whether these hodometers, which might afford "both use and pleas-ure," were much used in the Mediterranean. In the biography of the Emperor Pertinax by Julius Capitolinus, mention is made of the sale of the effects left by the Emperor Commodus, among which was a trav-eling carriage provided with a similar hodometric apparatus (cap. 8 in *Hist. Augustæ Script.*, ed. Lugd. Bat., 1671, t. i., p. 554). The wheels indicated both "the measure of the distance passed over, and the dura-

The influence exercised by Arabian civilization through the astronomical schools of Cordova, Seville, and Granada, on the

tion of the journey" in hours. A much more perfect way-measurer, used both on the water and on land, has been described by Hero of Alexandria, the pupil of Ctesibius, in his still inedited Greek manuscript on the Dioptra. (See Venturi, *Comment supra la Storia dell' Ottica*, Bologna, 1814, t. i., p. 134–139.) There is nothing to be found on the subject we are considering in the literature of the Middle Ages until we come to the period of several " books of Nautical Instruction," written or printed in quick succession by Antonio Pigafetta (*Trattato di Navigazione*, probably before 1530); Francisco Falero (1535, a brother of the astronomer Ruy Falero, who was to have accompanied Magellan on his voyage round the world, and left behind him a " Regimiento para observar· la longitud en la mar"); Pedro de Medina of Seville (*Arte de Navegar*, 1545); Martin Cortes of Bujalaroz (*Breve Compendio de la esfera, y de la arte de Navegar*, 1551); and Andres Garcia de Cespedes (*Regimiento de Navigacion y Hidrografia*, 1606). From almost all these works, some of which have become extremely rare, as well as from the *Suma de Geografia*, which Martin Fernandez de Enciso had published in 1519, we learn, most distinctly, that the " distance sailed over" is learned, in Spanish and Portuguese ships, not by any distinct measurement, but only by estimation by the eye, according to certain established principles. Medina says (libro iii., cap. 11 and 12), " in order to know the course of the ship, as to the length of distance passed over, the pilot must set down in his register how much distance the vessel has made according to hours (*i, e.*, guided by the hour-glass, *ampolleta*); and for this he must know that the most a ship advances in an hour is four miles, and with feebler breezes, three, or only two." Cespedes (*Regimiento*, p. 99 and 156) calls this mode of proceeding " echar punto por fantasia." This fantasia, as Enciso justly remarks, depends, if great errors are to be avoided, on the pilot's knowledge of the qualities of his ship: on the whole, however, every one who has been long at sea will have remarked, with surprise, when the waves are not very high, how nearly the mere estimation of the ship's velocity accords with the subsequent result obtained by the log. Some Spanish pilots call the old, and, it must be admitted, hazardous method of mere estimation (cuenta de estima) sarcastically, and certainly very incorrectly, " la corredera de los Holandeses, corredera de los perezosos." In Columbus's ship's journal, reference is frequently made to the dispute with Alonso Pinzon as to the distance passed over since their departure from Palos. The hour or sand glasses, *ampolletas*, which they made use of, ran out in half an hour, so that the interval of a day and night was reckoned at 48 ampolletas. We find in this important journal of Columbus (as, for example, on the 22d of January, 1493): " andaba 8 millas por hora hasta pasadas 5 ampolletas, y 3 antes que comenzase la guardia, que eran 8 ampolletas." (Navarrete, t. i., p. 143.) No mention is ever made of the log (la corredera). Are we to assume that Columbus was acquainted with and employed it, and that he did not think it necessary to name it, owing to its being already in very general use, in the same way that Marco Polo has not mentioned tea, or the great wall of China? Such an assumption appears to me very improbable, because I find in the proposals made by the pilot, Don Jayme Ferrer, 1495, for the exact determination of the position of the papal line of demarkation, that when there is a question regarding the

navigation of the Spaniards and Portuguese, can not be overlooked. The great instruments of the schools of Bagdad and
Cairo were imitated, on a small scale, for nautical purposes.
Their names even were transferred ; thus, for instance, that
of "astrolabon," given by Martin Behaim to the main-mast,
belongs originally to Hipparchus. When Vasco de Gama
landed on the eastern coast of Africa, he found that the Indian
pilots at Melinde were acquainted with the use of astrolabes
and ballestilles.* Thus, by the more general intercourse consequent on increasing cosmical relations, by original inventions,
and by the mutual fructification afforded by the mathematical
and astronomical sciences, were all things gradually prepared
for the discovery of tropical America; the rapid determination
of its configuration ; the passage round the southern point of
Africa to India ; and, finally, the first circumnavigation of
the globe—great and glorious events, which, in the space of
thirty years (from 1492 to 1522), contributed so largely in extending the general knowledge of the regions of the earth.
The minds of men were rendered more acute and more capable of comprehending the vast abundance of new phenomena
presented to their consideration, of analyzing them, and, by
comparing one with another, of employing them for the foundation of higher and more general views regarding the universe.

It will be sufficient here to touch upon the more prominent
elements of these higher views, which were capable of lead-

distance sailed over, the appeal is made only to the accordant judgment
(juicio) of twenty very experienced seamen ("que apunten en su carta de 6 en 6 horas el camino que la nao fará segun su juicio"). If the
log had been in use, no doubt Ferrer would have indicated how often
it should be thrown. I find the first mention of the application of the
log in a passage of Pigafetta's Journal of Magellan's voyage of circum
navigation, which long lay buried among the manuscripts in the Am-
brosian Library at Milan. It is there said that, in the month of January, 1521, when Magellan had already arrived in the Pacific, "Secondo
la misura che facevamo del viaggio colla catena a poppa, noi percorre-
vamo da 60 in 70 leghe al giorno" (Amorelli, *Primo Viaggio intorno
al Globo Terracqueo, ossia Navigazione fatta dal Cavaliere Antonio
Pigafetta sulla squadra del Cap. Magaglianes*, 1800, p. 46). What
can this arrangement of a chain at the hinder part of a ship (catena a
poppa), "which we used throughout the entire voyage to measure the
way," have been, except an apparatus similar to our log ? No special
mention is made of the log-line divided into knots, the ship's log, and
the half-minute or log-glass, but this silence need not surprise us when
reference is made to a long-known matter. In the part of the *Trattato
di Navigazione* of the Cavalier Pigafetta, given by Amoretti in extracts,
amounting, indeed, only to ten pages, the "catena della poppa" is not
again mentioned.   * Barros, Dec. i., liv. iv., p. 320.

ing men to a clearer insight into the connection of phenomena. On entering into a serious consideration of the original works of the earliest writers of the history of the Conquista, we are surprised so frequently to discover the germ of important physical truths in the Spanish writers of the sixteenth century. At the sight of a continent in the vast waste of waters which appeared separated from all other regions in creation, there presented themselves to the excited curiosity, both of the earliest travelers themselves and of those who collected their narratives, many of the most important questions which occupy us in the present day. Among these were questions regarding the unity of the human race, and its varieties from one common original type ; the migrations of nations, and the affinity of languages, which frequently manifest greater differences in their radical words than in their inflections or grammatical forms ; the possibility of the migration of certain species of plants and animals ; the cause of the trade winds, and of the constant oceanic currents ; the regular decrease of temperature on the declivities of the Cordilleras, and in the superimposed strata of water in the depths of the ocean ; and the reciprocal action of the volcanoes occurring in chains, and their influence on the frequency of earthquakes, and on the extent of circles of commotion. The ground-work of what we at present term physical geography, independently of mathematical considerations, is contained in the Jesuit Joseph Acosta's work, entitled *Historia natural y moral de las Indias*, and in the work by Gonzalo Hernandez de Oviedo, which appeared hardly twenty years after the death of Columbus. At no other period since the origin of society had the sphere of ideas been so suddenly and so wonderfully enlarged in reference to the external world and geographical relations ; never had the desire of observing nature at different latitudes and at different elevations above the sea's level, and of multiplying the means by which its phenomena might be investigated, been more powerfully felt.

We might, perhaps, as I have already elsewhere remarked,* be led to adopt the erroneous idea that the value of these great discoveries, each one of which reciprocally led to others, and the importance of these two-fold conquests in the physical and the intellectual world, would not have been duly appreciated before our own age, in which the history of civilization has happily been subjected to a philosophical mode of treatment. Such an assumption is, however, refuted by the cotem-

* *Examen Crit.*, t. i., p. 3–6 and 290.

poraries of Columbus. The most talented among them fore-saw the influence which the events of the latter years of the fifteenth century would exercise on humanity. "Every day," writes Peter Martyr de Anghiera,* in his letters written in the years 1493 and 1494, "brings us new wonders from a new world—from those antipodes of the West—which *a certain Genoese* (*Christophorus quidam, vir Ligur*) has discovered. Although sent forth by our monarchs Ferdinand and Isabella, he could with difficulty obtain three ships, since what he said was regarded as fabulous. Our friend Pomponius Lætus (one of the most distinguished promoters of classical learning, and persecuted at Rome for his religious opinions) could scarcely refrain from tears of joy when I communicated to him the first tidings of so unhoped-for an event." Anghiera, from whom we take these words, was an intelligent statesman at the court of Ferdinand the Catholic and of Charles V., once em-bassador at Egypt, and the personal friend of Columbus, Amer-igo Vespucci, Sebastian Cabot, and Cortez. His long life embraced the discovery of Corvo, the westernmost island of the Azores, the expeditions of Diaz, Columbus, Gama, and Magellan. Pope Leo X. read to his sister and to the car-dinals, "until late in the night," Anghiera's *Oceanica*. "I would wish never more to quit Spain," writes Anghiera, "since I am here at the fountain head of tidings of the new-ly-discovered lands, and where I may hope, as the historian of such great events, to acquire for my name some renown with posterity."† Thus clearly did cotemporaries appreciate the

* Compare *Opus Epistolarum Petri Martyris Anglerii Mediolanensis*, 1670, ep. cxxx. and clii. "Præ lætitia prosiliisse te vixque à lachry-mis præ gaudio temperasse quando literas adspexisti meas, quibus de Antipodium Orbe, latenti hactenus, te certiorem feci, mi suavissime Pomponi, insinuasti. Ex tuis ipse literis colligo, quid senseris. Sen-sisti autem, tantique rem fecisti, quanti virum summa doctrina insigni-tum decuit. Quis namque cibus sublimibus præstari potest ingeniis isto suavior? quod condimentum gratius? à me facio conjecturam. Beari sentio spiritus meos, quando accitos alloquor prudentes aliquos ex his qui ab ea redeunt provincia (Hispaniola insula)." The expression, "Christophorus quidam Colonus," reminds us, I will not say of the too often and unjustly cited "nescio quis Plutarchus" of Aulus Gellius (*Noct. Atticæ*, xi., 16), but certainly of the "quodam Cornelio scri-bente," in the answer written by the King Theodoric to the Prince of the Æstyans, who was to be informed of the true origin of amber, as recorded in Tacitus, *Germ.*, cap. 45.

† *Opus Epistol.*, No. ccccxxxvii. and dlxii. The remarkable and in-telligent Hieronymus Cardanus, a magician, a fantastic enthusiast, and, at the same time, an acute mathematician, also draws attention, in his "physical problems," to how much of our knowledge of the earth was

glory of events which will survive in the memory of the latest ages.

Columbus, in sailing westward from the meridian of the Azores, through a wholly unexplored ocean, and applying the newly-improved astrolabe for the determination of the ship's place, sought Eastern Asia by a western course, not as a mere adventurer, but under the guidance of a systematic plan. He certainly had with him the sea chart which the Florentine physician and astronomer, Paolo Toscanelli, had sent him in 1477, and which, fifty-three years after his death, was still in the possession of Bartholomew de las Casas.* It would ap-

derived from facts, to the observation of which one man has led.— *Cardani Opera*, ed. Lugdun., 1663, t. ii., probl. p. 630 and 659, at nunc quibus te laudibus afferam Christophore Columbi, non familiæ tantum, non Genuensis urbis, non Italiæ Provinciæ, non Europæ, partis orbis solum, sed humani generis decus. I have been led to compare the "problems" of Cardanus with those of the latter Aristotelian school, because it appears to me remarkable, and characteristic of the sudden enlargement of geography at that epoch, that, amid the confusion and the feebleness of the physical explanations which prevail almost equally in both collections, the greater part of these problems relate to comparative meteorology. I allude to the considerations on the warm insular climate of England contrasted with the winter at Milan; on the dependence of hail on electric explosions; on the cause and direction of oceanic currents; on the maxima of atmospheric heat and cold occurring after the summer and winter solstices; on the elevation of the region of snow under the tropics; on the temperature dependent on the radiation of heat from the sun and from all the heavenly bodies; on the greater intensity of light in the southern hemisphere, &c. "Cold is merely absence of heat. Light and heat are only different in name, and are in themselves inseparable." *Cardani Opp.*, t. i., *De Vita Propria*, p. 40; t. ii., Probl. 621, 630–632, 653, and 713; t. iii., *De Subtilitate*, p. 417.

* See my *Examen Crit.*, t. ii., p. 210–249. According to the manuscript, *Historia General de las Indias*, lib. i., cap. 12, " la carta de marear que Maestro Paulo Fisico (Toscanelli) envio á Colon" was in the hands of Bartholomé de las Casas when he wrote his work. Columbus's ship's journal, of which we possess an extract (Navarrete, t. i., p. 13), does not entirely agree with the relation which I find in a manuscript of Las Casas, for a communication of which I am indebted to M. Ternaux Compans. The ship's journal says, " Iba hablando el Almirante (martes 25 de Setiembre, 1492), con Martin Alonso Pinzon, capitan de la otra carabela Pinta, sobra una carta que le habia enviado tres dias hacia á la carabela, donde segun parece *tenia pintadas el Almirante* ciertas islas por aquella mar . . . . ." In the manuscript of Las Casas (lib. i., cap. 12), we find, on the other hand, as follows: " La carta de marear que embió (Toscanelli al Almirante), yo que esta historia escrivo la tengo en mi poder. Creo que todo su viage sobre esta carta fundó" (lib. i., cap. 38); " asi fué que el martes 25 de Setiembre, llegase Martin Alonso Pinzon con su caravela Pinta á hablar con Christobal Colon, sobre una carta de marear que Christobal Colon le via embiado . . . .

pear from Las Casas's manuscript history, which I have examined, that this was the same "carta de marear" which the
admiral showed to Martin Alonso Pinzon on the 25th of September, 1492, and on which many prominent islands were delineated. Had Columbus, however, alone followed the chart
of his counselor and adviser, Toscanelli, he would have kept
a more northern course in the parallel of Lisbon; but instead
of this, he steered half the way in the latitude of Gomera,
one of the Canaries, in the hope of more speedily reaching
Zipangu (Japan); and subsequently keeping a less high latitude, he found himself, on the 7th of October, 1492, in the
parallel of 25° 30'.   Uneasy at not discovering the coast of
Zipangu, which, according to his reckoning, ought to lie 216
nautical miles further to the east, he yielded, after long contention, to the commander of the caravel Pinta, Martin Alonso Pinzon, of whom we have already spoken (one of three
wealthy and influential brothers, hostile to him), and steered
toward the southwest.   This change of direction led, on the
12th of October, to the discovery of Guanahani.

   We must here pause to consider the wonderful concatena-
tion of trivial circumstances which undeniably exercised an
influence on the course of the world's destiny.   The talented
and ingenious Washington Irving has justly observed, that if
Columbus had resisted the counsel of Martin Alonso Pinzon,
and continued to steer westward, he would have entered the
Gulf Stream, and been borne to Florida, and from thence
probably to Cape Hatteras and Virginia—a circumstance of
incalculable importance, since it might have been the means
of giving to the United States of North America a Catholic
Spanish population in the place of the Protestant English one
by which those regions were subsequently colonized.   "It
seems to me like an inspiration," said Pinzon to the admiral,
"that my heart dictates to me (*el corazon me da*) that we
ought to steer in a different direction."   It was on the strength
of this circumstance that in the celebrated lawsuit which Pinzon carried on against the heirs of Columbus between 1513
and 1515, he maintained that the discovery of America was
alone due to him.   This inspiration, emanating from the heart,

Esta carta es la que le embio Paulo Fisico el Florentin la qual yo tengo
en mi poder con otras cosas del Almirante y escrituras de su misma mano
que traxéron á mi poder.   En ella le pinto muchas islas. : . . . ." Are
we to assume that the admiral had drawn upon the map of Toscanelli
the islands which he expected to reach, or would "tenia pintadas"
merely mean that "the admiral had a map on which these were painted . . . ?"

Pinzon owed, as was related by an old sailor of Moguez, at the same trial, to the flight of a flock of parrots which he had observed in the evening flying toward the southwest, in order, as he might well have conjectured, to roost on trees on the land. Never has a flight of birds been attended by more important results. It may even be said that it has decided the first colonization in the New Continent, and the original distribution of the Roman and Germanic races of man.*

The course of great events, like the results of natural phenomena, is ruled by eternal laws, with few of which we have any perfect knowledge. The fleet which Emanuel, king of Portugal, sent to India, under the command of Pedro Alvarez Cabral, on the course discovered by Gama, was unexpectedly driven on the coast of Brazil on the 22d of April, 1500. From the zeal which the Portuguese had manifested, since the expedition of Diaz in 1487, to circumnavigate the Cape of Good Hope, a recurrence of fortuitous circumstances similar to those exercised by oceanic currents on Cabral's ships could hardly fail to manifest itself. The African discoveries would thus probably have brought about that of America south of the equator ; and thus Robertson was justified in saying that it was decreed in the destinies of mankind that the New Continent should be made known to European navigators before the close of the fifteenth century.

Among the characteristics of Christopher Columbus we must especially notice the penetration and acuteness with which, without intellectual culture, and without any knowledge of physical and natural science, he could seize and combine the phenomena of the external world. On his arrival in a new world and under a new heaven,† he examined with care the form of continental masses, the physiognomy of vegetation, the habits of animals, and the distribution of heat and the variations in terrestrial magnetism. While the old admiral strove to discover the spices of India, and the rhubarb (*rui-barba*), which had already acquired a great celebrity through

* Navarrete, *Documentos*, No. 69, in t. iii. of the *Viages y Discubr.*, p. 565–571 ; *Examen Crit.*, t. i., p. 234–249 and 252 ; t. iii., p. 158–165 and 224. On the contested spot of the first landing in the West Indies, see t. iii., p. 186–222. The map of the world of Juan de la Cosa, made six years before the death of Columbus, which was discovered by Valck-enaer and myself in the year 1832, during the cholera epidemic, and has since acquired so much celebrity, has thrown new light on these mooted questions.

† On the graphical and often poetical descriptions of nature found in Columbus, see *ante*, p. 66, 67.

the Arabian and Jewish physicians, and through the account of Rubruquis and the Italian travelers, he also examined with the greatest attention the roots, fruits, and leaves of the different plants. In drawing attention to the influence exercised by this great age of nautical discoverers on the extension of natural views, we impart more animation to our descriptions, by associating them with the individuality of one great man. In the journal of his voyage, and in his reports, which were first published from 1825 to 1829, we find almost all those circumstances touched upon to which scientific enterprise was directed in the latter half of the fifteenth and throughout the whole of the sixteenth centuries.

We need only revert generally and cursorily to the extension imparted to the geography of Western nations from the period when the Infante Dom Henrique the navigator, at his country seat of Terça Naval, on the lovely bay of Sagres, sketched his first plan of discovery, to the expeditions of Gaetano and Cabrillo to the South Sea.   The daring expeditions of the Portuguese, Spaniards, and English evince the suddenness with which a new sense, as it were, was opened for the appreciation of the grand and the boundless.   The advance of nautical science and the application of astronomical methods to the correction of the ship's reckoning favored the efforts which gave to this age its peculiar character, and revealed to men the image of the earth in all its completeness of form. The discovery of the main-land of tropical America (on the 1st of August, 1498) occurred seventeen months after Cabot reached the Labrador coast of North America.   Columbus did not see the terra firma of South America on the mountainous shores of Paria, as has generally been supposed, but at the Delta of the Orinoco, to the east of Caño Macareo.*   Sebastian Cabot† landed on the 24th of June, 1497, on the coast of Labrador, between 56° and 58° north latitude.   It has already been noticed that this inhospitable region had been visited by the Icelander Leif Ericksson, five hundred years earlier.

Columbus attached more importance on his third voyage to the circumstance of finding pearls in the islands of Margarita and Cabagua than to the discovery of the *tierra firme*, for he continued firmly persuaded to the day of his death that he had

* See the results of my investigations, in the *Relation Hist. du Voyage aux Régions Equinoxiales du Nouveau Continent*, t. ii., p. 702 ; and in the *Examen Crit. de l'Hist. de la Géographie*, t. i., p. 309.

† Biddle, *Memoir of Sebastian Cabot*, 1831, p. 52–61 ; *Examen Crit.*, t. iv., p. 231.

already touched a portion of the continent of Asia when on
his first voyage he reached Cuba, in November, 1492.* From
this point, as his son Don Fernando, and his friend the Cura
de los Palacios, relate, he proposed, if he had provisions enough,
" to continue his course westward, and to return to Spain
either by water, by way of Ceylon (Taprobane) *rodeando todo
la tierra de los Negros*, or by land, through Jerusalem and
Jaffa."† Such were the projects by which the admiral, in
1494, proposed to circumnavigate the globe, four years before
Vasco de Gama, and twenty-seven years before Magellan and
Sebastian de Elcano. The preparations for Cabot's second
voyage, in which he penetrated through blocks of ice to 67°
30′ north latitude, and endeavored to find a northwest passage
to Cathai (China), led him to think at "some future time of
an expedition to the north pole" (*á lo del polo arctico*).‡ The
more it became gradually recognized that the newly-discovered land constituted one connected tract, extending from Labrador to the promontory of Paria, and as the recently-found
map of Juan de la Cosa (1500) testified, beyond the equator,
far into the southern hemisphere, the more intense became the
desire of finding some passage either in the south or in the
north. Next to the rediscovery of the continent of America
and the knowledge of the extension of the new hemisphere
southward from Hudson's Bay to Cape Horn, discovered by
Garcia Jofre de Loaysa,§ the knowledge of the South Pacific,

* In a portion of Columbus's Journal, Nov. 1, 1492, to which but little
attention has been directed, it is stated, "I have (in Cuba) opposite,
and near to me, Zayto y Guinsay (*Zaitun* and *Quinsay*, Marco Polo, ii.,
77) of the Gran Can."—Navarrete, *Viages y Descubrim. de los Españoles*, t. i., p. 46. The curvature toward the south, which Columbus,
on his second voyage, remarked in the most western part of the coast
of Cuba, had an important influence, as I have elsewhere observed, on
the discovery of South America, and on that of the Delta of the Orinoco
and Cape Paria. See *Examen Crit.*, t. iv., p. 246–250. Anghiera
(*Epist.*, clxviii., ed. Amst., 1670, p. 96) writes as follows: "Putat
(Colonus) regiones has (Pariæ) esse Cubæ contiguas et adhærentes :
ita quod utræque sint Indiæ Gangetidis continens ipsum. . . . . ."
† See the important manuscript of Andres Bernaldez, Cura de la villa
de los Palacios (*Historia de los Reyes Catolicos*, cap. 123). This history
comprises the years from 1488 to 1513. Bernaldez had received Columbus into his house, in 1496, on his return from his second voyage.
Through the special kindness of M. Ternaux-Compans, to whom the
History of the Conquista owes much important elucidation, I was enabled at Paris, in Dec., 1838, to make a free use of this manuscript, which
was in the possession of my distinguished friend the historiographer
Don Juan Bautista Muñoz. (Compare Fern. Colon, *Vida del Almirante*,
cap. 56.)   ‡ *Examen Crit.*, t. iii., p. 244–248.
§ Cape Horn was discovered by Francisco de Hoces in February, 1526,

which bathes the western shores of America, was the most important cosmical event of the great epoch which we are here describing.

Ten years before Balboa, on the 25th of September, 1513, first caught sight of the Pacific from the heights of the Sierra de Quarequa at the Isthmus of Panama, Columbus distinctly learned, when he was coasting along the eastern shores of Veragua, that to the west of this land there was a sea " which in less than nine days' sail would bear ships to the *Chersonesus aurea* of Ptolemy and to the mouth of the Ganges." In the same *Carta rarissima*, which contains the beautiful and poetic narration of a dream, the admiral says, that " the opposite coasts of Veragua, near the Rio de Belen, are situated relatively to one another as Tortosa on the Mediterranean, and Fuenterrabia in Biscay, or as Venice and Pisa." The great ocean, the South Pacific, was even at that time regarded as merely a continuation of the *Sinus magnus* (μέγας κόλπος) of Ptolemy, situated before the golden Chersonesus, while Cattigara and the land of the Sines (Thinæ) were supposed to constitute its eastern boundary. The fanciful hypothesis of Hipparchus, according to which this eastern shore of the great gulf was connected with the portion of the African continent which extended far toward the east,* and thus supposed to make a closed inland sea of the Indian Ocean, was but little regarded in the Middle Ages, notwithstanding the partiality to the views of Ptolemy—a fortunate circumstance,

in the expedition of the Commendador Garcia de Loaysa, which, following that of Magellan, was destined to proceed to the Moluccas. While Loaysa was passing through the Straits of Magellan, Hoces, with his caravel, the *San Lesmes*, was separated from the flotilla, and driven as far as 55° south latitude. "Dijeron los del buque, que les parecia que era alli acabamiento de tierra." (Navarrete, *Viages de los Españoles*, t. v., p. 28 and 404–488.) Fleurieu maintains that Hoces only saw the Cabo del Buen Succeso, west of Staten Island. Toward the end of the sixteenth century, such a strange uncertainty again prevailed respecting the form of the land, that the author of the *Araucana* (canto i., oct. 9) believed that the Magellanic Straits had closed by an earthquake, and by the upheaval of the bottom of the sea, while, on the other hand, Acosta (*Historia Natural y Moral de las Indias*, lib. iii., cap. 10) regarded the Terra del Fuego as the beginning of a great south polar land. (Compare, also, *ante*, p. 72.)

* Whether the isthmus hypothesis, according to which Cape Prasum, on the eastern shore of Africa, was connected with the eastern Asiatic isthmus of Thinæ, is to be traced to Marinus of Tyre, or to Hipparchus, or to the Babylonian Seleucus, or rather to Aristotle, *De Cælo* (ii., 14), is a question treated in detail in another work, *Examen Crit.*, t. i., p. 144, 161, and 329; t. ii., p. 370–372.

when we consider the unfavorable influence which it would doubtlessly have exercised on the direction of great maritime enterprises.

The discovery and navigation of the Pacific indicate an epoch which was so much the more important with respect to the recognition of great cosmical relations, since it was owing to these events, and therefore scarcely three centuries and a half ago, that not only the configuration of the western coast of the New, and the eastern coast of the Old Continent were determined ; but also, what is far more important to meteorology, that the numerical relations of the area of land and water upon the surface of our planet first began to be freed from the highly erroneous views with which they had hitherto been regarded. The magnitude of these areas, and their relative distribution, exercise a powerful influence on the quantity of humidity contained in the atmosphere, the alternations in the pressure of the air, the force and vigor of vegetation, the greater or lesser distribution of certain species of animals, and on the action of many other general phenomena and physical processes. The larger area apportioned to the fluid over the solid parts of the earth's crust (in the ratio of $2\frac{4}{5}$ths to 1), does certainly diminish the habitable surface for the settlements of the human race, and for the nourishment of the greater portion of mammalia, birds, and reptiles ; but it is nevertheless, in accordance with the existing laws of organic life, a beneficent arrangement, and a necessary condition for the preservation of all living beings inhabiting continents.

When, at the close of the fifteenth century, a keen desire was awakened for discovering the shortest route to the Asiatic spice lands, and when the idea of reaching the east by sailing to the west simultaneously awoke in the minds of two intellectual men of Italy—the navigator Christopher Columbus, and the physician and astronomer Paul Toscanelli*—the opinion established in Ptolemy's Almagest still prevailed, that the Old Continent occupied a space extending over 180 equatorial degrees from the western shore of the Iberian peninsula to the meridian of Eastern Sinæ, or that it extended from east

* Paolo Toscanelli was so greatly distinguished as an astronomer, that Behaim's teacher, Regiomontanus, dedicated to him, in 1463, his work *De Quadratura Circuli*, directed against the Cardinal Nicolaus de Cusa. He constructed the great gnomon in the church of Santa Maria Novella at Florence, and died in 1482, at the age of 85, without having lived long enough to enjoy the pleasure of learning the discovery of the Cape of Good Hope by Diaz, and that of the tropical part of the New Continent by Columbus.

to west over half of the globe. Columbus, misled by a long series of false inferences, extended this space to 240 degrees, and in his eyes the desired eastern shores of Asia appeared to advance as far as the meridian of San Diego in New California. He therefore hoped that he should only have to sail 120 degrees instead of the 231 degrees at which the wealthy Chinese commercial city of Quinsay is actually situated to the west of the extremity of the Spanish peninsula. Toscanelli, in his correspondence with the admiral, diminished the expanse of the fluid element in a manner still more remarkable and more favorable to his designs. According to his calculations, the extent of the sea between Portugal and China was limited to 52 degrees, so that, in conformity with the expression of the Prophet Esdras, six sevenths of the earth were dry. Columbus, at a subsequent period, in a letter which he addressed to Queen Isabella from Haiti, immediately after the completion of his third voyage, showed himself the more inclined to these views, because they had been defended in the *Imago Mundi* by Cardinal d'Ailly, whom he regarded as the highest authority.*

* As the Old Continent, from the western extremity of the Iberian peninsula to the coast of China, comprehends almost 130° of longitude, there remain about 230° for the distance which Columbus would have had to traverse if he wished to reach Cathai (China), but less if he only desired to reach Zipangu (Japan). This difference of 230°, which I have here indicated, depends on the position of the Portuguese Cape St. Vincent (11° 20′ W. of Paris), and the far projecting part of the Chinese coast, near the then so celebrated port of Quinsay, so often named by Columbus and Toscanelli (lat. 30° 28′, long. 117° 47′ E. of Paris). The synonyms for Quinsay, in the province of Tschekiang, are Kanfu, Hangtscheufu, Kingszu. The East Asiatic general commerce was shared in the thirteenth century between Quinsay and Zaitun (Pinghai or Sseuthung), opposite to the island of Formosa (then Tungfan), in 25° 5′ N. lat. (see Klaproth, *Tableaux Hist. de l'Asie*, p. 227). The distance of Cape St. Vincent from Zipangu (Niphon) is 22° of longitude less than from Quinsay, therefore about 209° instead of 230° 53′. It is striking that the oldest statements, those of Eratosthenes and Strabo (lib. i., p. 64), come, through accidental compensations, within 10° of the above-mentioned result of 129° for the difference of longitude of the οἰκουμένη. Strabo, in the same passage in which he alludes to the possible existence of two great habitable continents in the northern hemisphere, says that our οἰκουμένη, in the parallel of Thinæ, Athens (see p. 189), constitutes more than one third of the earth's circumference. Marinus the Tyrian, misled by the length of the time occupied in the navigation from Myos Hormos to India, by the erroneously assumed direction of the major axis of the Caspian from west to east, and by the over-estimation of the length of the land route to the country of the Seres, gave to the Old Continent a breadth of 225° instead of 129°. The Chinese coast was thus advanced to the Sandwich Islands. Colum-

Six years after Balboa, sword in hand, and wading to his knees through the waves, claimed the possession of the Pacific for Castile, and two years after his head had fallen by the hand of the executioner in the revolt against the tyrannical Pedrarias Davila,* Magellan appeared in the Pacific (27th of November, 1520), and, traversing the vast ocean from south-

bus naturally preferred this result to that of Ptolemy, according to which Quinsay should have been found in the meridian of the eastern part of the archipelago of the Carolinas. Ptolemy, in the *Almagest* (II., 1), places the coast of Sinæ at 180°, and in his *Geography* (lib. i., cap. 12) at 177¼°. As Columbus estimated the navigation from Iberia to Sinæ at 120°, and Toscanelli at only 52°, they might certainly, estimating the length of the Mediterranean at about 40°, have called this apparently hazardous enterprise a "brevissimo camino." Martin Behaim, also, on his "*World Apple*," the celebrated globe which he completed in 1492, and which is still preserved in the Behaim house at Nuremberg, places the coast of China (or the throne of the King of Mango, Cambalu, and Cathai) at only 100° west of the Azores—*i. e.*, as Behaim lived four years at Fayal, and probably calculated the distance from that point—119° 40' west of Cape St. Vincent. Columbus was probably acquainted with Behaim at Lisbon, where both lived from 1480 to 1484. (See my *Examen Crit. de l'Hist. de la Géographie*, t. ii., p. 357–369.) The many wholly erroneous numbers which we find in all the writings on the discovery of America, and the then supposed extent of Eastern Asia, have induced me more carefully to compare the opinions of the Middle Ages with those of classical antiquity.

* The eastern portion of the Pacific was first navigated by white men in a boat, when Alonso Martin de Don Benito (who had seen the sea horizon with Vasco Nuñez de Balboa on the 25th of September, 1513, from the little Sierra de Quarequa) descended a few days afterward to the Gulf de San Miguel, before Balboa enacted the strange ceremony of taking possession of the ocean. Seven months before, in the month of January, 1513, Balboa had announced to his court that the South Sea, of which he had heard from the natives, was very easy to navigate: "mar muy mansa y que nunca anda brava como la mar de nuestra banda" (de las Antillas). The name *Oceano Pacifico* was, however, as Pigafetta tells us, first given by Magellan to the Mar del Sur (Balboa). Before Magellan's expedition (in August, 1519), the Spanish government, which was not wanting in watchful activity, had given secret orders, in November, 1514, to Pedrarias Davila, governor of the province of Castilla del Oro (the most northwestern part of South America), and to the great navigator Juan Diaz de Solis, for the former to have four caravels built in the Golfo de San Miguel, "to make discoveries in the newly-discovered South Sea;" and to the latter, to seek for an opening ("abertura de la tierra") from the eastern coast of America, with the view of arriving at the back ("á espel das") of the new country, *i. e.*, of the western portion of Castilla del Oro, which was surrounded by the sea. The expedition of Solis (October, 1515, to August, 1516) led him far to the south, and to the discovery of the Rio de la Plata, long called the Rio de Solis. (Compare, on the little known first discovery of the Pacific, Petrus Martyr, *Epist.*, dxl., p. 296, with the documents of 1513–1515, in Navarrete, t. iii., p. 134 and 357 ; also my *Examen Crit.*, t. i., p. 320 and 350.)

east to northwest, in a course of more than ten thousand geographical miles, by a singular chance, before he discovered the Marianas (his *Islas de los Ladrones*, or *de las Velas Latinas*) and the Philippines, saw no other land but two small uninhabited islands (the *Desventuradas*, or unfortunate islands), one of which, if we may believe his journal and his ship's reckoning, lies east of the Low Islands, and the other somewhat to the southwest of the Archipelago of Mendaña.* Sebastian de Elcano completed the first circumnavigation of the earth in the *Victoria* after Magellan's murder on the island of Zebu, and obtained as his armorial bearings a globe, with the glorious inscription, *Primus circumdedisti me.* He entered the harbor of San Lucar in the month of September, 1522, and scarcely had a year elapsed before the Emperor Charles, stimulated by the suggestions of cosmographers, urged, in a letter to Hernan Cortez, the discovery of a passage " by which the distance to the spice lands would be shortened by two thirds." The expedition of Alvaro de Saavedra was dispatched to the Moluccas from a port of the province Zacatula, on the western coast of Mexico. Hernan Cortez writes in 1527 from the recently-conquered Mexican capital, Tenochtitlan, " to the Kings of Zebu and Tidor in the Asiatic island world." So rapidly did the sphere of cosmical views enlarge, and with it the animation of general intercourse !

Subsequently, the conqueror of New Spain himself entering upon a course of discoveries in the Pacific, proceeded from thence in search of a northeast passage. Men could not habituate themselves to the idea that the continent extended uninterruptedly from such high southern to such high north-

---

* On the geographical position of the Desventuradas (San Pablo, S. lat. 16¼°, long. 135¾° west of Paris; Isla de Tiburones, S. lat. 10¼°, W. long. 145°), see my *Examen Crit.*, t. i., p. 286, and Navarrete, t. iv., p. lix., 52, 218, and 267. The great period of geographical discoveries gave occasion to many illustrious heraldic bearings, similar to the one mentioned in the text as bestowed on Sebastian de Elcano and his descendants (the terrestrial globe, with the inscription " Primus circumdedisti me"). The arms which were given to Columbus as early as May, 1493, to honor his person, " para sublimarlo," with posterity, contain the first map of America—a range of islands in front of a gulf (Oviedo, *Hist. General de las Indias*, ed. de 1547, lib. ii., cap. 7, fol. 10 a.; Navarrete, t. ii., p. 37 ; *Examen Crit.*, t. iv., p. 236). The Emperor Charles V. gave to Diego de Ordaz, who boasted of having ascended the volcano of Orizaba, the drawing of that conical mountain; and to the historian Oviedo (who lived in tropical America uninterruptedly for thirty-four years, from 1513 to 1547), the four beautiful stars of the Southern Cross, as armorial bearings. (Oviedo, lib. ii., cap. 11, fol. 16, b.)

ern latitudes. When tidings arrived from the coast of California that the expedition of Cortez had perished, the wife of
the hero, Juana de Zuñiga, the beautiful daughter of the
Count d'Aguilar, caused two ships to be fitted out and sent
forth to ascertain its fate.* California was already, in 1541,
recognized to be an arid, woodless peninsula—a fact that was
forgotten in the seventeenth century. We moreover gather
from the narratives of Balboa, Pedrarias Davila, and Hernan
Cortez, that hopes were entertained at that period of finding
in the Pacific, then considered to be a portion of the Indian
Ocean, groups of islands, rich in spices, gold, precious stones,
and pearls. Excited fancy urged men to undertake great enterprises, and the daring of these undertakings, whether successful or not, reacted on the imagination, and excited it still
more powerfully. Thus, notwithstanding the thorough absence of political freedom, many circumstances concurred at
this remarkable age of the Conquista—a period of overwrought
excitement, violence, and of a mania for discoveries by sea and
land—to favor individuality of character, and to enable some
highly-gifted minds to develop many noble germs drawn from
the depths of feeling. They err who believe that the *Conquistadores* were incited by love of gold and religious fanaticism alone. Perils always exalt the poetry of life; and, moreover, the remarkable age, whose influence on the development
of cosmical ideas we are now depicting, gave to all enterprises,
and to the natural impressions awakened by distant travels,
the charm of novelty and surprise, which is beginning to fail
us in the present well-instructed age, when so many portions
of the earth are opened to us. Not only one hemisphere, but
almost two thirds of the earth, were then a new and unexplored world, as unseen as that portion of the moon's surface
which the law of gravitation constantly averts from the glance
of the inhabitants of the earth. Our deeply-inquiring age
finds in the increasing abundance of ideas presented to the
human mind a compensation for the surprise formerly induced
by the novelty of grand, massive, and imposing natural phenomena—a compensation which will, it is true, long be denied to the many, but is vouchsafed to the few familiar with
the condition of science. To them the increasing insight into
the silent operation of natural forces, whether in electro-magnetism or in the polarization of light, in the influence of dia-

* See my *Essai Politique sur le Royaume de la Nouvelle Espagne*, t.
ii., 1827, p. 259; and Prescott, *History of the Conquest of Mexico* (New
York, 1843), vol. iii., p. 271 and 336.

thermal substances or in the physiological phenomena of vital organisms, gradually unvails a world of wonders, of which we have scarcely reached the threshold.

The Sandwich Islands, Papua or New Guinea, and some portions of New Holland, were all discovered in the early half of the sixteenth century.* These discoveries prepared the way for those of Cabrillo, Sebastian Vizcaino, Mendaña, and Quiros, whose Sagittaria is Tahiti, and whose Archipelago del Espiritu Santo is the same as the New Hebrides of Cook.† Quiros was accompanied by the bold navigator who subsequently gave his name to the Torres Straits. The Pacific no longer appeared as it had done to Magellan, a desert waste ; it was now animated by islands, which, however, for want of exact astronomical observations, appeared to have no fixed position, but floated from place to place over the charts. The Pacific remained for a long time the exclusive theater of the enterprises of the Spaniards and Portuguese. The important South Indian Malayan Archipelago, dimly described by Ptolemy, Cosmas, and Polo, unfolded itself in more distinct outlines after Albuquerque had established himself in 1511 in Malacca, and after the expedition of Anton Abreu. It is the special merit of the classical Portuguese historian, Barros, the cotemporary of Magellan and Camoens, to have so truly recognized the physical and ethnological character of this archipelago, as to be the first to propose that the Australian Polynesia should be distinguished as a fifth portion of the earth. It was not until the Dutch power acquired the ascendency in the Moluccas

---

* Gaetano discovered one of the Sandwich Islands in 1542. Respecting the voyage of Don Jorge de Menezes (1526), and that of Alvaro de Saavedra (1528), to the Ilhas de Papuas, see Barros, *Da Asia,* Dec. iv., liv. i., cap. 16; and Navarrete, t. v., p. 125. The " *Hydrography*" of Joh. Rotz (1542), which is preserved in the British Museum, and has been examined by the learned Dalrymple, contains outlines of New Holland, as does also the collection of maps of Jean Valard of Dieppe (1552), for the first knowledge of which we are indebted to M. Coquebert Monbret.

† After the death of Mendaña, his wife, Doña Isabela Baretos, a woman distinguished for personal courage and great mental endowments, undertook in the Pacific the command of the expedition, which did not terminate until 1596 (*Essai Polit. sur la Nouv. Esp.*, t. iv., p. 111).

Quiros practiced in his ships the distillation of fresh from salt water on a considerable scale, and his example was followed in several instances (Navarrete, t. i., p. liii.). The entire operation, as I have elsewhere shown on the testimony of Alexander of Aphrodisias, was known as early as the third century of our era, although it was not then practiced in ships.

that Australia began to emerge from its former obscurity, and to assume a definite form in the eyes of geographers.* Now began the great epoch of Abel Tasman. We do not purpose here to give the history of individual geographical discoveries, but simply to refer to the principal events by which, in a short space of time and in continuous connection, two thirds of the earth's surface were opened to the apprehension of men, in consequence of the suddenly awakened desire to reach the wide, the unknown, and the remote regions of our globe.

An enlarged insight into the nature and the laws of physical forces, into the distribution of heat over the earth's surface, the abundance of vital organisms and the limits of their distribution, was developed simultaneously with this extended knowledge of land and sea. The advance which the different branches of science had made toward the close of the Middle Ages (a period which, in a scientific point of view, has not been sufficiently estimated), facilitated and furthered the sensuous apprehension and the comparison of an unbounded mass of physical phenomena now simultaneously presented to the observation of men. The impressions were so much the deeper and so much the more capable of leading to the establishment of cosmical laws, because the nations of Western Europe, even before the middle of the sixteenth century, had explored the New Continent, at least along its coasts, in the most different degrees of latitude in both hemispheres; and because it was here that they first became firmly settled in the region of the equator, and that, owing to the singular configuration of the earth's surface, the most striking contrasts of vegetable organizations and of climate were presented to them at different elevations within very circumscribed limits of space. If I again take occasion to allude to the advantages presented by the mountainous districts of the equinoctial zone, I would observe, in justification of my reiteration of the same sentiment, that to the inhabitants of these regions alone it is granted to behold all the stars of the heaven, and almost all families and forms of vegetation ; but to behold is not to observe by a mental process of comparison and combination.

Although in Columbus, as I hope I have succeeded in showing in another work, a capacity for exact observation was developed in manifold directions, notwithstanding his entire deficiency of all previous knowledge of natural history, and solely by contact with great natural phenomena, we must by no

* See the excellent work of Professor Meinecke of Prenzlau, entitled *Das Festland Australien, eine Geogr. Monographie*, 1837, th. i., s. 2–10.

means assume a similar development in the rough and war-like body of the Conquistadores. Europe owes to another and more peaceful class of travelers, and to a small number of distinguished men among municipal functionaries, ecclesiastics, and physicians, that which it has unquestionably acquired by the discovery of America, in the gradual enrichment of its knowledge regarding the character and composition of the atmosphere, and its action on the human organization; the distribution of climates on the declivities of the Cordilleras; the elevation of the line of perpetual snow in accordance with the different degrees of latitude in both hemispheres; the succession of volcanoes; the limitation of the circles of commotion in earthquakes; the laws of magnetism; the direction of oceanic currents; and the gradations of new animal and vegetable forms. The class of travelers to whom we have alluded, by residing in native Indian cities, some of which were situated twelve or thirteen thousand feet above the level of the sea, were enabled to observe with their own eyes, and, by a continued residence in those regions, to test and to combine the observations of others, to collect natural products, and to describe and transmit them to their European friends. It will suffice here to mention Gomara, Oviedo, Acosta, and Hernandez. Columbus brought home from his first voyage of discovery some natural products, as, for instance, fruits, and the skins of animals. In a letter written from Segovia (August, 1494), Queen Isabella enjoins on the admiral to persevere in his collections; and she especially requires of him that he should bring with him specimens of " all the coast and forest birds peculiar to countries which have a different climate and different seasons." Little attention has hitherto been given to the fact that Martin Behaim's friend Cadamosto procured for the Infante Henry the Navigator black elephants' hair a palm and a half in length, from the same western coast of Africa whence Hanno, almost two thousand years earlier, had brought the " tanned skins of wild women " (of the large Gorilla apes), in order to suspend them in a temple. Hernandez, the private physician of Philip II., and sent by that monarch to Mexico, in order to have all the vegetable and zoological curiosities of the country depicted in accurate and finished drawings, was able to enlarge his collection by copies of many very carefully executed historical pictures, which had been painted at the command of Nezahualcoyotl, a king of Tezcuco,* half a century before the arrival of the

* This king died in the time of the Mexican king Axayacatl, who

Spaniards. Hernandez also availed himself of a collection of medicinal plants which he found still growing in the celebrated old Mexican garden of Huaxtepec, which, owing to its vicinity to a newly-established Spanish hospital,* the Conquistadores had not laid waste. Almost at this time the fossil mastodon bones on the elevated plateaux of Mexico, New Granada, and Peru, which have since become so important with respect to the theory of the successive elevation of mountain chains, were collected and described. The designations of giant bones and fields of giants (*Campos de Gigantes*) sufficiently testified the fantastic character of the early interpretation applied to these fossils.

One circumstance which specially contributed to the extension of cosmical views at this enterprising period was the immediate contact of a numerous mass of Europeans with the free and grand exotic forms of nature, on the plains and mountainous regions of America, and (in consequence of the voyage of Vasco de Gama) on the eastern shores of Africa and Southern India. Even in the beginning of the sixteenth century, a Portuguese physician, Garcia de Orta, under the protection of the noble Martin Alfonso de Sousa, established, on the present site of Bombay, a botanical garden, in which he cultivated the medicinal plants of the neighborhood. The muse of Camoens has paid Garcia de Orta the tribute of patriotic praise. The impulse to direct observation was now every where awakened, while the cosmographical writings of

---

reigned from 1464 to 1477. The learned native historian, Fernando de Alva Jxtlilxochitl, whose manuscript chronicle of the Chichimeque I saw in 1803, in the place of the Viceroy of Mexico, and of which Mr. Prescott has so ably availed himself in his work (*Conquest of Mexico*, vol. i., p. 61, 173, and 206; vol. iii., p. 112), was a descendant of the poet king Nezahualcoyotl. The Aztec name of the historian, Fernando de Alva, means Vanilla face. M. Ternaux Compans, in 1840, caused a French translation of this manuscript to be printed in Paris. The notice of the long elephants' hair collected by Cadamosto occurs in Ramusio, vol. i., p. 109, and in Grynæus, cap. 43, p. 33.

* Clavigero, *Storia antica del Messico* (Cesena, 1780), t. ii., p. 153. There is no doubt, from the accordant testimonies of Hernan Cortez in his reports to the Emperor Charles V., of Bernal Diaz, Gomara, Oviedo, and Hernandez, that, at the time of the conquest of Montezuma's empire, there were no menageries and botanic gardens in any part of Europe which could be compared with those of Huaxtepec, Chapoltapec, Iztapalapan, and Tezcuco. (Prescott, op. cit., vol. i., p. 178; vol. ii., p. 66 and 117–121; vol. iii., p. 42.) On the early attention which is mentioned in the text as having been paid to the fossil bones in the "fields of giants," see Garcilaso, lib. ix., cap. 9; Acosta, lib. iv., cap. 30; and Hernandez (ed. of 1556), t. i., cap. 32, p. 105.

the Middle Ages were to be regarded less as the result of actual observation than as mere compilations, reflecting the opinions of classical antiquity. Two of the greatest men of the sixteenth century, Conrad Gesner and Andreas Cæsalpinus, have the high merit of having opened a new path to zoology and botany.

In order to give a more vivid idea of the early influence exercised by oceanic discoveries on the enlarged sphere of the physical and astronomical sciences connected with navigation, I will call attention, at the close of this description, to some luminous points, which we may already see glimmering through the writings of Columbus. Their first faint light deserves to be traced with so much the more care, because they contain the germs of general cosmical views. I will not pause here to consider the proofs of the results which I have enumerated, since I have given them in detail in another work, entitled *Examen Critique de l'Histoire de la Géographie du Nouveau Continent et des Progrès de l'Astronomie Nautique aux* xve *et* xvie *Siècles*. But, in order to avoid the imputation of undervaluing the views of modern physical knowledge, in comparison with the observations of Columbus, I will give the literal translation of a few lines contained in a letter which the admiral wrote from Haiti in the month of October, 1498. He writes as follows: " Each time that I sail from Spain to India, as soon as I have proceeded about a hundred nautical miles to the west of the Azores, I perceive an extraordinary alteration in the movement of the heavenly bodies, in the temperature of the air, and in the character of the sea. I have observed these alterations with especial care, and I notice that the mariner's compass (*agujas de marear*), whose declination had hitherto been northeast, was now changed to northwest; and when I had crossed this line (*raya*), as if in passing the brow of a hill (*como quien traspone una cuesta*), I found the ocean covered by such a mass of sea weed, similar to small branches of pine covered with pistachio nuts, that we were apprehensive that, for want of a sufficiency of water, our ships would run upon a shoal. Before we reached the line of which I speak, there was no trace of any such sea weed. On the boundary line, one hundred miles west of the Azores, the ocean becomes at once still and calm, being scarcely ever moved by a breeze. On my passage from the Canary Islands to the parallel of Sierra Leone, we had to endure a frightful degree of heat, but, as soon as we had crossed the above-mentioned line (to the west of the meridian of the Azores), the climate changed, the air became temperate, and the freshness increased the further we advanced."

This passage, which is elucidated by many others in the writings of Columbus, contains views of physical geography, observations on the influence of geographical longitude on the declination of the magnetic needle, on the inflection of the isothermal lines between the western shores of the Old and the eastern shores of the New Continent, on the position of the Great Saragossa bank in the basin of the Atlantic Ocean, and on the relations existing between this part of the ocean and the superimposed atmosphere. Erroneous observations made in the vicinity of the Azores, on the movement of the polar star,[*] had misled Columbus during his first voyage, from the inaccuracy of his mathematical knowledge, to entertain a belief in the irregularity of the spheroidal form of the earth. In the western hemisphere, the earth, according to his views, " is more *swollen*, so that ships gradually arrive nearer the heavens on reaching the line (*raya*), where the magnetic needle points due north, and this elevation (*cuesta*) is the cause of the cooler temperature." The solemn reception of the admiral in Barcelona took place in April, 1493, and as early as the 4th of May of the same year, the celebrated bull was signed by Pope Alexander VI., which " establishes to all eternity" the line of demarkation[†] between the Spanish and Portuguese

---

[*] *Observations de Christophe Colomb sur le Passage de la Polaire par le Méridien*, in my *Relation Hist.*, t. i., p. 300, and in the *Examen Crit.*, t. iii., p. 17-20, 44-51, and 56-61. (Compare, also, Navarrete, in *Columbus's Journal* of 16th to 30th of September, 1492, p. 9, 15, and 254.)

[†] On the singular differences of the " Bula de concesion á los Reyes Catolicos de las Indias descubiertas y que se descubieren" of May 3, 1493, and the " Bula de Alexandro VI., sobre la particion del oceano" of May 4, 1493 (elucidated in the Bula de estension of the 25th of September, 1493), see *Examen Crit.*, t. iii., p. 52-54. Very different from this line of demarkation is that settled in the " Capitulacion de la particion del Mar Oceano entre los Reyes Catolicos y Don Juan, Rey de Portugal," of the 7th of June, 1494, 370 leagues (17½ to an equatorial degree) west of the Cape Verd Islands. (Compare Navarrete, *Coleccion de los Viages y Descub. de los Esp.*, t. ii., p. 28-35, 116-143, and 404; t. iv., p. 55 and 252.) This last-named line, which led to the sale of the Moluccas (de el Moluca) to Portugal, 1529, for the sum of 350,000 gold ducats, did not stand in any connection with magnetical or meteorological fancies. The papal lines of demarkation deserve, however, more careful consideration in the present work, because, as I have mentioned in the text, they exercised great influence on the endeavors to improve nautical astronomy, and especially on the methods attempted for the determination of the longitude. It is also very deserving of notice, that the *capitulacion* of June 7, 1494, affords the first example of a proposal for the establishment of a meridian in a permanent manner by

possessions, at a distance of one hundred miles to the west of the Azores. If we consider further that Columbus, immediately after his return from his first voyage of discovery, proposed to go to Rome, in order, as he said, to "give the pope notice of all that he had discovered," and if the importance attached by the cotemporaries of Columbus to the discovery of the line of no variation be further borne in mind, it will be admitted that I was justified in advancing the historical proposition that the admiral, at the moment of his highest court favor, strove to have a "*physical line of demarkation* converted into a *political one.*"

The influence which the discovery of America and the oceanic enterprises connected with that event so rapidly exercised on the combined mass of physical and astronomical science, is rendered most strikingly manifest when we recall the earliest impressions of those who lived at this period, and the extended range of those scientific efforts, of which the more important are comprehended in the first half of the sixteenth century. Christopher Columbus has not only the merit of being the first to discover *a line without magnetic variation*, but also of having excited a taste for the study of terrestrial magnetism in Europe, by means of his observations on the progressive increase of western declination in receding from that line. The fact that almost every where the ends of a freely-moving magnetic needle do not point exactly to the geographical north and south poles, must have repeatedly been recognized, even with very imperfect instruments, in the Mediterranean, and at all places where, in the twelfth century, the declination amounted to more than eight or ten degrees. But it is not improbable that the Arabs or the Crusaders, who were brought in contact with the East between the years 1096 and 1270, might, while they spread the use of the Chinese and Indian mariner's compass, also have drawn attention to the northeast and northwest pointing of the magnetic needle in different regions of the earth as to a long-known phenomenon. We learn positively from the Chinese *Penthsaoyan*, which was written under the dynasty of Song,*

marks graven in rocks, or by the erection of towers. It is commanded, "que se haga alguna señal ó torre," that some signal or tower be erected wherever the dividing meridian, whether in the eastern or the western hemisphere, intersects an island or a continent in its course from pole to pole. In the continents, the *rayas* were to be marked at proper intervals by a series of such marks or towers, which would indeed have been no slight undertaking.

* It appears to be a remarkable fact, that the earliest classical writer

between 1111 and 1117, that the mode of measuring the amount of western declination had long been understood. The merit due to Columbus is not to have made the first observation of the existence of magnetic variation, since we find, for example, that this is set down on the chart of Andrea Bianco in 1436, but that he was the first who remarked, on the 13th of September, 1492, that " $2\frac{1}{2}°$ east of the island of Corvo the magnetic variation changed and passed from N.E. to N.W."

This discovery of a *magnetic line without variation* marks a memorable epoch in nautical astronomy. It was celebrated with just praise by Oviedo, Las Casas, and Herrera. We can not assume, with Livio Sanuto, that this discovery is due to the celebrated navigator, Sebastian Cabot, without entirely losing sight of the fact that Cabot's first voyage, made at the expense of some merchants of Bristol, and distinguished for its success in reaching the continent of America, was not accomplished until five years after the first expedition of Columbus. The great Spanish navigator has not only the merit of having discovered a region in the Atlantic Ocean where at that period the magnetic meridian coincided with the geographical, but also that of having made the ingenious observation that magnetic variation might likewise serve to determine the ship's place with respect to longitude. In the journal of the second voyage (April, 1496) we find that the admiral actually determined his position by the observed declination. The difficulties were, it is true, at that period still unknown, which oppose this method of determining longitude, especially where the magnetic lines of declination are so much curved as to follow the parallels of latitude for considerable distances, instead of coinciding with the direction of the meridian. Mag-

on terrestrial magnetism, William Gilbert, who can not be supposed to have had the slightest knowledge of Chinese literature, should regard the mariner's compass as a Chinese invention, which had been brought to Europe by Marco Polo. " Illa quidem pyxide nihil unquam humanis excogitatum artibus humano generi profuisse magis, constat. Scientia nauticæ pyxidulæ traducta videtur in Italiam per Paulum Venetum, qui circa annum mcclx. apud Chinas artem pyxidis didicit." (*Gulielmi Gilberti Colcestrensis, Medici Londinensis de Magnete Physiologia nova*, Lond., 1600, p. 4.) The idea of the introduction of the compass by Marco Polo, whose travels occurred in the interval between 1271 and 1295, and who therefore returned to Italy after the mariner's compass had been mentioned as a long-known instrument by Guyot de Provins in his poem, as well as by Jacques de Vitry and Dante, is not supported by any evidence. Before Marco Polo set out on his travels in the middle of the thirteenth century, Catalans and Basques already made use of the compass. (See Raymond Lully, in the Treatise *De Contemplatione*, written in 1272.)

netic and astronomical methods were anxiously sought, in order to determine, on land and at sea, those points which are intersected by the ideal line of demarkation. The imperfect condition of science, and of all the instruments used at sea in 1493 to measure space and time, were unequal to afford a practical solution to so difficult a problem. Under these circumstances, Pope Alexander VI. actually rendered, without knowing it, an essential service to nautical astronomy and the physical science of terrestrial magnetism by his presumption in dividing half the globe between two powerful states. From that time forth the maritime powers were continually beset by a host of impracticable proposals. Sebastian Cabot, as we learn from his friend, Richard Eden, boasted on his death-bed of having had a "divine revelation made to him of an infallible method of finding geographical longitude." This revelation consisted in a firm conviction that magnetic declination changed regularly and rapidly with the meridian. The cosmographer Alonso de Santa Cruz, one of the instructors of Charles V., undertook, although certainly from very imperfect observations, to draw up the first general *variation chart** in the year 1530, and, therefore, one hundred and fifty years before Halley.

The advance or *movement* of the magnetic lines, the knowledge of which has generally been ascribed to Gassendi, was not even conjectured by William Gilbert, although Acosta, "from the instruction of Portuguese navigators," had at a much earlier period assumed that there were four lines without declination over the earth's surface.† No sooner was the

* In corroboration of this statement regarding Sebastian Cabot on his death-bed, see the well-written and critically-historical work by Biddle, entitled *A Memoir of Sebastian Cabo* (p. 222). "We do not know with certainty," says Biddle, "either the year of the death or the burying-place of the great navigator who gave to Great Britain almost an entire continent, and without whom (as without Sir Walter Raleigh) the English language would perhaps not have been spoken by many millions who now inhabit America." On the materials according to which the variation chart of Alonso de Santa Cruz was compiled, as well as on the variation compass, whose construction allowed altitudes of the sun to be taken at the same time, see Navarrete, *Noticia biografica del cosmografo Alonso de Santa Cruz*, p. 3–8. The first variation compass was constructed before 1525, by an ingenious apothecary of Seville, Felipe Guillen. The endeavors to learn more exactly the direction of the curves of magnetic declination were so earnest, that in 1585 Juan Jayme sailed with Francisco Gali from Manilla to Acapulco merely for the purpose of trying in the Pacific a declination instrument which he had invented. See my *Essai Politique sur la Nouvelle Espagne*, t. iv., p. 110.

† Acosta, *Hist. Natural de las Indias*, lib. i., cap. 17. These four

dipping-needle invented in England, in 1576, by Robert Norman, than Gilbert boasted that, by means of this instrument, he could determine a ship's place in dark, starless nights (*aëre calignoso*).* Immediately after my return to Europe, I showed from my own observations in the Pacific that, under certain local relations, as, for instance, during the season of the constant mist (*garua*) on the coasts of Peru, the latitude might be determined from the magnetic inclination with sufficient accuracy for the purposes of navigation. I have purposely dwelt at length on these individual points, in order to show, in our consideration of an important cosmical event, that, with the exception of measuring the intensity of magnetic force, and the horary variations of the declination, all those questions were broached in the sixteenth century, with which the physicists of the present day are still occupied. On the remarkable chart of America appended to the edition of the geography of Ptolemy, published at Rome in 1508, we find the magnetic pole marked as an insular mountain north of Gruentlant (Greenland), which is represented as a part of Asia. Martin Cortez in the *Breve Compendio de la Sphera* (1545), and Livio Sanuto in the *Geographia di Tolomeo* (1588), place it further to the south. The latter writer entertained a prejudice, which has unfortunately survived to the present time, that "if we were so fortunate as to reach the magnetic pole (*il calamitico*), we should there experience some miraculous effects (*alcun miraculoso stupendo effetto*").

Attention was directed at the close of the fifteenth and the beginning of the sixteenth century, in reference to the distribution of heat and meteorology, to the decrease of heat with the increase of western longitude† (the curvature of the isothermal lines); to the law of rotation of the winds, generalized by Lord

magnetic lines without variation led Halley, by the contests between Henry Bond and Beckborrow, to the theory of four magnetic poles.

* Gilbert, *De Magnete Physiologia nova*, lib. v., cap. 8, p. 200.

† In the temperate and cold zones, this inflection of the isothermal lines is general between the west coast of Europe and the east coast of North America, but within the tropical zone the isothermal lines run almost parallel to the equator; and in the hasty conclusions into which Columbus was led, no account was taken of the difference between sea and land climates, or between east and west coasts, or of the influence of latitudes and winds, as, for instance, those blowing over Africa. (Compare the remarkable considerations on climates which are brought together in the *Vida del Almirante*, cap. 66.) The early conjecture of Columbus regarding the curvature of the isothermal lines in the Atlantic Ocean was well founded, if limited to the extra-tropical (temperate and cold) zones.

Bacon ;* to the decrease of humidity in the atmosphere, and of the quantity of rain owing to the destruction of forests ;† to the decrease of heat with the increase of ·elevation above the level of the sea ; and to the lower limit of the line of perpetual snow. The fact of this limit being a *function of geographical latitude* was first recognized by Peter Martyr Anghiera in 1510. Alonso de Hojeda and Amerigo Vespucci had seen the snowy mountains of Santa Marta (*Tierras nevadas de Citarma*) as early as the year 1500 ; Rodrigo Bastidas and Juan de la Cosa examined them more closely in 1501 ; but it was not until the pilot Juan Vespucci, nephew of Amerigo, had communicated to his friend and patron Anghiera an account of the expedition of Colmenares, that the *tropical snow region* visible on the mountainous shore of the Caribbean Sea acquired a great, and, we might say, a cosmical importance. A connection was now established between the lower limit of perpetual snow and the general relations of the decrease of heat and the differences of climate. Herodotus (ii., 22), in his investigations on the rising of the Nile, wholly denied the existence of snowy mountains south of the tropic of Cancer. Alexander's campaigns indeed led the Greeks to the Nevados of the Hindoo-Coosh range (ὄρη ἀγάννιφα), but this is situated between 34° and 36° north latitude. The only notice of snow in the equatorial region with which I am acquainted, before the discovery of America, and prior to the year 1500, and which has been but little regarded by physicists, is contained in the celebrated inscription of Adulis, which is considered by Niebuhr to be later than Juba and Augustus. The knowledge of the dependence of the lower limit of snow on the latitude of the place,‡ the first insight into the law of the vertical decrease of temperature, and the sinking of an

---

* An observation of Columbus. (*Vida del Almirante*, cap. 55 ; *Examen Crit.*, t. iv., p. 253 ; and see, also, vol. i., p. 316.)

† The admiral, says Fernando Colon (*Vida del Alm.*, cap. 58), ascribed the extent and denseness of the forests which clothed the ridges of the mountains to the many refreshing falls of rain, which cooled the air while he continued to sail along the coast of Jamaica. He remarks in his ship's journal on this occasion, that "formerly the quantity of rain was equally great in Madeira, the Canaries, and the Azores; but since the trees which shaded the ground have been cut down, rain has become much more rare." This warning has remained almost unheeded for three centuries and a half.

‡ See vol. i., p. 329 ; *Examen Crit.*, t. iv., p. 294 ; *Asie Centrale*, t. iii., p. 235. The inscription of Adulis, which is almost fifteen hundred years older than Anghiera, speaks of "Abyssinian snow, in which the traveler sinks up to the knees."

almost equally cold upper stratum of air from the equator toward the poles, designate an important epoch in the history of our physical knowledge.

If, on the one hand, accidental observations, having a wholly unscientific origin, favored this knowledge in the suddenly enlarged spheres of natural investigation, the age we are describing was, on the other hand, from an unfortunate combination of circumstances, singularly deficient in the advantages arising from a purely scientific impulse. Leonardo da Vinci, the greatest physicist of the fifteenth century, who combined an enviable insight into nature with distinguished mathematical knowledge, was the cotemporary of Columbus, and died three years after him. Meteorology, as well as hydraulics and optics, had occupied the attention of this celebrated artist. The influence which he exercised during his life was made manifest by his great works in painting, and by the eloquence of his discourse, and not by his writings. Had the physical views of Leonardo da Vinci not remained buried in his manuscripts, the field of observation opened by the new world would in a great degree have been worked out in many departments of science before the great epoch of Galileo, Pascal, and Huygens. Like Francis Bacon, and a whole century before him, he regarded induction as the only sure method of treating natural science (" *dobbiamo cominciare dall' esperienza, e per mezzo di questa scoprirne la regione*").[*]

As we find, notwithstanding the want of instruments of measurement, that the questions of climatic relations in the tropical mountainous regions—the distribution of heat, the extremes of atmospheric dryness, and the frequency of electric explosions—were frequently discussed in the accounts of the first land journeys, so also it appears that mariners very early acquired correct views of the direction and rapidity of the currents which traverse the Atlantic Ocean, like rivers of very variable breadth. The actual *equatorial current*, the movement of the waters between the tropics, was first described by Columbus. He expresses himself most positively and gener-

[*] Leonardo da Vinci correctly observes of this proceeding, " questo è il methodo daosservarsi nella ricerca de' fenomeni della natura." See Venturi, *Essai sur les Ouvrages Physico-mathematiques de Leonardo da Vinci*, 1797, p. 31 ; Amoretti, *Memorie Storiche sù la Vita di Lionardo da Vinci*, Milano, 1804, p. 143 (in his edition of *Trattato della Pittura*, t. xxxiii. of the Classici Italiani) ; Whewell, *Philos. of the Inductive Sciences*, 1840, vol. ii., p. 368–370 ; Brewster, *Life of Newton*, p. 332. Most of Leonardo da Vinci's physical works bear the date of the year 1498.

ally on the subject on his third voyage, saying, " the waters
move with the heavens (con los cielos) from east to west."
Even the direction of separate floating masses of sea weed
confirmed this view.* A small pan of tinned iron, which he
found in the hands of the natives of the island of Guadaloupe,
confirmed Columbus in the idea that it might be of European
origin and obtained from the remains of a shipwrecked vessel,
borne by the equatorial current from Spain to the coasts of
America. In his geognostic fancies, he regarded the existence of the series of the smaller Antilles and the peculiar configuration of the larger islands, or, in other words, the correspondence in the direction of their coasts with that of their
parallels of latitude, as the long-continued action of the movement of the sea between the tropics from east to west.

When the admiral, on his fourth and last voyage, discovered the inclination from north to south of the coasts of the
continent from Cape Gracias á Dios to the Laguna de Chiri-
qui, he felt the action of the violent current which runs N.
and N.N.W., and is induced by the contact of the equatorial
current with the opposite dike-like projecting coast-line. An-

---

* The great attention paid by the early navigators to natural phenomena may be seen in the oldest Spanish accounts. Diego de Lepe,
for instance, found, in 1499 (as we learn from a witness in the lawsuit
against the heirs of Columbus), by means of a vessel having valves,
which did not open until it had reached the bottom, that at a distance
from the mouth of the Orinoco, a stratum of fresh water of six fathoms
depth flowed above the salt water (Navarrete, Viages y Descubrim., t.
iii., p. 549). Columbus drew milk-white sea water (" white as if meal
had been mixed with it") on the south coast of Cuba, and carried it
to Spain in bottles (Vida del Almirante, p. 56). I have myself been at
the same spots for the purpose of determining longitudes, and it surprised me to think that the milk-white color of sea water, so common on
shoals, should have been regarded by the experienced admiral as a new
and unexpected phenomenon. With reference to the Gulf Stream itself, which must be regarded as an important cosmical phenomenon,
many effects had been observed long before the discovery of America,
produced by the sea washing on shore at the Canaries and the Azores
stems of bamboos, trunks of pines, corpses of strange aspect from the
Antilles, and even living men in canoes "which could never sink."
These effects were, however, then attributed solely to the strength of
the westerly gales (Vida del Almirante, cap. 8; Herrera, Dec. i., lib.
i., cap. 2; lib. ix., cap. 12), while the movement of the waters, which
is wholly independent of the direction of the winds—the returning
stream of the oceanic current, which brings every year tropical fruits
from the West Indian Islands to the coasts of Ireland and Norway, was
not accurately recognized. Compare the memoir of Sir Humphrey
Gilbert, On the Possibility of a Northwest Passage to Cathay, in Hakluyt, Navigations and Voyages, vol. iii., p. 14; Herrera, Dec. i., lib. ix.,
cap. 12; and Examen Crit., t. ii., p. 247–257; t. iii., p. 99–108.

ghiera survived Columbus sufficiently long to become acquainted with the deflection of the waters of the Atlantic throughout their whole course, and to recognize the existence of the
rotatory movement in the Mexican Gulf, and the propagation
of this movement to the Tierra de los Bacallaos (Newfoundland) and the mouth of the St. Lawrence. I have elsewhere
circumstantially considered how much the expedition of Ponce
de Leon, in the year 1512, contributed to the establishment
of more exact ideas, and have shown that in a treatise written by Sir Humphrey Gilbert between the years 1567 and
1576, the movement of the waters of the Atlantic Ocean from
the Cape of Good Hope to the Banks of Newfoundland is
treated according to views which coincide almost entirely with
those of my excellent deceased friend, Major Rennell.

At the same time that the knowledge of oceanic currents
was generally diffused, men also became acquainted with those
great banks of sea weed (*Fucus natans*)—the oceanic meadows which presented the singular spectacle of the accumulation of a social plant over an extent of space almost seven times
greater than the area of France. The *great Fucus Bank*, the
Mar de Sargasso, extends between 19° and 34° north latitude.
The major axis is situated about 7° west of the island of Corvo. The *lesser Fucus Bank* lies in the space between the
Bermudas and the Bahamas. Winds and partial currents
variously affect, according to the character of the season, the
length and circumference of these Atlantic fucoid meadows,
for the first description of which we are indebted to Columbus.
No other sea in either hemisphere presents an accumulation
of social plants on so large a scale."*

The important era of geographical discoveries and of the
sudden opening of an unknown hemisphere not only extended
our knowledge of the earth, but it also expanded our views of
the whole universe, or, in other words, of the visible vault of
heaven. Since man, to borrow a fine expression of Garcilaso
de la Vega, in his wanderings to distant regions sees " lands
and stars simultaneously change,"† the advance to the equator on both coasts of Africa, and even beyond the southern
extremity of the New Continent, must have presented to travelers, by sea and land, the glorious aspect of the southern constellations longer and more frequently than could have been

* *Examen Crit.*, t. iii., p. 26 and 66–99 ; and see, also, *Cosmos*, vol.
i., p. 308.
† Alonso de Ercilla has imitated the passage of Garcilaso in the *Araucana :* " Climas passè, mudè constelaciones."—See *Cosmos. ante*, p. 72.

the case at the time of Hiram and the Ptolemies, or during
the Roman dominion, and the period in which the Arabs
maintained commercial intercourse with the nations dwelling
on the shores of the Red Sea or of the Indian Ocean, between
the Straits of Bab-el-Mandeb and the western peninsula of
India.   Amerigo Vespucci, in his letters, Vicente Yañez Pinzon, Pigafetta, the companion of Magellan and Elcano, and
Andrea Corsali, in his voyage to Cochin in the East Indies,
in the beginning of the sixteenth century, gave the first and
most animated accounts of the southern sky (beyond the feet
of the Centaur and the glorious constellation Argo).   Amerigo, who had higher literary acquirements, and whose style
was also more redundant than that of the others, extols, not
ungracefully, the glowing richness of the light, and the picturesque grouping and strange aspect of the constellations that
circle round the southern pole, which is surrounded by so few
stars.   He maintains, in his letters to Pierfrancesco de' Medici, that he had carefully devoted his attention, on his third
voyage, to the southern constellations, having made drawings
of them and measured their polar distances.   His communications regarding these observations do not, indeed, leave
much cause to regret that any portion of them should have
been lost.

I find that the first mention of the mysterious black specks
(coal-bags) was made by Anghiera in the year 1510.   They
had already been observed in 1499 by the companions of Vicente Yañez Pinzon, on the expedition dispatched from Palos,
and which took possession of the Brazilian Cape San Augustin.*   The *Canopo fosco* (*Canopus niger*) of Amerigo is probably also one of these *coal-bags*.   The intelligent Acosta compares them to the darkened portion of the moon's disk (in partial eclipses), and appears to ascribe them to a void in the
heavens, or to an absence of stars.   Rigaud has shown how
the reference to the coal-bags, of which Acosta says positively
that they are visible in Peru (and not in Europe), and move
round the south pole, has been regarded by a celebrated astronomer as the first notice of spots on the sun.†   The knowledge of the two Magellanic clouds has been unjustly ascribed
to Pigafetta, for I find that Anghiera, on the observations of
Portuguese seamen, mentions these clouds fully eight years

* Petr. Mart., *Ocean.*, Dec. i., lib. ix., p. 96 ; *Examen Crit.*, t. iv., p.
221 and 317.

† Acosta, *Hist. Natural de las Indias*, lib. i., cap. 2 ; Rigaud, *Account
of Harriot's Astron. Papers*, 1833, p. 37.

before the termination of Magellan's voyage of circumnavigation. He compares their mild effulgence to that of the Milky
Way. The larger cloud did not, however, escape the vigilance
of the Arabs, and it is probably the white ox (*El Bakar*) of
their southern sky, the *white spot* of which the astronomer
Abdurrahman Sofi says that it could not be seen at Bagdad
or in northern Arabia, but at Tchama, and in the parallel of
the Straits of Bab-el-Mandeb. The Greeks and Romans,
who followed the same path under the Lagides and later, did
not observe, or, at least, make no mention, in their extant
writings, of a cloud of light, which, nevertheless, between 11°
and 12° north latitude, rose three degrees above the horizon
at the time of Ptolemy, and more than four degrees in that of
Abdurrahman, in the year 1000.* At the present day, the
altitude of the central part of the *Nubecula major* may be
about 5° at Aden. The reason that seamen usually first see
the Magellanic clouds in much more southern latitudes, as,
for instance, near the equator, or even far to the south of it,
is probably to be ascribed to the character of the atmosphere,
and to the vapors near the horizon, which reflect white light.
In Southern Arabia, especially in the interior of the country,
the deep azure of the sky and the great dryness of the atmosphere must favor the recognition of the Magellanic clouds, as
we see exemplified by the visibility of comets' tails at daylight
between the tropics and in very southern latitudes.

The arrangement of the stars near the antarctic pole into
new constellations was made in the seventeenth century. The
observations made with imperfect instruments by the Dutch
navigators Petrus Theodori of Embden, and Friedrich Houtmann, who was a prisoner in Java and Sumatra to the King
of Bantam and Atschin (1596–1599), were incorporated in
the celestial charts of Hondius Bleaw (Jansonius Cæsius) and
Bayer.

---

* Pigafetta, *Primo Viaggio intorno al Globo Terracqueo*, publ. da C.
Amoretti, 1800, p. 46 ; *Ramusio*, vol. i., p. 355, c. ; Petr. Mart., *Ocean.*,
Dec. iii., lib. i., p. 217. (According to the events referred to by Anghiera, Dec. ii., lib. x., p. 204, and Dec. iii., lib. x., p. 232, the passage
in the *Oceanica* which speaks of the Magellanic clouds must have been
written between 1514 and 1516.) Andrea Corsali (*Ramusio*, vol. i.,
p. 177) also describes, in a letter to Giuliano de' Medici, the rotatory
and translatory movement of "*due nugolette di ragionevol grandezza.*"
The star which he represents between *Nubecula major* and *minor* appears to me to be β Hydræ (*Examen Crit.*, t. v., p. 234–238). Regarding Petrus Theodori of Embden, and Houtmann, the pupil of the mathematician Plancius, see an historical article by Olbers, in Schumacher's
Jahrbuch für 1840, s. 249.

The less regular distribution of masses of light gives to the zone of the southern sky situated between the parallels of 50° and 80°, which is so rich in crowded nebulous spots and starry masses, a peculiar, and, one might almost say, picturesque character, depending on the grouping of the stars of the first and second magnitudes, and their separation by intervals, which appear to the naked eye desert and devoid of radiance. These singular contrasts—the Milky Way, which presents numerous portions more brilliantly illumined than the rest, and the insulated, revolving, rounded Magellanic clouds, and the coal-bags, the larger of which lies close upon a beautiful constellation—all contribute to augment the diversity of the picture of nature, and rivet the attention of the susceptible mind to separate regions on the confines of the southern sky. One of these, the constellation of the Southern Cross, has acquired a peculiar character of importance from the beginning of the sixteenth century, owing to the religious feelings of Christian navigators and missionaries who have visited the tropical and southern seas and both the Indies. The four principal stars of which it is composed are mentioned in the Almagest, and, therefore, were regarded in the time of Adrian and Antoninus Pius as parts of the constellation of the Centaur.* It seems singular that, since the figure of this constellation is so striking, and is so remarkably well defined and individualized, in the same way as those of the Greater and Lesser Bear, the Scorpion, Cassiopeia, the Eagle, and the Dolphin, these four stars of the Southern Cross should not have been earlier separated from the large ancient constellation of the Centaur; and this is so much the more remarkable, since the Persian Kazwini, and other Mohammedan astronomers, took pains to discover crosses in the Dolphin and the Dragon. Whether the courtly flattery of the Alexandrian literati, who converted Canopus into a *Ptolemæon*, likewise included the stars of our Southern Cross, for the glorification of Augustus, in a *Cæsaris thronon*, never visible in Italy, is a question that can not now be very readily answered.† At the time of Claudius Ptolemæus, the beautiful star at the base of the Southern Cross had still an altitude of 6° 10′ at its meridian passage at Alexandria, while in the present day it culminates there several degrees below the horizon. In order at this time (1847) to

* Compare the researches of Delambre and Encke with Ideler, *Ursprung der Sternnamen*, s. xlix., 263 und 277; also my *Examen Crit.*, t. iv., p. 319–324; t. v., p. 17–19, 30, and 230–234.
† Plin., ii., 70; Ideler, *Sternnamen*, s. 260 und 295.

see *a* Crucis at an altitude of 6° 10′, it is necessary, taking the refraction into account, to be ten degrees south of Alexandria, in the parallel of 21° 43′ north latitude. In the fourth century the Christian anchorites in the Thebaid desert might have seen the Cross at an altitude of ten degrees. I doubt, however, whether its designation is due to them, for Dante, in the celebrated passage of the *Purgatorio,*

> Io mi volsi a man destra, e posi mente
> All'altro polo, e vidi quattro stelle
> Non viste mai fuor ch' alla prima gente;

and Amerigo Vespucci, who, at the aspect of the starry skies of the south, first called to mind this passage on his third voyage, and even boasted that he now "looked on the four stars never seen till then by any save the first human pair," were both unacquainted with the denomination of the *Southern Cross.* Amerigo simply observes that the four stars form a rhomboidal figure (*una mandorla*), and this remark was made in the year 1501. The more frequently the maritime expeditions on the routes opened by Gama and Magellan round the Cape of Good Hope and through the Pacific were multiplied, and as Christian missionaries penetrated into the newly-discovered tropical lands of America, the fame of this constellation continually increased. I find it mentioned first by the Florentine, Andrea Corsali, in 1517, and subsequently, in 1520, by Pigafetta, as a wonderful cross (*croce maravigliosa*), more glorious than all the constellations in the heavens. The learned Florentine extols Dante's "prophetic spirit," as if the great poet had not as much erudition as creative imagination, and as if he had not seen Arabian celestial globes, and conversed with many learned Oriental travelers of Pisa.* Acos-

---

* I have elsewhere attempted to dispel the doubts which several distinguished commentators of Dante have advanced in modern times respecting the "*quattro stelle.*" To take this problem in all its completeness, we must compare the passage, "Io mi volsi," &c. (*Purgat.*, l., v. 22–24), with the other passages: *Purg.*, l., v. 37 ; viii., v. 85–93 ; xxix., v. 121 ; xxx., v. 97 ; xxxi., v. 106 ; and *Inf.*, xxvi., v. 117 and 127. The Milanese astronomer, De Cesaris, considers the three "*facelle*" ("Di che il polo di quà tutto quanto arde," and which set when the four stars of the Cross rise) to be Canopus, Achernar, and Fomalhaut. I have endeavored to solve these difficulties by the following considerations. "The philosophical and religious mysticism which penetrates and vivifies the grand composition of Dante, assigns to all objects, besides their real or material existence, an ideal one. It seems almost as if we beheld two worlds reflected in one another. The four stars represent, in their moral order, the *cardinal virtues,* prudence, justice, strength, and temperance ; and they, therefore, merit the name of the

ta, in his *Historia Natural y Moral de las Indias*,* remarks, that in the Spanish settlements of tropical America, the first settlers were accustomed, even as is now done, to use, as a celestial clock, the Southern Cross, calculating the hours from its inclined or vertical position.

In consequence of the precession of the equinoxes, the starry heavens are continually changing their aspect from every portion of the earth's surface. The early races of mankind beheld in the far north the glorious constellation of our southern hemisphere rise before them, which, after remaining long invisible, will again appear in those latitudes after the lapse of thousands of years. Canopus was fully 1° 20′ below the horizon at Toledo (39° 54′ north latitude) in the time of Columbus, and now the same star is almost as much above the horizon at Cadiz. While at Berlin and in the northern latitudes the stars of the Southern Cross, as well as a and β Centauri, are receding more and more from view, the Magellanic clouds are slowly approaching our latitudes. Canopus was at its greatest northern approximation during the last century, and is now moving nearer and nearer to the south, although

holy lights, '*luci sante.*' The three stars which light the pole represent the *theological virtues*, faith, hope, and charity. The first of these beings themselves reveals their double nature, chanting, ' Here we are nymphs, in heaven we are stars ;' *Noi sem qui ninfe, e nel cielo semo stelle.* In the land of truth, in the terrestrial paradise there are seven nymphs. *In cerchio faceran di se claustro le sette ninfe.* This is the union of all the cardinal and theological virtues. Under these mystic forms we can scarcely recognize the real objects of the firmament separated from each other, according to the eternal laws of the *celestial mechanism.* The ideal world is a free creation of the soul, the product of poetic inspiration." (*Examen Crit.*, t. iv., p. 324–332.)

* Acosta, lib. i., cap. 5. Compare my *Relation Historique*, t. i., p. 209. As the stars a and γ of the Southern Cross have almost the same right ascension, the Cross appears perpendicular when passing the meridian ; but the natives too often forget that this celestial clock marks the hour each day 3′ 56″ earlier. I am indebted to the communications of my friend, Dr. Galle, by whom Le Verrier's planet was first discovered in the heavens, for all the calculations respecting the visibility of southern stars in northern latitudes. " The inaccuracy of the calculation, according to which the star a of the Southern Cross, taking refraction into account, would appear to have begun to be invisible in 52° 25′ north latitude, about the year 2900 before the Christian era, may perhaps amount to more than 100 years, and could not be altogether set aside, even by the strictest mode of calculation, as the proper motion of the fixed stars is probably not uniform for such long intervals of time. The proper motion of a Crucis is about one third of a second annually, chiefly in right ascension. It may be presumed that the uncertainty produced by neglecting this does not exceed the above-mentioned limit."

very slowly, owing to its vicinity to the south pole of the ecliptic. The Southern Cross began to become invisible in 52° 30′ north latitude 2900 years before our era, since, according to Galle, this constellation might previously have reached an altitude of more than 10°. When it disappeared from the horizon of the countries on the Baltic, the great pyramid of Cheops had already been erected more than five hundred years. The pastoral tribe of the Hyksos made their incursion seven hundred years earlier. The past seems to be visibly nearer to us when we connect its measurement with great and memorable events.

The progress made in nautical astronomy, that is to say, in the improvement of methods of determining the ship's place (its geographical latitude and longitude), was simultaneous with the extension of a knowledge of the regions of space, although this knowledge was more the result of sensuous observation than of scientific induction. All that was able in the course of ages to favor advance in the art of navigation—the compass and the more correct acquaintance with magnetic declination; the measurement of a ship's speed by a more careful construction of the log, and by the use of chronometers and lunar observations; the improved construction of ships; the substitution of another force for that of the wind; and lastly and most especially, the skillful application of astronomy to the ship's reckoning—must all be regarded as powerful means toward the opening of the different portions of the earth, the more rapid and animated furtherance of general intercourse, and the acquirement of a knowledge of cosmical relations. Assuming this as one point of view, we would again observe, that even in the middle of the thirteenth century, nautical instruments capable of determining the time by the altitude of the stars were in use among the seamen of Catalonia and the island of Majorca, and that the astrolabe described by Raymond Lully in his *Arte de Navegar* was almost two hundred years older than that of Martin Behaim. The importance of astronomical methods was so thoroughly appreciated in Portugal, that toward the year 1484 Behaim was nominated president of a *Junta de Mathematicos*, who were to form tables of the sun's declination, and, as Barros observes, to teach pilots the method of navigating by the sun's altitude, *maniera de navegar por altura del Sol.** This mode of navigating by the meridian altitude of the sun was even at that

---

* Barros, *Da Asia*, Dec. i., liv. iv., cap. 2 (1788), p. 282.

time clearly distinguished from that by the determination of the longitude, *por la altura del Este-Oeste.**

The importance of determining the position of the papal *line of demarkation,* and of thus fixing the limits between the possessions of the Portuguese and Spanish crowns in the newly-discovered land of Brazil, and in the group of islands in the South Indian Ocean, increased, as we have already observed, the desire for ascertaining a practical method for determining the longitude.   Men perceived how rarely the ancient and imperfect method of lunar eclipses employed by Hipparchus could be applied, and the use of lunar distances was recommended as early as 1514 by the Nuremberg astronomer, Johann Werner, and soon afterward by Orontius Finæus and Gemma Frisius.   Unfortunately, however, these methods also remained impracticable until, after many fruitless attempts with the instruments of Peter Apianus (Bienewitz) and Alonso de Santa Cruz, the mirror sextant was invented by the ingenuity of Newton in 1700, and was brought into use among seamen by Hadley in 1731.

The influence of the Arabian astronomers acted, through the Spaniards, on the general progress of nautical astronomy. Many methods were certainly attempted for determining the longitude, which did not succeed ; and the fault of the want of success was less rarely ascribed to the incorrectness of the observation, than to errors of printing in the astronomical ephemerides of Regiomontanus which were then in use.   The Portuguese even suspected the correctness of the astronomical data as given by the Spaniards, whose tables they accused of being falsified from political grounds.†   The suddenly-awakened desire for the auxiliaries which nautical astronomy promised, at any rate theoretically, is most vividly expressed in the narrations of the travels of Columbus, Amerigo Vespucci, Pigafetta, and of Andreas de San Martin, the celebrated pilot of the Magellanic expedition, who was in possession of the methods of Ruy Falero for determining the longitude.   Oppositions of planets, occultations of the stars, differences of altitude between the moon and Jupiter, and changes in the moon's declination, were all tried with more or less success.   We possess observations of conjunction by Columbus on the night of the 13th of January, 1493, at Haiti.   The necessity for at-

* Navarrete, *Coleccion de los Viages y Descubrimientos que Hiciéron por mar los Españoles,* t. iv., p. xxxii. (in the *Noticia Biographica de Fernando de Magellanes*).

† Barros, Dec. iii., parte ii., p. 650 and 658-662.

taching a special and well-informed astronomer to every great expedition was so generally felt, that Queen Isabella wrote to Columbus on the 5th of September, 1493, " that although he had shown in his undertakings that he knew more than any other living being (*que ninguno de los nacidos*), she counseled him, nevertheless, to take with him Fray Antonio de Marchena, as being a learned and skillful astronomer." Columbus writes, in the narrative of his fourth voyage, that " there was only one infallible method of taking a ship's reckoning, viz., that employed by astronomers. He who understands it may rest satisfied, for that which it yields is like unto a prophetic vision (*vision profetica.*)* Our ignorant pilots, when they

* The queen writes to Columbus : " Nosotros mismos *y no otro alguno, habemos visto algo del libro que nos dejústes,*" " we ourselves, and no one else, have seen the book you have sent us" (a journal of his voyage, in which the distrustful navigator had omitted all numerical data of degrees of latitude and of distances) : " quanto mas en esto platicamos y vemos, conocemos cuan gran cosa ha seido este negocio vuestro, y que habeis sabido en ello mas que nunca se pensó que pudiera saber ninguno de los nacidos. Nos parece que seria bien que llevásedes con vos un buen Estrologo, y nos parescia que seria bueno para esto Fray Antonio de Marchena, porque es buen Estrologo, y siempre, nos pareció que se conformaba con vuestro parecer." " The more we have examined it, the more we have appreciated your undertaking, and the more we have felt that you have shown by it that you know more than any human being could be supposed to know. It appears to us that it would be well for you to take with you some astrologer, and that Fray Antonio de Marchena would be a very suitable person for such a purpose." Respecting this Marchena, who is identical with Fray Juan Perez, the guardian of the Convent de la Rabida, where Columbus, in his poverty, in 1484, " asked the monks for bread and water for his child," see Navarrete, t. ii., p. 110 ; t. iii., p. 597 and 603 (Muñoz, *Hist. del Nuevo Mundo,* lib. iv., § 24.) Columbus, in a letter from Jamaica to the *Christianisimos Monarcas,* July 7, 1503, calls the astronomical ephemerides " *una vision profetica.*" (Navarrete, t. i., p. 306.) The Portuguese astronomer, Ruy Falero, a native of Cubilla, nominated by Charles V., in 1519, Caballero de la Orden de Santiago, at the same time as Magellan, played an important part in the preparations for Magellan's voyage of circumnavigation. He had prepared expressly for him a treatise on determinations of longitude, of which the great historian Barros possessed some chapters in manuscript (*Examen Crit.*, t. i., p. 276 and 302 ; t. iv., p. 315), probably the same which were printed at Seville by John Escomberger in 1535. Navarrete (*Obra póstuma sobre la Hist. de la Nautica y de las ciencias Matematicas,* 1846, p. 147) had not been able to find the book even in Spain. Respecting the four methods of determining the longitude which Falero had received from the suggestions of his " *Demonio familiar,*" see Herrera, Dec. ii., lib. ii., cap. 19, and Navarrete, t. v., p. lxxvii. Subsequently the cosmographer Alonso de Santa Cruz, the same who (like the apothecary of Seville, Felipe Guillen, 1525) attempted to determine the longitude by means of the variation of the magnetic needle, made impracticable pro-

have lost sight of land for several days, know not where they are. They would not be able to find the countries again which I have discovered. To navigate a ship requires the compass (*compas y arte*), and the knowledge or art of the astronomer."

I have given these characteristic details in order more clearly to show the manner in which nautical astronomy—the powerful instrument for rendering navigation more secure, and thereby of facilitating access to all portions of the earth—was first developed in the period of time under consideration, and how, in the general intellectual activity of the age, men perceived the possibility of establishing methods which could not be made practically applicable until improvements were effected in solar and lunar tables, and in the construction of time-pieces and instruments for measuring angles. If the character of an age be " the manifestation of the human mind in any definite epoch," the age of Columbus and of the great nautical discoveries must be regarded as having given a new and higher impetus to the acquirements of succeeding centuries, while it increased in an unexpected manner the objects of science and contemplation. It is the peculiar attribute of important discoveries at once to extend the domain of our possessions, and the prospect into the new territories which yet remain open to conquest. Weak minds complacently believe that in their own age humanity has reached the culminating point of intellectual progress, forgetting that by the internal connection existing among all natural phenomena, in proportion as we advance, the field to be traversed acquires additional extension, and that it is bounded by a horizon which incessantly recedes before the eyes of the inquirer.

Where, in the history of nations, can we find an epoch similar to that in which events so fraught with important results as the discovery and first colonization of America, the passage to the East Indies round the Cape of Good Hope, and Magellan's first circumnavigation, occurred simultaneously with the highest perfection of art, with the attainment of intellectual

posals for accomplishing the same object by the conveyance of time; but his chronometers were sand-and-water clocks, wheel-works moved by weights, and even by wicks " dipped in oil," which were consumed in very equal intervals of time ! Pigafetta (*Transunto del Trattato di Navigazione*, p. 219) recommends altitudes of the moon at the meridian. Amerigo Vespucci, speaking of the method of determining longitude by lunar distances, says, with great naïveté and truth, that its advantages arise from the "*corso più leggier de la luna*." (Canovai, *Viaggi*, p. 57.)

and religious freedom, and with the sudden enlargement of the knowledge of the earth and the heavens ? Such an age owes a very inconsiderable portion of its greatness to the distance at which we contemplate it, or to the circumstance of its appearing before us amid the records of history, and free from the disturbing reality of the present. But here too, as in all earthly things, the brilliancy of greatness is dimmed by the association of emotions of profound sorrow. The advance of cosmical knowledge was bought at the price of the violence and revolting horrors which conquerors—the so-called civilizers of the earth—spread around them. But it were irrational and rashly bold to decide dogmatically on the balance of blessings and evils in the interrupted history of the development of mankind. It becomes not man to pronounce judgment on the great events of the world's history, which, slowly developed in the womb of time, belong but partially to the age in which we place them.

The first discovery of the central and southern portions of the United States of America by the Northmen coincides very nearly with the mysterious appearance of Manco Capac in the elevated plateaux of Peru, and is almost two hundred years prior to the arrival of the Azteks in the Valley of Mexico. The foundation of the principal city (Tenochtitlan) occurred fully three hundred and twenty-five years later. If these Scandinavian colonizations had been attended by permanent results, if they had been maintained and protected by a powerful mother country, the advancing Germanic races would still have found many unsettled hordes of hunters in those regions where the Spanish conquerors met with only peacefully-settled agriculturists.*

* The American race, which was the same from 65° north latitude to 55° south latitude, passed directly from the life of hunters to that of cultivators of the soil, without undergoing the intermediate gradation of a pastoral life. This circumstance is so much the more remarkable, because the bison, which is met with in enormous herds, is susceptible of domestication, and yields an abundant supply of milk. Little attention has been paid to an account given in Gomara (*Hist. Gen. de las Indias*, cap. 214), according to which it would appear that in the sixteenth century there was a race of men living in the northwest of Mexico, in about 40° north latitude, whose greatest riches consisted in herds of tamed bisons (*bueyes con una giba*). From these animals the natives obtained materials for clothing, food, and drink, which was probably the blood (Prescott, *Conquest of Mexico*, vol. iii., p. 416), for the dislike to milk, or, at least, its non-employment, appears, before the arrival of Europeans, to have been common to all the natives of the New Continent, as well as to the inhabitants of China and Cochin China. There were certainly, from the earliest times, herds of domesticated llamas in

The age of the *Conquista*, which comprises the end of the fifteenth and the beginning of the sixteenth century, indicates a remarkable concurrence of great events in the political and social life of the nations of Europe. In the same month in which Hernan Cortez, after the battle of Otumba, advanced upon Mexico, with the view of besieging it, Martin Luther burned the pope's bull at Wittenberg, and laid the foundation of the Reformation, which promised to the human mind both freedom and progress on paths which had hitherto been almost wholly untrodden.* Still earlier, the noblest forms of ancient Hellenic art, the Laocoon, the Torso, the Apollo de Belvidere, and the Medicean Venus, had been resuscitated, as it were, from the tombs in which they had so long been buried. There flourished in Italy, Michael Angelo, Leonardo da Vinci, Titian, and Raphael; and in Germany, Holbein and Albert Durer. The Copernican system of the universe was discovered, if not made generally known, in the year in which Columbus died, and fourteen years after the discovery of the New Continent.

The importance of this discovery, and of the first colonization of Europeans, involves a consideration of other fields of inquiry besides those to which these pages are devoted, and closely bears upon the intellectual and moral influences exercised on the improvement of the social condition of mankind by the sudden enlargement of the accumulated mass of new ideas. We would simply draw attention to the fact that,

the mountainous parts of Quito, Peru, and Chili. These herds constituted the riches of the nations who were settled there, and were engaged in the cultivation of the soil; in the Cordilleras of South America there were no "pastoral nations," and "pastoral life" was not known. What are the "tame deer," near the Punta de St. Helena, which are mentioned in Herrera, Dec. ii., lib. x., cap. 6 (t. i., p. 471, ed. Amberes, 1728)? These deer are said to have given milk and cheese, "*ciervos que dan leche y queso y se crian en casa!*" From what source is this notice taken? It can not have arisen from a confusion with the llamas (having neither horns nor antlers) of the cold mountainous region, of which Garcilaso affirms that in Peru, and especially on the plateau of Callao, they were used for plowing. (*Comment reales*, Part i., lib. v., cap. 2, p. 133. Compare, also, Pedro de Cieça de Leon, *Chronica del Peru*, Sevilla, 1553, cap. 110, p. 264.) This employment of llamas appears, however, to have been a rare exception, and a merely local custom. In general, the American races were remarkable for their deficiency of domesticated animals, and this had a profound influence on family life.

* On the hope which Luther, in the execution of his great and free-minded work, placed especially on the younger generation, the youth of Germany, see the remarkable expressions in a letter written in June, 1518. (Neander, *De Vicelio*, p. 7.)

since this period, a new and more vigorous activity of the mind and feelings, animated by bold aspirations and hopes which can scarcely be frustrated, has gradually penetrated through all grades of civil society; that the scanty population of one half of the globe, especially in the portions opposite to Europe, has favored the settlements of colonies, which have been converted by their extent and position into independent states, enjoying unlimited power in the choice of their mode of free government ; and, finally, that religious reform—the precursor of great political revolutions—could not fail to pass through the different phases of its development in a portion of the earth which had become the asylum of all forms of faith, and of the most different views regarding divine things. The daring enterprise of the Genoese seaman is the first link in the immeasurable chain of these momentous events. Accident, and not fraud and dissensions, deprived the continent of America of the name of Columbus.* The New World continuously

* I have shown elsewhere how a knowledge of the period at which Vespucci was named royal chief pilot alone refutes the accusation first brought against him by the astronomer Schoner, of Nuremberg, in 1533, of having artfully inserted the words " *Terra di Amerigo*" in charts which he altered. The high esteem which the Spanish court paid to the hydrographical and astronomical knowledge of Amerigo Vespucci is clearly manifested in the instructions (*Real titulo con todas sas facultades*) which were given to him when he was appointed *piloto mayor* on the 22d of March, 1508. (Navarrete, t. iii., p. 297–302.) He was placed at the head of a true *Deposito hydrografico*, and was to prepare for the *Casade Contratacion* in Sevilla (the central point of all oceanic expedition) a general description of coasts and account of positions (*Padron general*), in which all new discoveries were to be annually entered. But even as early as 1507 the name of " Americi terra" had been proposed for the New Continent by a person whose existence even was undoubtedly unknown to Vespucci, the geographer Waldseemüller (Martinus Hylacomylus) of Freiburg, in the Breisgau (the director of a printing establishment at St. Dié in Lorraine), in a small work entitled *Cosmographiæ Introductio, insuper quatuor Americi Vespucii Navigationes* (impr. in oppido S. Deodati, 1507). Ringmann, professor of cosmography at Basle (better known under the name of Philesius), Hylacomylus, and Father Gregorius Reisch, who edited the *Margarita Philosophica*, were intimate friends. In the last-named work we find a treatise written in 1509 by Hylacomylus on architecture and perspective. (*Examen Crit.*, t. iv., p. 112.) Laurentius Phrisius of Metz, a friend of Hylacomylus, and, like him, patronized by Duke René of Lorraine, who maintained a correspondence with Vespucci, in the Strasburg edition of Ptolemy, 1522, speaks of Hylacomylus as deceased. In the map of the New Continent contained in this edition, and drawn by Hylacomylus, the name of America occurs for the first time *in the editions of Ptolemy's Geography*. According to my investigations, a map of the world by Petrus Apianus, which was once included in Cramer's edition of Solinus, and a second time in the Va-

brought nearer to Europe during the last half century, by
means of commercial intercourse and the improvement of nav-

dian edition of Mela, and represented, like more modern Chinese maps,
the Isthmus of Panama broken through, had appeared two years earlier. (*Examen Crit.*, t. iv., p. 99–124; t. v., p. 168–176.) It is a great
error to regard the map of 1527, obtained from the Ebner library at
Nuremberg, now in Weimar, and the map of 1529 of Diego Ribero,
which differs from the former, and is engraved by Gussefeld, as the
oldest maps of the New Continent (op. cit., t. ii., p. 184; t. iii., p. 191).
Vespucci had visited the coasts of South America in the expedition of
Alonso de Hojeda, a year after the third voyage of Columbus, in 1499,
in company with Juan de la Cosa, whose map, drawn at Puerto de
Santa Maria in 1500, fully six years before Columbus's death, was first
made known by myself. Vespucci could not have had any motive for
feigning a voyage in the year 1497, for he, as well as Columbus, was
firmly persuaded, until his death, that only parts of Eastern Asia had
been reached. (Compare the letter of Columbus, February, 1502, to
Pope Alexander VII., and another, July, 1506, to Queen Isabella, in
Navarrete, t. i., p. 304; t. ii., p. 280; and Vespucci's letter to Pierfrancesco de' Medici, in Bandini's *Vita e Lettere di Amerigo Vespucci*, p. 66
and 83.) Pedro de Ledesma, the pilot of Columbus on his third voyage, says, even in 1513, in the lawsuit against the heirs, " that Paria is
regarded as a part of Asia, *la tierra firme que dicese que es de Asia.*"—
Navarrete, t. iii., p. 539. The frequent periphrases, *Mondo nouvo, alter
Orbis, Colonus novit Orbis repertor*, are not at variance with this, as
they only denote regions not before seen, and are so used by Strabo,
Mela, Tertullian, Isidore of Seville, and Cadamosto. (*Examen Crit.*,
t. i., p. 118; t. v., p. 182–184.) For more than twenty years after the
death of Vespucci, which occurred in 1512, and until the calumnious
charges of Schoner, in the *Opusculum Geographicum*, 1533, and of
Servet, in the Lyons edition of Ptolemy's *Geography* of 1535, we find
no complaint against the Florentine navigator. Christopher Columbus, a year before his death, calls him *mucho hombre de bien*, a man of
worth, " worthy of all confidence," and " always inclined to render
him service." (*Carta a mi muy caro fijo D. Diego*, in Navarrete, t. i.,
p. 351.) Fernando Colon expresses the same good will toward Vespucci. He wrote the life of his father in 1535, in Seville, four years
before his death, and with Juan Vespucci, a nephew of Amerigo's, attended the astronomical junta of Badajoz, and the proceedings respecting the possession of the Moluccas. Similar feelings were entertained
by Petrus Martyr de Anghiera, the personal friend of the admiral,
whose correspondence goes down to 1525; by Oviedo, who seeks for
every thing which can lessen the fame of Columbus; by Ramusio; and
by the great historian Guicciardini. If Amerigo had intentionally falsified the dates of his voyage, he would have brought them into agree-
ment with each other, and not have made the first voyage terminate
five months after the second began. The confusion of dates in the
many different translations of his voyages is not to be attributed to him,
as he did not himself publish any of these accounts. Such confusions
of figures were, besides, very frequently to be met with in writings
printed in the sixteenth century. Oviedo had been present, as one of
the queen's pages, at the audience at which Ferdinand and Isabella, in
1493, received Columbus with much pomp on his return from his first
voyage of discovery. Oviedo has three times stated in print that this

igation, has exercised an important influence on the political
institutions, the ideas and feelings of those nations who occu-

audience took place in the year 1496, and even that America was discovered in 1491. Gomara had the same printed, not in numerals, but
in words, and placed the discovery of the *tierra firme* of America in
1497, in the very year, therefore, which proved so fatal to Amerigo
Vespucci's reputation. (*Examen Crit.*, t. v., p. 196–202.) The wholly
irreproachable conduct of the Florentine (who never attempted to attach his name to the New Continent, but who, in the grandiloquent
accounts which he addressed to the Gonfalionere Piero Goderini, to
Pierfrancesco de' Medici, and to Duke René II. of Lorraine, had the
misfortune of drawing upon himself the attention of posterity more
than he deserved) is most positively proved by the lawsuit which the
fiscal authorities carried on from 1508 to 1527 against the heirs of Christopher Columbus, for the purpose of withdrawing from them the rights
and privileges which had been granted by the crown to the admiral in
1492. Amerigo entered the service of the state as *Piloto mayor* in the
same year that the lawsuit began. He lived at Seville during four
years of this suit, in which it was to be decided what parts of the New
Continent had been first reached by Columbus. The most miserable
reports found a hearing, and were converted into subjects of accusation
by the fiscal; witnesses were sought for at St. Domingo, and all the
Spanish ports, at Moguer, Palos, and Seville, and even under the eyes
of Amerigo Vespucci and his nephew Juan. The *Mundus Novus*, printed by Johann Otmer, at Augsburg, in 1504; the *Raccolta di Vicenza*
(*Mondo Novo e paesi novamente retrovati da Alberico Vespuzio Fioren-
tino*), by Alessandro Zorzi, in 1507, and generally ascribed to Fracan-
zio di Montalboddo; and the *Quatuor Navigationes* of Martin Waldseemüller (Hylacomylus), had already appeared. Since 1520, maps had
been constructed, on which was marked the name of *America*, which
had been proposed by Hylacomylus in 1507, and praised by Joachim
Vadius in a letter addressed to Rudolphus Agricola from Vienna in 1512;
and yet the person to whom widely-circulated writings in Germany,
France, and Italy attributed a voyage of discovery in 1497, to the *tierra firme* of Paria, was neither cited by the fiscal as a witness in the
lawsuit which had been begun in 1508, and was continued during
nineteen years, nor was he even spoken of as the predecessor or the
opponent of Columbus. Why, after the death of Amerigo Vespucci
(22d February, 1512, in Seville), was not his nephew, Juan Vespucci,
called upon to show (as Martin Alonso, Vicente Yañez Pinzon, Juan de
la Cosa, and Alonso de Hojeda had done) that the coast of Paria, which
did not derive its importance from its being " part of the main land of
Asia," but on account of the productive pearl fishery in its vicinity,
had been already reached by Amerigo, before Columbus landed there
on the 1st of August, 1498? The disregard of this most important testimony is inexplicable if Amerigo Vespucci had ever boasted of having
made a voyage of discovery in 1497, or if any serious import had been
attached at that time to the confused dates and mistakes in the printing
of the " *Quatuor Navigationes.*" The great and still unprinted work
of a friend of Columbus, Fra Bartholomé de las Casas (the *Historia
general de las Indias*), was written, as we know with certainty, at
very different periods. It was not begun until fifteen years after the
death of Amerigo in 1527, and was finished in 1559, seven years before the death of the aged author, in his 92d year. Praise and bitter

py the eastern shores of the Atlantic, the boundaries of which
appear to be constantly brought nearer and nearer to one an-

blame are strangely mingled in it. We see that dislike and suspicion of
fraud augmented in proportion as the fame of the Florentine navigator
spread. In the preface (*Prolongo*) which was written first, Las Casas
says, " Amerigo relates what he did in two voyages to our Indies, but
he appears to have passed over many circumstances, whether designedly (*á saviendas*), or because he did not attend to them. This circumstance has led some to attribute to him that which is due to others, and
which ought not to be taken from them." The judgment pronounced
in the 1st book (chap. 140) is equally moderate: " Here I must speak
of the injustice which Amerigo, or perhaps those who printed (*ó los
que imprimiéron*) the *Quatuor Navigationes*, appear to have committed
toward the admiral. To Amerigo alone, without naming any other, the
discovery of the continent is ascribed. He is also said to have placed
the name of America in maps, thus sinfully failing toward the admiral.
As Amerigo was learned, and had the power of writing eloquently (*era
latino y eloquente*), he represented himself in the letter to King René
as the leader of Hojeda's expedition ; yet he was only one of the seamen, although experienced in seamanship and learned in cosmography
(*hombre entendido en las cosas de la mar y docto en Cosmographia*). . . .
In the world the belief prevails that he was the first to set foot on the
main land. If he purposely gave currency to this belief, it was great
wickedness; and if it was not done intentionally, it looks like it (*clara
pareze la falsedad : y si fué de industria hecha maldad grande fué ; y
ya que no lo fuese, al menos parezelo*). . . . Amerigo is represented as
having sailed in the year 7 (1497): a statement that seems, indeed, to
have been only an oversight in writing, and not an intentional false
statement (*pareze aver avido yerro de pendola y no malicia*), because he
is stated to have returned at the end of eighteen months. The foreign
writers call the country America ; it ought to be called Columba."
This passage shows clearly that up to that time Las Casas had not accused Amerigo of having himself brought the name *America* into usage.
He says, *an tomado los escriptores estrangeros de nombrar la nuestra
Tierra firme America, como si Americo solo y no otro con él y antes que
todos la oviera descubierto.* In lib. i., cap. 164–169, and in lib. ii., cap.
2, of the work, his hatred is fully expressed; nothing is now attributed
to erroneous dates, or to the partiality of foreigners for Amerigo ; all is
intentional deceit, of which Amerigo himself is guilty (*de industria lo
hizo . . . persisitó en el engaño . . . . de falsedad está claramente con-
vencido*). Bartholomé de las Casas takes pains, moreover, in two passages, to show especially that Amerigo, in his accounts, falsified the
succession of the occurrences of his first two voyages, placing many
things which belonged to the second voyage in the first, and *vice versà*.
It seems very strange to me that the accuser does not appear to have
felt how much the weight of his accusations is diminished by the circumstance that he himself speaks of the opposite opinion, and of the
indifference of the person who would have been most interested in at-
tacking Vespucci, if he had believed him guilty and hostilely disposed
against his father and himself. " I can not but wonder," says Las Casas
(cap. 164), " that Hernando Colon, a clear-sighted man, who, as I certainly know, had in his hands Amerigo's accounts of his travels, should
not have remarked in them any deceit or injustice toward the admiral." As I had a fresh opportunity, a few months ago, of examining the

other. (See my *Examen Crit. de l' Hist. de la Géographie,*
t. iii., p. 154–158 and 225–227.)

GREAT DISCOVERIES IN THE HEAVENS BY THE APPLICATION OF THE
TELESCOPE.—PRINCIPAL EPOCHS IN THE HISTORY OF ASTRONOMY
AND MATHEMATICS, FROM GALILEO AND KEPLER TO NEWTON AND
LEIBNITZ.—LAWS OF THE PLANETARY MOTIONS AND GENERAL
THEORY OF GRAVITATION.

AFTER having endeavored to enumerate the most distinctly
defined periods and stages of development in the history of the
contemplation of the universe, we have proceeded to delineate
the epoch in which the civilized nations of one hemisphere became acquainted with the inhabitants of the other. The periods of the greatest discoveries in space over the surface of
our planet was immediately succeeded by the revelations of

rare manuscript of Bartholomé de las Casas, I would wish to embody in
this long note what I did not employ in 1839 in my *Examen Critique,*
t. v., p. 178–217. The conviction which I then expressed, in the same
volume, p. 217 and 224, has remained unshaken. "Where the designation of a large continent, generally adopted as such, and consecrated
by the usage of many ages, presents itself to us as a monument of human injustice, it is natural that we should at first sight attribute the
cause to the person who would appear most interested in the matter.
A careful study of the documentary evidence has, however, shown
that this supposition in the present instance is devoid of foundation, and
that the name of America has originated in a distant region (as, for instance, in France and Germany), owing to many concurrent circumstances which appear to remove all suspicion from Vespucci. Here
historical criticism stops, for the field of *unknown* causes and *possible*
moral contingencies does not come within the domain of positive history. We here find a man who, during a long life, enjoyed the esteem
of his cotemporaries, raised by his attainments in nautical astronomy
to an honorable employment. The concurrence of many fortuitous
circumstances gave him a celebrity which has weighed upon his memory, and helped to throw discredit on his character. Such a position is
indeed rare in the history of human misfortunes, and affords an instance
of a moral stain deepened by the glory of an illustrious name. It seems
most desirable to examine, amid this mixture of success and adversity,
what is owing to the navigator himself, to the accidental errors arising
from a hasty supervision of his writings, or to the indiscretion of dangerous friends." Copernicus himself contributed to this dangerous
celebrity, for he also ascribes the discovery of the new part of the globe
to Vespucci. In discussing the "*centrum gravitatis*" and "*centrum
magnitudinis*" of the continent, he adds, "magis id erit clarum, si ad-
dentur insulæ ætate nostra sub Hispaniarum Lusitaniæque principibus
repertæ et præsertim America ab inventore denominata navium præ-
fecto, quem, ob incompertam ejus adhuc magnitudinem, alterum orbem
terrarum putent." (*Nicolai Copernici de Revolutionibus Orbium Cœlestium,* libri sex, 1543, p. 2, a.)

the telescope, through which man may be said to have taken possession of a considerable portion of the heavens. The application of a newly-created organ—an instrument possessed of the power of piercing the depths of space—calls forth a new world of ideas. Now began a brilliant age of astronomy and mathematics ; and in the latter, the long series of profound inquirers, leading us on to the " all transforming" Leonhard Euler, the year of whose birth (1707) is so near that of the death of Jacques Bernouilli.

A few names will suffice to give an idea of the gigantic strides with which the human mind advanced in the seventeenth century, especially in the development of mathematical induction, under the influence of its own subjective force rather than from the incitement of outward circumstances. The laws which control the fall of bodies and the motions of the planets were now recognized. The pressure of the atmosphere ; the propagation of light, and its refraction and polarization, were investigated. Mathematical physics were created, and based on a firm foundation. The invention of the infinitesimal calculus characterizes the close of the century ; and, strengthened by its aid, human understanding has been enabled, during the succeeding century and a half, successfully to venture on the solution of the problems presented by the perturbations of the heavenly bodies ; by the polarization and interference of the waves of light ; by the radiation of heat ; by electro-magnetic re-entering currents ; by vibrating chords and surfaces ; by the capillary attraction of narrow tubes ; and by many other natural phenomena.

Henceforward the work in the world of thought progresses uninterruptedly, each portion continually contributing its aid to the remainder. None of the earlier germs are stifled. With the abundance of the materials to be elaborated, strictness in the methods and improvements in the instruments of observation are simultaneously increased. We will here limit ourselves more especially to the seventeenth century, the age of Kepler, Galileo, and Bacon, of Tycho Brahe, Descartes, and Huygens, of Fermat, Newton, and Leibnitz. The labors of these distinguished inquirers are so generally known, that slight references will be sufficient to point out those portions by which they have most brilliantly contributed to the enlargement of cosmical views.

We have already shown* how the discovery of telescopic vision gave to the eye—the organ of the sensuous contempla-

* See *Cosmos*, vol. i., p. 83.

tion of the universe—a power from whose limits we are still far removed, and which, in its first feeble beginning, when scarcely magnifying thirty-two linear diameters,* was yet enabled to penetrate into depths of space which until then had remained closed to the eyes of man.   The exact knowledge of many of the heavenly bodies which belong to our solar system, the eternal laws which regulate their revolution in their orbits, and the more perfect insight into the true structure of the universe, are the characteristics of the age which I am here delineating.   The results produced by this epoch determine the principal outlines of the great natural picture of the Cosmos, and add to the earlier investigated contents of terrestrial space the newly-acquired knowledge of the contents of the celestial regions, at least with reference to the well-organized arrangement of one planetary group.   In my desire of assuming only general views, I will confine myself to the consideration of the most important objects of the astronomical labors of the seventeenth century.   I would here refer to their influence in powerfully inciting to great and unexpected mathematical discoveries, and to more comprehensive and grander views of the universe.

I have already remarked that the age of Columbus, Gama, and Magellan—the age of great maritime enterprises—coincided in a most wonderful manner with many great events, with the awakening of a feeling of religious freedom, with the development of nobler sentiments for art, and with the diffusion of the Copernican views regarding the system of the universe.   Nicolaus Copernicus (who, in two letters still extant, calls himself Koppernik) had already attained his twenty-first year, and was engaged in making observations with the astronomer Albert Brudzewski, at Cracow, when Columbus discovered America.   Hardly a year after the death of the great discoverer, and after a six years' residence at Padua, Bologna, and Rome, we find him returned to Cracow, and busily engaged in bringing about a thorough revolution in the astronomical views of the universe.   By the favor of his uncle, Lucas Waisselrode of Allen, bishop of Ermland, he was nominated, in 1510, canon of Frauenburg, where he labored

---

* " The telescopes which Galileo constructed, and others of which he made use for observing Jupiter's satellites, the phases of Venus, and the solar spots, possessed the gradually increasing powers of magnifying four, seven, and thirty-two linear diameters, but they never had a higher power."   (Arago, in the *Annuaire du Bureau des Longitudes pour l'an.* 1842, p. 268.)

for thirty-three years on the completion of his work, entitled *De Revolutionibus Orbium Cœlestium.** The first printed copy was brought to him when, shattered in mind and body, he was preparing himself for death. He saw it and touched it, but his thoughts were no longer fixed on earthly things, and he died—not, as Gassendi says, a few hours, but several days afterward (on the 24th of May, 1543†). Two years

* Westphal, in his *Biographie des Copernicus* (1822, s. 33), dedicated to the great astronomer of Königsberg, Bessel, calls the Bishop of Ermland Lucas Watzelrodt von Allen, as does also Gassendi. According to explanations which I have very recently obtained, through the kindness of the learned historian of Prussia, Voigt, director of the Archives, "the family of the mother of Copernicus is called in original documents Weiselrodt, Weisselrot, Weisselrodt, and most commonly Waisselrode. His mother was undoubtedly of German descent, and the family of Waisselrode, who were originally distinct from that of Von Allen, which had flourished at Thorn from the beginning of the 15th century, probably took the latter name in addition to their own, through adoption, or from family connections." Sniadecki and Czynski (*Kopernik et ses Travaux*, 1847, p. 26) call the mother of the great Copernicus Barbara Wasselrode, and state that she was married at Thorn, in 1464, to his father, whose family they believe to be of Bohemian origin. The name of the astronomer, which Gassendi writes Tornæus Borussus, Westphal and Czynksi write Kopernik, and Krzyzianowski, Kopirnig. In a letter of the Bishop of Ermland, Martin Cromer of Heilsberg, dated Nov. 21, 1580, it is said, " Cum Jo. (Nicolaus) Copernicus vivens ornamento fuerit, atque etiam nunc post fata sit, non solum huic ecclesiæ, verùm etiam toti Prussiæ patriæ suæ, iniquam esse puto, eum post obitum carere honor esepulchri sive monumenti."

† Thus Gassendi, in *Nicolai Copernici Vita,* appended to his biography of Tycho (*Tychonis Brahei Vita,* 1655, Hagæ Comitum, p. 320): " eodem die et horis non multis priusquam animam efflaret." It is only Schubert, in his *Astronomy,* th. i., s. 115, and Robert Small, in the very learned *Account of the Astronomical Discoveries of Kepler,* 1804, p. 92, who maintain that Copernicus died "a few days after the appearance of his work." This is also the opinion of Voigt, the director of the Archives at Königsberg; because, in a letter which George Donner, canon of Ermland, wrote to the Duke of Prussia shortly after the death of Copernicus, it is said that "the estimable and worthy Doctor Nicolaus Koppernick sent forth his work, like the sweet song of the swan, a short time before his departure from this life of sorrows." According to the ordinarily received opinion (Westphal, *Nikolaus Kopernikus,* 1822, s. 73 und s. 82), the work was begun in 1507, and was so far completed in 1530 that only a few corrections were subsequently added. The publication was hastened by a letter from Cardinal Schonberg, written from Rome in 1536. The cardinal wishes to have the manuscript copied and sent to him by Theodor von Reden. We learn from Copernicus himself, in his dedication to Pope Paul III., that the performance of the work has lingered on into the *quartum novennium.* If we remember how much time was required for printing a work of 400 pages, and that the great man died in May, 1543, it may be conjectured that the dedication was not written in the last-named year; which, reckoning backward thirty-six years, would not give us a later, but an earlier year

earlier an important part of his theory had been made known by the publication of a letter of one of his most zealous pupils and adherents, Joachim Rhæticus to Johann Schoner, professor at Nuremberg. It was not, however, the propagation of the Copernican doctrines, the renewed opinion of the existence of one central sun, and of the diurnal and annual movement of the earth, which somewhat more than half a century after its first promulgation led to the brilliant astronomical discoveries that characterize the commencement of the seventeenth century ; for these discoveries were the result of the accidental invention of the telescope, and were the means of at once perfecting and extending the doctrine of Copernicus. Confirmed and extended by the results of physical astronomy (by the discovery of the satellite-system of Jupiter and the phases of Venus), the fundamental views of Copernicus have indicated to theoretical astronomy paths which could not fail to lead to sure results, and to the solution of problems which of necessity demanded, and led to a greater degree of perfection in the analytic calculus. While George Peuerbach and Regiomontanus (Johann Müller, of Königsberg, in Franconia) exercised a beneficial influence on Copernicus and his pupils Rhæticus, Reinhold, and Möstlin, these, in their turn, influenced in a like manner, although at longer intervals of time, the works of Kepler, Galileo, and Newton. These are the ideal links which connect the sixteenth and seventeenth centuries ; and we can not delineate the extended astronomical views of the latter of these epochs without taking into consideration the incitements yielded to it by the former.

An erroneous opinion unfortunately prevails, even in the present day,* that Copernicus, from timidity and from apprehension of priestly persecution, advanced his views regarding the planetary movement of the earth, and the position of the sun in the center of the planetary system, as mere hypotheses, which fulfilled the object of submitting the orbits of the heavenly bodies more conveniently to calculation, "but which need

than 1507. Herr Voigt doubts whether the aqueduct and hydraulic works at Frauenburg, generally ascribed to Copernicus, were really executed in accordance with his designs. He finds that, so late as 1571, a contract was concluded between the Chapter and the "skillful master Valentine Lendel, manager of the water-works at Breslau," to bring the water to Frauenburg, from the mill-ponds to the houses of the canons. Nothing is said of any previous water-works, and those which exist at present can not have been commenced until twenty-eight years after the death of Copernicus.

* Delambre, *Histoire De l'Astronomie Moderne*, t. i., p. 140.

not necessarily either be true or even probable." These singular words certainly do occur in the anonymous preface* attached to the work of Copernicus, and inscribed *De Hypothesibus hujus Operis*, but they are quite contrary to the opinions expressed by Copernicus, and in direct contradiction with his dedication to Pope Paul III. The author of these prefatory remarks was, as Gassendi most expressly says, in his Life of the great astronomer, a mathematician then living at Nuremberg, and named Andreas Osiander, who, together with Scho-

---

* " Neque enim necesse est, eas hypotheses esse veras, imo ne verisimiles quidem, sed sufficit hoc unum, si calculum observationibus congruentem exhibeant," says the preface of Osiander. " The Bishop of Culm, Tidemann Gise, a native of Dantzic, who had for years urged Copernicus to publish his work, at last received the manuscript, with the permission of having it printed fully in accordance with his own free pleasure. He sent it first to Rhæticus, professor at Wittenberg, who had, until recently, been living for a long time with his teacher at Frauenburg. Rhæticus considered Nuremberg as the most suitable place for its publication, and intrusted the superintendence of the printing to Professor Schoner and to Andreas Osiander." (Gassendi, *Vita Copernici*, p. 319.) The expressions of praise pronounced on the work at the close of the preface might be sufficient to show, without the express testimony of Gassendi, that the preface was by another hand. Osiander has used an expression on the title of the first edition (that of Nuremberg, 1543) which is always carefully avoided in all the writings of Copernicus, " motus stellarum novis insuper ac admirabilibus hypothesibus ornati," together with the very ungentle addition, " Igitur studiose lector, eme, lege, fruere." In the second Basle edition of 1566, which I have very carefully compared with the first Nuremberg edition, there is no longer any reference in the title of the book to the "admirable hypothesis;" but Osiander's *Præfatiuncula de Hypothesibus hujus Operis*," as Gassendi calls the intercalated preface, is preserved. That Osiander, without naming himself, meant to show that the *Præfatiuncula* was by a different hand from the work itself, appears very evident, from the circumstance of his designating the dedication to Paul III. as the *Præfatio Authoris*." The first edition has only 196 leaves; the second 213, on account of the *Narratio Prima* of the astronomer George Joachim Rhæticus, and a letter addressed to Schoner, which, as I have remarked in the text, was printed in 1541 by the intervention of the mathematician Gassarus of Basle, and gave to the learned world the first accurate knowledge of the Copernican system. Rhæticus had resigned his professional chair at Wittenberg, in order that he might enjoy the instructions of Copernicus at Frauenburg itself. (Compare, on these subjects, Gassendi, p. 310–319.) The explanation of what Osiander was induced to add from timidity is given by Gassendi: "Andreas porro Osiander fuit, qui non modo operarum inspector (the superintendent of the printing) fuit, sed Præfatiunculam quoque ad lectorem (tacito licet nomine) de Hypothesibus operis adhibuit. Ejus in ea consilium fuit, ut, tametsi Copernicus Motum Terræ habuisset, non solum pro Hypothesi, sed pro vero etiam placito, ipse tamen ad rem, ob illos, qui hinc offenderentur, leniendam, excusatum eum faceret, quasi talem motum non pro dogmate, sed pro Hypothesi mera assumpsisset."

ner, superintended the printing of the work *De Revolutionibus*, and who, although he makes no express declaration of any religious scruples, appears nevertheless to have thought it expedient to speak of the new views as of an hypothesis, and not, like Copernicus, as of demonstrated truth.

The founder of our present system of the universe (for to him incontestably belong the most important parts of it, and the grandest features of the design) was almost more distinguished, if possible, by the intrepidity and confidence with which he expressed his opinions, than for the knowledge to which they owed their origin. He deserves to a high degree the fine eulogium passed upon him by Kepler, who, in the introduction to the Rudolphine Tables, says of him, "*Vir fuit maximo ingenio et quod in hoc exercitio* (combating prejudices) *magni momenti est, animo liber.*" When Copernicus is describing, in his dedication to the pope, the origin of his work, he does not scruple to term the opinion generally expressed among theologians of the immobility and central position of the earth "an absurd acroama," and to attack the stupidity of those who adhere to so erroneous a doctrine. "If even," he writes, "any empty-headed babblers (ματαιολόγοι), ignorant of all mathematical science, should take upon themselves to pronounce judgment on his work through an intentional distortion of any passage in the Holy Scriptures (*propter aliquem locum scripturæ male ad suum propositum detortum*), he should despise so presumptuous an attack. It was, indeed, universally known that the celebrated Lactantius, who, however, could not be reckoned among mathematicians, had spoken childishly (*pueriliter*) of the form of the earth, deriding those who held it to be spherical. On mathematical subjects one should write only to mathematicians. In order to show that, deeply penetrated with the truth of his own deductions, he had no cause to fear the judgment that might be passed upon him, he turned his prayers from a remote corner of the earth to the head of the Church, begging that he would protect him from the assaults of calumny, since the Church itself would derive advantage from his investigations on the length of the year and the movements of the moon." Astrology and improvements in the calendar long procured protection for astronomy from the secular and ecclesiastical powers, as chemistry and botany were long esteemed as purely subservient auxiliaries to the science of medicine.

The strong and free expressions employed by Copernicus sufficiently refute the old opinion that he advanced the sys-

tem which bears his immortal name as an hypothesis convenient for making astronomical calculations, and one which
might be devoid of foundation. "By no other arrangement,"
he exclaims with enthusiasm, "have I been able to find so admirable a symmetry of the universe, and so harmonious a connection of orbits, as by placing the lamp of the world (*lucernam mundi*), the Sun, in the midst of the beautiful temple of
nature as on a kingly throne, ruling the whole family of circling stars that revolve around him (*circumagentem gubernans
astrorum familiam*)."* Even the idea of universal gravitation or attraction (*appetentia quædam naturalis partibus in-
dita*) toward the sun as the center of the world (*centrum
mundi*), and which is inferred from the force of gravity in
spherical bodies, seems to have hovered before the mind of
this great man, as is proved by a remarkable passage in the
9th chapter of the 1st book *De Revolutionibus*.†

* Quis enim in hoc pulcherrimo templo lampadem hanc in alio vel
meliori loco poneret, quam unde totum simul possit illuminare? Siqui-
dem non inepte quidam lucernam mundi, alii mentem, alii rectorem
vocant. Trismegistus visibilem Deum, Sophoclis Electra intuentem
omnia. Ita profecto tanquam in solio regali Sol residens circumagentem gubernat Astrorum familiam : Tellus quoque minime fraudatur lu-
nari ministerio, sed ut Aristoteles de animalibus ait, maximam Luna
cum terra cognationem habet. Concepit interea a Sole terra, et im-
pregnatur annuo partu. Invenimus igitur sub hac ordinatione admi-
randam mundi symmetriam ac certum harmoniæ nexum motus et magnitudinis orbium; qualis alio modo reperiri non potest. (Nicol. Copern.,
De Revol. Orbium Cælestium, lib. i., cap. 10, p. 9, b.) In this passage,
which is not devoid of poetic grace and elevation of expression, we recognize, as in all the works of the astronomers of the seventeenth century, traces of long acquaintance with the beauties of classical antiquity.
Copernicus had in his mind Cic., *Somn. Scip.*, c. 4 ; Plin., ii., 4 ; and
Mercur. Trismeg., lib. v. (ed. Cracov., 1586), p. 195 and 201. The allusion to the *Electra* of Sophocles is obscure, as the sun is never any
where expressly termed "all-seeing," as in the *Iliad* and the *Odyssey*,
and also in the *Choephoræ* of Æschylus (v. 980), which Copernicus
would not probably have called *Electra*. According to Böckh's conjecture, the allusion is to be ascribed to an imperfect recollection of
verse 869 of the *Œdipus Coloneus* of Sophocles. It very singularly
happens that quite lately, in an otherwise instructive memoir (Czynski,
Kopernik et ses Travaux, 1847, p. 102), the *Electra* of the tragedian is
confounded with *electric currents*. The passage of Copernicus, quoted
above, is thus rendered : "If we take the sun for the torch of the universe, for its spirit and its guide—if Trismegistes call it a god, and if
Sophocles consider it to be an electrical power which animates and
contemplates all that is contained in creation—"

† Pluribus ergo existentibus centris, de centro quoque mundi non
temere quis dubitabit, an videlicet fuerit istud gravitatis terrenæ, an
aliud. Equidem existimo, *gravitatem* non aliud esse, quam appeten-
tiam quandam naturalem partibus inditam a divina providentia officis

On considering the different stages of the development of cosmical contemplation, we are able to trace from the earliest ages faint indications and presentiments of the attraction of masses and of centrifugal forces. Jacobi, in his researches on the mathematical knowledge of the Greeks (unfortunately still in manuscript), justly comments on "the profound consideration of nature evinced by Anaxagoras, in whom we read with astonishment a passage asserting that the moon, if its centrifugal force ceased, would fall to the earth like a stone from a sling."*

I have already, when speaking of aërolites, noticed similar expressions of the Clazomenian and of Diogenes of Apollonia on the "cessation of the rotatory force."† Plato truly had a clearer idea than Aristotle of the *attractive force* exercised by the earth's center on all heavy masses removed from it, for the Stagirite was indeed acquainted, like Hipparchus, with the acceleration of falling bodies, although he did not correctly understand the cause. In Plato, and according to Democritus, *attraction* is limited to bodies having an affinity for one an-

universorum, ut in unitatem integritatemque suam sese conferant in formam globi coëuntes. Quam affectionem credibile est etiam Soli, Lunæ, cæterisque errantium fulgoribus inesse, ut ejus efficacia in ea qua se repræsentant rotunditate permaneant, quom nihilominus multis modis suos efficiunt circuitus. Si igitur et terra faciat alios, utpote secundum centrum (mundi), necesse erit eos esse qui similiter extrinsecus in multis apparent, in quibus invenimus annuum circuitum. Ipse denique Sol medium mundi putabitur possidere, quæ omnia ratio ordinis, quo illa sibi invicem succedunt, et mundi totius harmonia nos docet, si modo rem ipsam ambobus (ut aiunt) oculis inspiciamus." (Copern., *De Revol. Orb. Cœl.*, lib. i., cap. 9, p. 7, b.)

* Plut., *De Facie in Orbe Lunæ*, p. 923. (Compare Ideler, *Meteorologia veterum Græcorum et Romanorum*, 1832, p. 6.) In the passage of Plutarch, Anaxagoras is not named; but that the latter applied the same theory of "falling where the force of rotation had been intermitted" to all (the material) celestial bodies, is shown in Diog. Laert., ii., 12, and by the many passages which I have collected (p. 122). Compare, also, Aristot., *De Cœlo*, ii., 1, p. 284, a. 24, Bekker, and a remarkable passage of Simplicius, p. 491, b., in the *Scholia*, according to the edition of the Berlin Academy, where the "non-falling of heavenly bodies" is noticed "when the rotatory force predominates over the actual falling force or downward attraction." With these ideas, which also partially belong to Empedocles and Democritus, as well as to Anaxagoras, may be connected the instance adduced by Simplicius (l. c.), "that water in a vial is not spilled when the movement of rotation is more rapid than the downward movement of the water," τῆς ἐπὶ τὸ κάτω τοῦ ὕδατος φαρᾶς.

† See *Cosmos*, vol. i., p. 134. (Compare Letronne, *Des Opinions Cosmographiques des Pères de l'Eglise*, in the *Revue des Deux Mondes*, 1834. *Cosmos*. t. i., p. 621.)

other, or, in other words, to those in which there exists a tendency of the *homogeneous* elementary substances to combine together.* John Philoponus, the Alexandrian, a pupil of Ammonius, the son of Hermias, who probably lived in the sixth century, was the first who ascribed the movement of the heavenly bodies to a primitive impulse, connecting with this idea that of the fall of bodies, or the tendency of all substances, whether heavy or light, to reach the ground.† The idea conceived by Copernicus, and more clearly expressed by Kepler, in his admirable work *De Stella Martis,* who even applied it to the ebb and flow of the ocean, received in 1666 and 1674 a new impulse and a more extended application through the sagacity of the ingenious Robert Hooke ;‡ Newton's theory of gravitation, which followed these earlier advances, presented the grand means of converting the whole of physical astronomy into a true *mechanism of the heavens.*§

Copernicus, as we find not only from his dedication to the pope, but also from several passages in the work itself, had a tolerable knowledge of the ideas entertained by the ancients of the structure of the universe. He, however, only names in the period anterior to Hipparchus, Hicetas (or, as he always calls him, Nicetas) of Syracuse, Philolaüs the Pythagorean, the Timæus of Plato, Ecphantus, Heraclides of Pontus, and the great geometrician Apollonius of Perga. Of the two mathematicians, Aristarchus of Samos and Seleucus of Babylon, whose systems came most nearly to his own, he mentions only the first, making no reference to the second.‖ It has

* See, regarding all that relates to the ideas of the ancients on attraction, gravity, and the fall of bodies, the passages collected with great industry and discrimination, by Th. Henri Martin, *Etudes sur le Timée de Platon,* 1841, t. ii., p. 272–280, and 341.

† Joh. Philoponus, *De Creatione Mundi,* lib. i., cap. 12.

‡ He subsequently relinquished the correct opinion (Brewster, *Martyrs of Science,* 1846, p. 211) ; but the opinion that there dwells in the central body of the planetary system—the sun—a power which governs the movements of the planets, and that this solar force decreases either as the squares of the distances or in direct ratio, was expressed by Kepler in the *Harmonices Mundi,* completed in 1618.

§ See *Cosmos,* vol. i., p. 48 and 63.

‖ See op. cit., p. 177. The scattered passages to be found in the work of Copernicus, relating to the ante-Hipparchian system of the structure of the universe, are, exclusive of the dedication, the following : lib. i., cap. 5 and 10 ; lib. v., cap. 1 and 3 (ed. princ., 1543, p. 3, b. ; 7, b. ; 8, b. ; 133, b. ; 141 and 141, b. ; 179 and 181, b.). Every where Copernicus shows a predilection for, and a very accurate acquaintance with, the views of the Pythagoreans, or, to speak less definitely, with those which were attributed to the most ancient among them. Thus,

often been asserted that he was not acquainted with the views of Aristarchus of Samos regarding the central sun and the condition of the earth as a planet, because the *Arenarius*, and all the other works of Archimedes, appeared only one year after his death, and a whole century after the invention of the art of printing ; but it is forgotten that Copernicus, in his dedication to Pope Paul III., quotes 'a long passage on Philolaüs, Ecphantus, and Heraclides of Pontus, from Plutarch's work on *The Opinions of Philosophers* (III., 13), and therefore that he might have read in the same work (II., 24) that Aristarchus of Samos regards the sun as one of the fixed stars.

for instance, he was acquainted, as may be seen by the beginning of the dedication, with the letter of Lysis to Hipparchus, which, indeed, shows that the Italian school, in its love of mystery, intended only to communicate its opinions to friends, " as had also at first been the purpose of Copernicus." The age in which Lysis lived is somewhat uncertain ; he is sometimes spoken of as an immediate disciple of Pythagoras himself ; sometimes, and with more probability, as a teacher of Epaminondas (Böckh, *Philolaos*, s. 8–15). The letter of Lysis to Hipparchus, an old Pythagorean, who had disclosed the secrets of the sect, is, like many similar writings, a forgery of later times. It had probably become known to Copernicus from the collection of Aldus Manutius, *Epistola diversorum Philosophorum* (Romæ, 1494), or from a Latin translation by Cardinal Bessarion (Venet., 1516). In the prohibition of Copernicus's work, *De Revolutionibus*, in the famous decree of the *Congregazione dell' Indice* of the 5th of March, 1616, the new system of the universe is expressly designated as " falsa illa *doctrina Pythagorica*, Divinæ Scripturæ omnino adversans." The important passage on Aristarchus of Samos, of which I have spoken in the text, occurs in the *Arenarius*, p. 449 of the Paris edition of Archimedes of 1615, by David Rivaltus. The editio princeps is the Basle edition of 1544, apud Jo. Hervagium. The passage in the *Arenarius* says, very distinctly, that " Aristarchus had confuted the astronomers who supposed the earth to be immovable in the center of the universe. The sun, which constituted this center, was immovable like the other stars, while the earth revolved round the sun." In the work of Copernicus, Aristarchus is twice named, p. 69, b., and 79, without any reference being made to his system. Ideler, in Wolf and Buttmann's *Museum der Alterthumswissenschaft* (bd. ii., 1808, s. 452), asks whether Copernicus was acquainted with Nicolaus de Cusa's work, *De Docta Ignorantia*. The first Paris edition was indeed published in 1514, and the expression "jam nobis manifestum est terram in veritate moveri," from a Platonizing cardinal, might certainly have made some impression on the Canon of Frauenburg (Whewell, *Philosophy of the Inductive Sciences*, vol. ii., p. 343) ; but a fragment of Cusa's writing, discovered very recently (1843) by Clemens in the library of the Hospital at Cues, proves sufficiently, as does the work *De Venatione Sapientiæ*, cap. 28, that Cusa imagined that the earth did not move round the sun, but that they moved together, though more slowly, "round the constantly changing pole of the universe." (Clemens, in *Giordano Bruno*, and *Nicol. von Cusa*, 1847, s. 97–100.)

Among all the opinions of the ancients, those which appeared to exercise the greatest influence on the direction and gradual development of the ideas of Copernicus are expressed, according to Gassendi, in a passage in the encyclopædic work of Martianus Mineus Capella, written in a half-barbarous language, and in the *System of the World* of Apollonius of Perga. According to the opinions described by Martianus Mineus of Madaura, and which have been very confidently ascribed, sometimes to the Egyptians, and sometimes to the Chaldeans,*

---

* See the profound treatment of this subject in Martin, *Etudes sur Timée*, t. ii., p. 111, *Cosmographie des Egyptiens*); and p. 129–133) '*Antécédents du Système de Copernic*). The assertion of this learned philologist, that the original system of Pythagoras differed from that of Philolaüs, and that it regarded the earth as fixed in the center of the universe, does not appear to me to be entirely conclusive (t. ii., p. 103 and 107). I would here explain myself more fully respecting the remarkable statement of Gassendi regarding the similarity of the systems of Tycho Brahe and Apollonius of Perga, to which I have referred in the text. We find the following passage in Gassendi's biographies : " Magnam imprimis rationem habuit Copernicus duarum opinionum affinium, quarum unam Martiano Capellæ, alteram Apollonio Pergaco attribuit. Apollonius solem delegit, circa quem, ut centrum, non modo Mercurius et Venus, verum etiam Mars, Jupiter, Saturnus suas obirent periodos, dum Sol interim, uti et Luna, circa Terram, ut circa centrum, quod foret Affixarum mundique centrum, moverentur ; quæ deinceps quoque opinio Tychonis propemodum fuit. Rationem autem magnam harum opinionum Copernicus habuit, quod utraque eximie Mercurii ac Veneris circuitiones repræsentaret, eximieque causam retrogradationum, directionum, stationum in iis apparentium exprimeret et posterior (Pergæi) quoque in tribus Planetis superioribus præstaret." (Gassendi, *Tychonis Brahei Vita*, p. 296.) My friend the astronomer Galle, to whom I applied for information, agrees with me in thinking that nothing could justify Gassendi's decided statement. " In the passages," he writes to me, " to which you refer in Ptolemy's *Almagest* (in the commencement of book xii.), and in the works of Copernicus (lib. v., cap. 3, p. 141, a. ; cap. 35, p. 179, a. and b. ; cap. 36, p. 181, b.), the only questions considered are the retrogressions and stationary conditions of the planets, in which Apollonius's assumption of their revolution round the sun is indeed referred to (and Copernicus himself mentions expressly the assumption of the earth's standing still), but it can not be determined when he became acquainted with what he supposes to have been derived from Apollonius. We can only, therefore, conjecture that he assumed, on some later authority, that Apollonius of Perga had constructed a system similar to that of Tycho, although I do not find, even in Copernicus, any clear exposition of such a system, or any reference to ancient passages in which it may be spoken of. If lib. xii. of the Almagest should be the only source from whence the complete Tychonic view is ascribed to Apollonius, we may consider that Gassendi has gone too far in his suppositions, and that the case is precisely the same as that of the phases of Mercury and Venus, of which Copernicus spoke (lib. i., cap. 10, p. 7, b., and 8, a.), without decidedly applying them to his system. Apollonius may, perhaps, in a similar manner, have treat

the earth is immovably fixed in a central point, while the sun revolves around it as a circling planet, attended by two satellites, Mercury and Venus. Such a view of the structure of the world might, indeed, prepare the way for that of the central force of the sun. There is, however, nothing in the Almagest, or in the works of the ancients generally, or in the work of Copernicus, *De Revolutionibus*, which justifies the assertion so confidently maintained by Gassendi, of the perfect resemblance existing between the system of Tycho Brahe and that which has been ascribed to Apollonius of Perga. After Böckh's complete investigation, nothing further need be said of the confusion of the Copernican system with that of the Pythagorean, Philolaüs, according to which, the non-rotating earth (the Antichthon or opposite earth, being not in itself a planet, but merely the opposite hemisphere of our planet) moves like the sun itself round the focus of the world—the central fire, or vital flame of the whole planetary system.

The scientific revolution originated by Nicolaus Copernicus has had the rare fortune (setting aside the temporary retrograde movement imparted by the hypothesis of Tycho Brahe) of advancing without interruption to its object—the discovery of the true structure of the universe. The rich abundance of accurate observations furnished by Tycho Brahe himself, the zealous opponent of the Copernican system, laid the foundation for the discovery of those eternal laws of the planetary movements which prepared imperishable renown for the name of Kepler, and which, interpreted by Newton, and proved to be theoretically and necessarily true, have been transferred into the bright and glorious domain of thought as *the intellectual recognition of nature*. It has been ingeniously said, although, perhaps, with too feeble an estimate of the free and independent spirit which created the theory of gravitation, that "Kepler wrote a code of laws, and Newton the spirit of those laws.*"

ed mathematically the assumption of the retrogressions of the planets under the idea of a revolution round the sun, without adding any thing definite and general as to the truth of this assumption. The difference of the Apollonian system, described by Gassendi, from that of Tycho, would only be, that the latter likewise explained the *inequalities* of the movements. The remark of Robert Small, that the idea which forms the basis of Tycho's system was by no means unfamiliar to the mind of Copernicus, but had rather served him as a point of transition to his own system, appears to me well founded."

* Schubert, *Astronomie*, th. i., s. 124. In the *Philosophy of the Inductive Sciences*, vol. ii., p. 282, Whewell, in his Inductive Table of Astronomy, has given an exceedingly good and complete view of the

The figurative and poetical myths of the Pythagorean and Platonic pictures of the universe, changeable as the fancy from which they emanated,* may still be traced partially reflected in Kepler ; but while they warmed and cheered his often saddened spirit, they never turned him aside from his earnest course, the goal of which he reached in the memorable night of the 15th of May, 1618, twelve years before his death.†
Copernicus had furnished a satisfactory explanation of the ap-

astronomical contemplation of the structure of the universe, from the earliest ages to Newton's system of gravitation.

* Plato, in the *Phædrus*, adopts the system of Philolaüs, but in the *Timæus*, that according to which the earth is immovable in the center, and which was subsequently called the Hipparchian or the Ptolemaic. (Böckh, *De Platonico systemate cœlestium globorum, et de vera indole astronomiæ Philolaicæ*, p. xxvi.–xxxii. ; the same author in the *Philolaos*, s. 104–108. Compare, also, Fries, *Geschichte der Philosophie*, bd. i., s. 325–347, with Martin's *Etudes sur Timée*, t. ii., p. 64–92.) The astronomical vision, in which the structure of the universe is shrouded, at the end of the *Book of the Republic*, reminds us at once of the intercalated spherical systems of the planets, and of the concord of tones, " the voices of the Syrens moving in concert with the revolving spheres." (See, on the discovery of the true system of the universe, the fine and comprehensive work of Apelt, *Epochen der Gesch. der Menscheit*, bd. i., 1845, s. 205–305, and 379–445.)

† Kepler, *Harmonices Mundi, libri quinque*, 1619, p. 189. " On the 8th of March, 1618, it occurred to Kepler, after many unsuccessful attempts, to compare the squares of the times of revolution of the planets with the cubes of the mean distances ; but he made an error in his calculations, and rejected this idea. On the 15th of May, 1618, he again reverted to it, and calculated correctly. The third law of Kepler was now discovered." This discovery, and those related to it, coincide with the unhappy period when this great man, who had been exposed from early childhood to the hardest blows of fate, was striving to save from the torture and the stake his mother, who, at the age of seventy years, in a trial for witchcraft, which lasted six years, had been accused of poison-mixing, inability of shedding tears, and of sorcery. The suspicion was increased from the circumstance that her own son, the wicked Christopher Kepler, a worker in tin, was her accuser, and that she had been brought up by an aunt, who was burned at Weil as a witch. See an exceedingly interesting work, but little known in foreign countries, drawn from newly-discovered manuscripts by Baron von Breitschwert, entitled " *Johann Keppler's Leben und Wirken*," 1831, s. 12, 97–147, and 196. According to this work, Kepler, who in German letters always signed his name Keppler, was not born on the 21st of December, 1571, in the imperial town of Weil, as is usually supposed, but on the 27th of December, 1571, in the village of Magstadt, in Würtemberg. It is uncertain whether Copernicus was born on the 19th of January, 1472, or on the 19th of February, 1473, as Mostlin asserts, or (according to Czynski) on the 12th of February of the same year. The year of Columbus's birth was long undetermined within nineteen years. Ramusio places it in 1430, Bernaldez, the friend of the discoverer, in 1436 and the celebrated historian Muñoz in 1446.

parent revolution of the heaven of the fixed stars by the diurnal rotation of the earth round its axis ; and by its annual
movement round the sun he had afforded an equally perfect
solution of the most striking movements of the planets (their
stationary conditions and their retrogressions), and thus given
the true reason of the so-called *second inequality of the planets.*   The *first inequality,* or the unequal movement of the
planets in their orbits, he left unexplained.   True to the ancient Pythagorean principle of the perfectibility inherent in
circular movements, Copernicus thought that he required for
his structure of the universe some of the *epicycles* of Apollonius of Perga, besides the *eccentric* circles having a vacuum
in their center.   However bold was the path adventured on,
the human mind could not at once emancipate itself from all
earlier views.

The equal distance at which the stars remained, while the
whole vault of heaven seemed to move from east to west, had
led to the idea of a firmament and a solid crystal sphere, in
which Anaximenes (who was probably not much later than
Pythagoras) had conjectured that the stars were riveted like
nails.*   Geminus of Rhodes, the cotemporary of Cicero, doubt-
ed whether the constellations lay in one uniform plane, being
of opinion that some were higher and others lower than the
rest.   The idea formed of the heaven of the fixed stars was
extended to the planets, and thus arose the theory of the eccentric intercalated spheres of Eudoxus and Menæchmus, and
of Aristotle, who was the inventor of *retrograde* spheres.   The
theory of epicycles—a construction which adapted itself most
readily to the representation and calculation of the planetary
movements—was, a century afterward, made by the acute
mind of Apollonius to supersede solid spheres.   However much
I may incline to mere ideal abstraction, I here refrain from
attempting to decide historically whether, as Ideler believes,
it was not until after the establishment of the Alexandrian
Museum that " a free movement of the planets in space was
regarded as possible," or whether, before that period, the intercalated transparent spheres (of which there were twenty-
seven according to Eudoxus, and fifty-five according to Aristotle), as well as the epicycles which passed from Hipparchus
and Ptolemy to the Middle Ages, were regarded generally not

* Plat., *De plac. Philos.*, ii., 14; Aristot., *Meteorol.*, xi., 8; *De Cælo,*
ii., 8.   On the theory of spheres generally, and on the retrograding
spheres of Aristotle in particular, see Ideler's *Vorlesung. über Eudoxus,*
1828, s. 49–60.

as solid bodies of material thickness, but merely as ideal abstractions. It is more certain that in the middle of the sixteenth century, when the theory of the seventy-seven homocentric spheres of the learned writer, Girolamo Fracastoro, found general approval; and when, at a later period, the opponents of Copernicus sought all means of upholding the Ptolemaic system, the idea of the existence of *solid* spheres, circles, and epicycles, which was especially favored by the Fathers of the Church, was still very widely diffused. Tycho Brahe expressly boasts that his considerations on the orbits of comets first proved the impossibility of solid spheres, and thus destroyed the artificial fabrics. He filled the free space of heaven with air, and even believed that the resisting medium, when disturbed by the revolving heavenly bodies, might generate tones. The unimaginative Rothmann believed it necessary to refute this renewed Pythagorean myth of celestial harmony.

Kepler's great discovery that all the planets move round the sun in ellipses, and that the sun lies in one of the foci of these ellipses, at length freed the original Copernican system from eccentric circles and all epicycles.* The planetary structure of the world now appeared objectively, and as it were architecturally, in its simple grandeur; but it remained for Isaac Newton to disclose the play and connection of the internal forces which animate and preserve the system of the universe. We have already often remarked, in the history of the gradual development of human knowledge, that important but apparently accidental discoveries, and the simultaneous appearance of many great minds, are crowded together in a short period of time; and we find this phenomenon most strikingly manifested in the first ten years of the seventeenth century; for Tycho Brahe (the founder of modern astronomical calculations), Kepler, Galileo, and Lord Bacon, were cotemporaries. All these, with the exception of Tycho Brahe, were enabled, in the prime of life, to benefit by the labors of Descartes and Fermat. The elements of Bacon's *Instauratio Magna* appeared in the English language in 1605, fifteen years before

---

* A better insight into the free movement of bodies, and into the independence of the direction once given to the earth's axis, and into the rotatory and progressive movement of the terrestrial planet in its orbit, has freed the original system of Copernicus from the assumption of a *declination movement*, or a so-called third movement of the earth (*De Revolut. Orb. Cœl.*, lib. i., cap. 11, *triplex motus telluris*). The parallelism of the earth's axis is maintained in the annual revolution round the sun, in conformity with the law of inertia, without the application of a *correcting* epicycle.

the *Novum Organon.* The invention of the telescope, and the greatest discoveries in physical astronomy (viz., Jupiter's satellites, the sun's spots, the phases of Venus, and the remarkable form of Saturn), fall between the years 1609 and 1612. Kepler's speculations on the elliptic orbit of Mars* were began in 1601, and gave occasion, eight years after, to the completion of the work entitled *Astronomia nova seu Physica celestis.* " By the study of the orbit of Mars," writes Kepler, " we must either arrive at a knowledge of the secrets of astronomy, or forever remain ignorant of them. I have succeeded, by untiring and continued labor, in subjecting the inequalities of the movement of Mars to a natural law." The generalization of the same idea led the highly-gifted mind of Kepler to the great cosmical truths and presentiments which, ten years later, he published in his work entitled *Harmonices Mundi libri quinque.* " I believe," he well observes in a letter to the Danish astronomer Longomontanus, " that astronomy and physics are so intimately associated together, that neither can be perfected without the other." The results of his researches on the structure of the eye and the theory of vision appeared in 1604 in the *Paralipomena ad Vitellionem,* and in 1611† in the *Dioptrica.* Thus were the knowledge of the most important objects in the perceptive world and in the regions of space, and the mode of apprehending these objects by means of new discoveries, alike rapidly increased in the short period of the first ten or twelve years of a century which began with Galileo and Kepler, and closed with Newton and Leibnitz.

The accidental discovery of the power of the telescope to penetrate through space originated in Holland, probably in the closing part of the year 1608. From the most recent investigations it would appear that this great discovery may be claimed by Hans Lippershey, a native of Wesel and a spectacle maker at Middleburg ; by Jacob Adriaansz, surnamed Metius, who is said also to have made burning glasses of ice ; and by Zacharias Jansen.‡ The first-named is always called

* Delambre, *Hist. de l'Astronomie Ancienne,* t. ii., p. 381.

† See Sir David Brewster's judgment on Kepler's optical works, in the " *Martyrs of Science,*" 1846, p. 179–182. (Compare Wilde, *Gesch. der Optik,* 1838, th. i., s. 182–210.) If the law of the refraction of the rays of light belong to Willebrord Snellius, professor at Leyden (1626), who left it behind him buried in his papers, the publication of the law in a trigonometrical form was, on the other hand, first made by Descartes. See Brewster, in the *North British Review,* vol. vii., p. 207 ; Wilde, *Gesch. der Optik,* th. i., s. 227.

‡ Compare two excellent treatises on the discovery of the telescope, by Professor Moll, of Utrecht, in the *Journal of the Royal Institution,*

Laprey in the important letter of the Dutch embassador Boreel to the physician Borelli, the author of the treatise *De vero*

1831, vol. i., p. 319 ; and by Wilde, of Berlin, in his *Gesch. der Optik*, 1838, th. i., s. 138–172. The work referred to, and written in the Dutch language, is entitled *" Geschiedkundig Onderzoek naar de eerste Uitfinders der Vernkykers, uit de Aunekenningen van wyle den Hoogl. van Swinden zamengesteld door*, G. Moll," Amsterdam, 1831. Albers has given an extract from this interesting treatise in Schumacher's *Jahrbuch für* 1843, s. 56–65. The optical instruments with which Jansen furnished Prince Maurice of Nassau, and the Archduke Albert (the latter gave his to Cornelius Drebbel), were (as is shown by the letter of the embassador Boreel, who, when a child, had been often in the house of Jansen, the spectacle maker, and who subsequently saw the instruments in the shop) microscopes eighteen inches in length, " through which small objects were wonderfully magnified when one looked down at them from above." The confusion between the microscope and the telescope has rendered the history of the invention of both instruments obscure. The letter of Boreel (Paris, 1655), above alluded to, notwithstanding the authority of Tiraboschi, renders it improbable that the first invention of the compound microscope belonged to Galileo. Compare, on this obscure history of optical instruments, Vicenzio Antinori, in the *Saggi di Naturali Esperienze fatte nell' Accademia del Cimento*, 1841, p. 22–26. Even Huygens, who was born scarcely twenty-five years after the conjectural date of the invention of the telescope, does not venture to decide with certainty on the name of the first inventor (*Opera Reliqua*, 1728, vol. ii., p. 125). According to the researches made in public archives by Van Swiden and Mole, Lippershey was not only in possession of a telescope made by himself as early as the 2d of October, 1608, but the French embassador at the Hague, President Jeannin, wrote, on the 28th of December of the same year, to Sully, " that he was in treaty with the Middleburg spectacle maker for a telescope, which he wished to send to the king, Henry IV." Simon Marius (Mayor of Genzenhausen, one of the discoverers of Jupiter's satellites) even relates that a telescope was offered for sale in the autumn of 1608, at Frankfort-on-Maine, by a Belgian, to his friend Fuchs of Bimbach, Privy Counselor of the Margrave of Ansbach. Telescopes were made in London in February, 1610, therefore a year after Galileo had completed his own. (Rigaud, *On Hariot's Papers*, 1833, p. 23, 26, and 46.) They were at first called *cylinders*. Porta, the inventor of the *camera obscura*, like Francastero, the cotemporary of Columbus, Copernicus and Cardanus, at earlier periods, had merely spoken of the possibility " of seeing all things larger and nearer" by means of convex and concave glasses being placed on each other (duo specilla ocularia alterum alteri superposita) ; but we can not ascribe the invention of the telescope to them (Tiraboschi, *Storia della Letter.*, ital., t. xi., p. 467 ; Wilde, *Gesch. der Optik*, th. i., s. 121). Spectacles had been known in Haarlem since the beginning of the fourteenth century; and an epitaph in the church of Maria Maggiore, at Florence, names Salvino degli Armati, who died in 1317, as the inventor (inventore degli occhiali). Some apparently authentic notices of the use of spectacles by aged persons are to be met with as early as 1299 and 1305. The passages of Roger Bacon refer to the magnifying power of spherical segments of glass. See Wilde, *Gesch. der Optik*, th. i., s. 93–96 ; and *ante*, p. 245.

telescopii inventore (1655). If the claim of priority be determined by the periods at which offers were made to the General States, the honor belongs to Hans Lippershey ; for, on the 2d of October, 1608, he offered to the government three instruments " by which one might see objects at a distance." The offer of Metius was made on the 17th of October of the same year ; but he expressly says " that he has already, for two years, constructed similar instruments, through industry and thought." Zacharias Jansen (who, like Lippershey, was a spectacle maker at Middleburg) invented, in conjunction with his father Hans Jansen, toward the end of the sixteenth century, and probably after 1590, the *compound microscope*, the eye-piece of which is a concave lens ; but, as we learn from the embassador Boreel, it was not until 1610 that he discovered the telescope, which he and his friends directed to distant terrestrial, but not toward celestial objects. The influence which has been exercised by the microscope in giving us a more profound knowledge of the conformation and movement of the separate parts of all organic bodies, and by the telescope in suddenly opening to us the regions of space, has been so immeasurably great, that it seems requisite to enter somewhat circumstantially into the history of these discoveries.

When, in May, 1609, the news of the discovery made in Holland of telescopic vision reached Venice, Galileo, who was accidentally there, conjectured at once what must be the essential points in the construction of a telescope, and immediately completed one for himself at Padua.* This instrument

---

* The above-named physician and mathematician of the Margravate of Ansbach, Simon Marius, after receiving a description of the action of a Dutch telescope, is likewise believed to have constructed one himself as early as the year 1608. On Galileo's earliest observation of the mountainous regions in the moon, to which I have referred in the text, compare Nelli, *Vita di Galilei*, vol. i., p. 200–206 ; Galilei, *Opere*, 1744, t. ii., p. 60, 403, and *Lettera al Padre Cristoforo Grienberger, in materia delle Montuosità della Luna*, p. 409–424. Galileo found in the moon some circular districts, surrounded on all sides by mountains similar to the form of Bohemia. " Eundem facit aspectum Lunæ locus quidam, ac faceret in terris regio consimilis Boemiæ, si montibus altissimis, inque peripheriam perfecti circuli dispositis occluderetur undique" (t. ii., p. 8). The measurements of the mountains were made by the method of the tangents of the solar ray. Galileo, as Helvetius did still later, measured the distance of the summit of the mountains from the boundary of the illuminated portion, at the moment when the mountain summit was first struck by the solar ray. I find no observation of the lengths of the shadows of the mountains. He found the summits "in circa miglia quattro" in height, and " much higher than the mountains

he first directed toward the mountainous parts of the moon, and showed how their summits might be measured, while he, like Leonardo da Vinci and Möstlin, ascribed the ash-colored light of the moon to the reflection of solar light from the earth to the moon. He observed with low magnifying powers the group of the Pleiades, the starry cluster in Cancer, the Milky Way, and the group of stars in the head of Orion. Then followed, in quick succession, the great discoveries of the four satellites of Jupiter, the two handles of Saturn (his indistinctly-seen rings, the form of which was not recognized), the solar spots, and crescent shape of Venus.

The moons of Jupiter, the first of all the secondary planets discovered by the telescope, were first seen, almost simultaneously and wholly independently, on the 29th of December, 1609, by Simon Marius at Ansbach, and on the 7th of January, 1610, by Galileo at Padua. In the publication of this discovery, Galileo, by the *Nuncius Siderius* (1610), preceded the *Mundus Jovialis* (1614) of Simon Marius,* who had

on our earth." The comparison is remarkable, since, according to Riccioli, very exaggerated ideas of the height of our mountains were then entertained, and one of the principal or most celebrated of these elevations, the Peak of Teneriffe, was first measured trigonometrically, with some degree of exactness, by Feuillée, in 1724. Galileo, like all other observers up to the close of the eighteenth century, believed in the existence of many seas and of a lunar atmosphere.

* I here again find occasion (*Cosmos*, vol. i., p. 185) to refer to the proposition laid down by Arago: "The only rational and just method of writing the history of science is to base it exclusively on works, the date of whose publication is certain. All beyond this must be confused and obscure." The singularly-delayed publication of the *Fränkische Kalender* or *Practica* (1612), and of the astronomically important memoir entitled "*Mundus Jovialis anno* 1609 *detectus ope perspicilli Belgici* (February, 1614)," may indeed have given occasion to the suspicion that Marius had drawn his materials from the *Nuncius Sidereus* of Galileo, the dedication of which is dated March, 1610, or even from earlier manuscript communications. Galileo, irritated by the still remembered lawsuit against Balthasar Capra, a pupil of Marius, calls him the usurper of the system of Jupiter, "Usurpatore del sistema di Giove," and he even accuses the heretical Protestant astronomer of Gunzenhausen of having founded his apparently earlier observation on a confusion between the calendars. "Tace il Mario di far cauto il lettore, come essendo egli separato della chiesa nostra, ne avendo accettato l'emendatione Gregoriana, il giorno 7 di gennaio del 1610, di noi Cattolici (the day on which Galileo discovered the satellites) è l'istesso, che il di 28 di Decembre del 1609, di loro eretici, e questa è tutta la precedenza delle sue finte osservationi" (Venturi, *Memoire è Lettere di G. Galilei*, 1818, Part i., p. 279 ; and Delambre, *Hist. de l'Astr. Mod.*, t. i., p. 696). According to a letter written by Galileo in 1614 to the *Accademia di Lincei*, it would appear that he attempted, somewhat unphilosophically, to direct his complaint against Marius to the Marchese

proposed to give to Jupiter's satellites the names of *Sidera Brandenburgica*, while Galileo preferred the names *Sidera Cosmica* or *Medicea*, of which the latter found most approval at the court of Florence. This collective appellation did not satisfy the yearnings of flattery. Instead of designating the satellites by numbers, as we do at present, Marius had named them Io, Europa, Ganymede, and Callisto; but for these mythological designations Galileo's nomenclature substituted the family names of the ruling house of Medici— Catharina, Maria, Cosimo the elder, and Cosimo the younger.

The knowledge of Jupiter's satellite-system, and of the phases of Venus, has exercised the most marked influence on the establishment and general diffusion of the Copernican system. The little world of Jupiter (*Mundus Jovialis*) presented to the intellectual contemplation of men a perfect image of the large planetary and solar systems. It was recognized that the secondary planets obeyed the laws discovered by Kepler; and it was now first observed that the squares of their

di Brandeburgo. On the whole, however, Galileo continued well disposed toward the German astronomers. He writes, in March, 1611, "Gli ingegni singolari, che in gran numero fioriscono nell' Alemagna, mi hanno lungo tempo tenuto in desiderio di vederla" (*Opere*, t. ii., p. 44). It has always appeared very remarkable to me, that if Kepler, in a conversation with Marius, was playfully adduced as a sponsor for these mythological designations of Io and Callisto, there should be no mention of his countryman either in the Commentary published in Prague, in April, 1610, to the *Nuncius Siderius, nuper ad mortales a Galilæo missus,* or in his letters to Galileo, or in those addressed to the Emperor Rudolph in the autumn of the same year; but that, on the contrary, Kepler should every where speak of "the glorious discovery of the Medicean stars by Galileo." In publishing his own observations on the satellites, from the 4th to the 9th of September, 1610, he gives to a little memoir which appeared at Frankfort in 1611, the title, "*Kepleri Narratio de observatis a se quatuor Jovis satellitibus erronibus quos Galilæus Mathematicus Florentinus jure inventionis Medicea Sidera nuncupavit.*" A letter from Prague, October 25, 1610, addressed to Galileo, concludes with the words "neminem habes, quem metuas amulum." Compare Venturi, Part i., p. 100, 117, 139, 144, and 149. Misled by a mistake, and after a very careless examination of the valuable manuscripts preserved at Petworth, the seat of Lord Egremont, Baron von Zach asserted that the distinguished astronomer and Virginian traveler, Thomas Hariot, had discovered the satellites of Jupiter simultaneously with, or even earlier than Galileo. A more careful examination of Hariot's manuscripts, by Rigaud, has shown that his observations began, not on the 16th of January, but only on the 17th of October, 1610, nine months after Galileo and Marius. (Compare Zach, *Corr. Astron.*, vol. vii., p. 105. Rigaud, *Account of Harriot's Astron. Papers*, Oxf., 1833, p. 37; Brewster, *Martyrs of Science*, 1846, p. 32.) The earliest original observations of Jupiter's satellites made by Galileo and his pupil Renieri were only discovered two years ago.

periodic times were as the cubes of the mean distances of the
satellites from the primary planets.   It was this which led
Kepler, in the *Harmonices Mundi*, to state, with the firm
confidence and security of a German spirit of philosophical
independence, to those whose opinions bore sway beyond the
Alps ; "eighty years have elapsed,* during which the doctrines
of Copernicus, regarding the movement of the earth, and the
immobility of the sun, have been promulgated without hinderance, because it is deemed allowable to dispute concerning
natural things, and to elucidate the works of God ; and now
that *new testimony is discovered in proof of the truth of those
doctrines*—testimony which was not known to the spiritual
judges—ye would prohibit the promulgation of the true system of the structure of the universe !"   Such a prohibition—
a consequence of the old contest between natural science and
the Church—Kepler had early encountered in Protestant Germany.†

The discovery of Jupiter's satellites marks an ever-memo-
rable epoch in the history and the vicissitudes of astronomy.‡
The occultations of the satellites, or their entrance into Jupiter's
shadow, led to a knowledge of the *velocity of light* (1675),
and, through this knowledge, to the explanation of the *aberration-ellipse* of the fixed stars (1727), in which the great orbit
of the earth, in its annual course round the sun, is, as it were,
reflected on the vault of heaven.   These discoveries of Römer
and Bradley have been justly termed "the keystone of the
Copernican system," the perceptible evidence of the translatory motion of the earth.

Galileo had also early perceived (September, 1612) the importance of the occultations of Jupiter's satellites for geographical determinations of longitude on land.   He proposed this
method, first to the Spanish court in 1616, and afterward to
the States-General of Holland, with a view of its being applied to nautical purposes,§ little aware, as it would appear,

* It should be seventy-three years ; for the prohibition of the Copernican system by the Congregation of the *Index* was promulgated on
the 5th of March, 1616.

† Freiherr von Breitschwert, *Keppler's Leben*, s. 36.

‡ Sir John Herschel, *Astron.*, s. 465.

§ Galilei, *Opere*, t. ii. (*Longitudine per via de' Pianeti Medicei*), p.
435–506 ; Nelli, *Vita*, vol. ii., p. 656–688 ; Venturi, *Memorie e Lettere
di G. Galilei*, Part i., p. 177.   As early as 1612, or scarcely two years
after the discovery of Jupiter's satellites, Galileo boasted, somewhat
prematurely indeed, of having completed tables of those secondary satellites "to within 1′ of time."   A long diplomatic correspondence was
carried on with the Spanish embassador in 1616, and with the Dutch

of the insuperable difficulties presented to its practical application on the unstable element.   He wished to go himself, or
to send his son Vicenzio, to Spain, with a hundred telescopes,
which he would prepare.   He required as a recompense " una
croce di San Jago," and an annual payment of 4000 scudi, a
small sum, he says, considering that hopes had been given to
him, in the house of Cardinal Borgia, of receiving 6000 ducats
annually.

The discovery of the secondary planets of Jupiter was soon
followed by the observations of the so-called triple form of
Saturn as a *planeta tergeminus*.   As early as November,
1610, Galileo informed Kepler that " Saturn consisted of three
stars, which were in mutual contact with one another."   In
this observation lay the germ of the discovery of Saturn's ring.
Hevelius, in 1656, described the variations in its form, the unequal opening of the handles (ansæ), and their occasional total
disappearance.   The merit of having given a scientific explanation of all the phenomena of Saturn's ring belongs, however, to the acute observer Huygens, who, in 1655, in accordance with the suspicious custom of the age, and like Galileo,
concealed his discovery in an anagram of eighty-eight letters.
Dominicus Cassini was the first who observed the black stripe
on the ring, and in 1684 he recognized that it is divided into
at least two concentric rings.   I have here collected together
what has been learned during a century regarding the most
wonderful and least anticipated of all the forms occurring in
the heavenly regions—a form which has led to ingenious conjectures regarding the original mode of formation of the secondary and primary planets.

embassador in 1636, but without leading to the desired object.   The
telescopes were to magnify from forty to fifty times.   In order more
easily to find the satellites when the ship is in motion, and (as he believed) to keep them in the field, he invented, in 1617 (Nelli, vol. ii.,
p. 663), the binocular telescope, which has generally been ascribed to
the Capucine monk Schyrleus de Rheita, who had much experience in
optical matters, and who endeavored to construct telescopes magnifying
four thousand times.   Galileo made experiments with his binocular
(which he also called a *celatone* or *testiera*) in the harbor of Leghorn,
while the ship was violently moved by a strong wind.   He also caused
a contrivance to be prepared in the arsenal at Pisa, by which the observer of the satellites might be protected from all motion, by seating
himself in a kind of boat, floating in another boat filled with water or
with oil (*Lettera al Picchena de'* 22 *Marzo*, 1617 ; Nelli, *Vita*, vol. i., p.
281 ; Galilei, *Opere*, t. ii., p. 473 ; *Lettera a Lorenzo Realio del* 5 *Giug-
no*, 1637).   The proof which Galileo (*Opere*, t. ii., p. 454) brought forward of the advantage to the naval service of his method over Morin's
method of lunar distances is very striking.

The *spots upon the sun* were first observed through telescopes by Johann Fabricius of East Friesland, and by Galileo
(at Padua or Venice, as is asserted). In the publication of
the discovery, in June, 1611, Fabricius incontestably preceded
Galileo by one year, since his first letter to the burgomaster,
Marcus Welser, is dated the 4th of May, 1612. The earliest
observations of Fabricius were made, according to Arago's
careful researches, in March, 1611,* and, according to Sir
David Brewster, even as early as toward the close of the year
1610 ; while Christopher Scheiner did not carry his own observations back to an earlier period than April, 1611, and it
is probable that he did not seriously occupy himself with the
solar spots until October of the same year. Concerning Galileo we possess only very obscure and discrepant data on this
subject. It is probable that he recognized the solar spots in
April, 1611, for he showed them publicly at Rome in Cardinal Bandini's garden on the Quirinal, in the months of April
and May of that year. Hariot, to whom Baron Zach ascribes
the discovery of the sun's spots (16th of January, 1610), certainly saw three of them on the 8th of December, 1610, and
noted them down in a register of observations ; but he was
ignorant that they were solar spots ; thus, too, Flamstead, on
the 23d of December, 1690, and Tobias Mayer, on the 25th
of September, 1756, did not recognize Uranus as a planet
when it passed across the field of their telescope. Hariot first
observed the solar spots on the 1st of December, 1611, five
months, therefore, after Fabricius had published his discovery.
Galileo had made the observation that the solar spots, " many
of which are larger than the Mediterranean, or even than
Africa and Asia," form a definite zone on the sun's disk. He
occasionally noticed the same spots return, and he was convinced that they belonged to the sun itself. Their differences
of dimension in the center of the sun, and when they disappeared on the sun's edge, especially attracted his attention,

---

* See Arago, in the *Annuaire* for 1842, p. 460–476 (*Découvertes des
taches Solaires et de la Rotation du Soleil*). Brewster (*Martyrs of
Science*, p. 36 and 39) places the first observation of Galileo in October
or November, 1610. Compare Nelli, *Vita*, vol. i., p. 324–384 ; Galilei,
Opere, t. i., p. lix. ; t. ii., p. 85–200 ; t. iv., p. 53. On Harriot's observations, see Rigaud, p. 32 and 38. The Jesuit Scheiner, who was sum-
moned from Gratz to Rome, has been accused of striving to revenge
himself on Galileo, on account of the literary contest regarding the discovery of the solar spots, by getting it whispered to Pope Urban VIII.,
through another Jesuit, Grassi, that he (the pope), in the *Dialoghi delle
Scienze Nuove*, was represented as the foolish and ignorant Simplicio.
(Nelli, vol. ii., p. 515)

but still I find nothing in his second remarkable letter of the 14th of August, 1612, to Marcus Welser, that would indicate his having observed an inequality in the ash-colored margin on both sides of the black nucleus when approaching the sun's edge (Alexander Wilson's accurate observation in 1773). The Canon Tarde in 1620, and Malapertus in 1633, ascribed all obscurations of the sun to small cosmical bodies revolving around it and intercepting its light, and named the Bourbon and Austrian stars* (*Borbonia et Austriaca Sidera*). Fabricius recognized, like Galileo, that the spots belonged to the sun itself;† he also noticed that the spots he had seen vanish all reappear ; and the observation of these phenomena taught him the rotation of the sun, which had already been conjectured by Kepler before the discovery of the solar spots. The most accurate determinations of the period of rotation were, however, made in 1630, by the diligent Scheiner. Since the strongest light ever produced by man, Drummond's incandescent lime-ball, appears inky black when thrown on the sun's disk, we can not wonder that Galileo, who undoubtedly first described the great solar *faculæ*, should have regarded the light of the nucleus of the sun's spots as more intense than that of the full moon, or the atmosphere near the sun's disk.‡ Fanciful conjectures regarding the many envelopes of air, clouds, and light, which surround the black, earth-like nucleus of the sun, may be found in the writings of Cardinal Nicholas of Cusa as early as the middle of the fifteenth century.§

To close our consideration of the cycle of remarkable discoveries, which scarcely comprised two years, and in which the great and undying name of the Florentine shines pre-eminent, it still remains for us to notice the observation of the phases of Venus. In February, 1610, Galileo observed the crescentic form of this planet, and on the 11th of December, 1610, in accordance with a practice already alluded to, he concealed this important discovery in an anagram, of which Kepler makes mention in the preface to his *Dioptrica*. We learn

---

* Delambre, *Hist. de l'Astronomie Moderne*, t. i., p. 690.

† The same opinion is expressed in Galileo's Letters to Prince Cesi (May 25, 1612); Venturi, Part i., p. 172.

‡ See some ingenious and interesting considerations on this subject by Arago, in the *Annuaire pour l'an* 1842, p. 481–488. Sir John Herschel, in his *Astronomy*, § 334, speaks of the experiments with Drummond's light projected on the sun's disk.

§ *Giordano Bruno und Nic. von Cusa verglichen*, von J. Clemens, 1847, s. 101. On the phases of Venus, see Galilei, *Opere*, t. ii., p. 53, and Nelli, *Vita*, vol. i., p. 213–215.

also, from a letter of his to Benedetto Castelli (30th of December, 1610), that he believed, notwithstanding the low magnifying power of his telescope, that he could recognize changes in the illumined disk of Mars. The discovery of the moonlike or crescent shape of Venus was the triumph of the Copernican system. The founder of that system could scarcely fail to recognize the necessity of the existence of these phases; and we find that he discusses circumstantially, in the tenth chapter of his first book, the doubts which the more modern adherents of the Platonic opinions advance against the Ptolemaic system on account of these phases. But, in the development of his own system, he does not speak expressly of the phases of Venus, as is stated by Thomas Smith in his *Optics*.

The enlargement of cosmical knowledge, whose description can not, unhappily, be wholly separated from unpleasant dissensions regarding the right of priority to discoveries, excited, like all that refers to physical astronomy, more general attention, from the fact that several great discoveries in the heavens had aroused the attention of the public mass at the respective periods of thirty-six, eight, and four years prior to the invention of the telescope in 1608, viz., the sudden apparition and disappearance of three new stars, one in Cassiopeia in 1572, another in the constellation of the Swan in 1600, and the third in the foot of Ophiuchus in 1604. All these stars were brighter than those of the first magnitude, and the one observed by Kepler in the Swan continued to shine in the heavens for twenty-one years, throughout the whole period of Galileo's discoveries. Three centuries and a half have now nearly passed since then, but no new star of the first or second magnitude has appeared; for the remarkable event witnessed by Sir John Herschel in the southern hemisphere (in 1837)* was a great increase in the intensity of the light of a long-known star of the second magnitude (η Argo), which had not until then been recognized as variable. The writings of Kepler, and our own experience of the effect produced by the appearance of comets visible to the naked eye, will teach us to understand how powerfully the appearance of new stars, between the years 1572 and 1604, must have arrested attention, increased the general interest in astronomical discoveries, and excited the minds of men to the combination of imaginative conjectures. Thus, too, terrestrial natural events, as earthquakes in regions where they have been but seldom experienced; the eruption of volcanoes that had long remained inactive; the

* Compare *Cosmos*, vol. i., p. 153 and 353.

sounds of aërolites traversing our atmosphere and becoming ignited within its confines, impart a new stimulus, for a certain time, to the general interest in problems, which appear to the people at large even more mysterious than to the dogmatizing physicist.

My reason for more particularly naming Kepler in these remarks on the influence of direct sensuous contemplation has been to point out how, in this great and highly-gifted man, a taste for imaginative combinations was combined with a remarkable talent for observation, an earnest and severe method of induction, a courageous and almost unparalleled perseverance in calculation, and a mathematical profoundness of mind, which, revealed in his *Stereometria Doliorum,* exercised a happy influence on Fermat, and, through him, on the invention of the theory of the infinitesimal calculus.* A man endowed with such a mind was pre-eminently qualified by the richness and mobility of his ideas,† and by the bold cosmical conjectures which he advanced, to animate and augment the movement which led the seventeenth century uninterruptedly forward to the exalted object presented in an extended contemplation of the universe.

The many comets visible to the naked eye from 1577 to the appearance of Halley's comet in 1607 (eight in number), and the sudden apparition already alluded to of three stars almost at the same period, gave rise to speculations on the origin of these heavenly bodies from a cosmical vapor filling the regions of space. Kepler, like Tycho Brahe, believed that the new stars had been conglomerated from this vapor, and that they were again dissolved in it.‡ Comets to which,

---

* Laplace says of Kepler's theory of the measurement of casks (*Stereometria Doliorum*), 1615, " which, like the sand-reckoning of Archimedes, develops elevated ideas on a subject of little importance;" " Kepler présente dans cet ouvrage des vues sur l'infini qui ont influé sur la révolution que la Géométrie a éprouvée à la fin du 17ᵐᵉ siècle; et Fermat, que l'on doit regarder comme le véritable inventeur du calcul différentiel, a fondé sur elles sa belle méthode *de maximis et minimis.* (*Précis de l'Hist. de l'Astronomie,* 1821, p. 95.)" On the geometrical power manifested by Kepler in the five books of his *Harmonices Mundi,* see Chasles, *Aperçu Hist. des Méthodes en Géométrie,* 1837, p. 482–487.

† Sir David Brewster elegantly remarks, in the account of Kepler's method of investigating truth, that " the influence of imagination as an instrument of research has been much overlooked by those who have ventured to give laws to philosophy. This faculty is of greatest value in physical inquiries; if we use it as a guide and confide in its indications, it will infallibly deceive us; but if we employ it as an auxiliary, it will afford us the most invaluable aid" (*Martyrs of Science,* p. 215).

‡ Arago, in the *Annuaire,* 1842, p. 434 (*De la Transformation des*

before the discovery of the elliptic orbit of the planets, he ascribed a rectilinear and not a closed revolving course, were
regarded by him, in 1608, in his "new and singular discourse
on the hairy stars," as having originated from "celestial air."
He even added, in accordance with ancient fancies on *spontaneous generation*, that comets arise "as an herb springs from
the earth without seed, and as fishes are formed in the sea by
a *generatio spontanea*."

Happier in his other cosmical conjectures, Kepler hazarded
the following propositions : that all the fixed stars are suns
like our own luminary, and surrounded by planetary systems ;
that our sun is enveloped in an atmosphere which appears
like a white *corona* of light during a total solar eclipse ; that
our sun is so situated in the great cosmical island as to constitute the center of the compressed stellar ring of the Milky
Way ;* that the sun itself, whose spots had not then been
discovered, together with all the planets and fixed stars, rotates
on its axis ; that satellites, like those discovered by Galileo
round Jupiter, will also be discovered round Saturn and Mars ;
and that in the much too great interval of space between
Mars and Jupiter,† where we are now acquainted with seven
asteroids (as between Venus and Mercury), there revolve
planets which, from their smallness alone, are invisible to the
naked eye.   Presentient propositions of this nature, felicitous
conjectures of that which was subsequently discovered, excited general interest, while none of Kepler's cotemporaries, including Galileo, conferred any adequate praise on the discovery of the three laws, which, since Newton and the promul-

---

Nébuleuses et de la Matière diffuse en Etoiles).   Compare *Cosmos*, vol.
i., p. 144 and 152.
  * Compare the ideas of Sir John Herschel on the position of our
planetary system, vol. i., p. 141 ; also Struve, *Etudes d'Astronomie Stel-
laire*, 1847, p. 4.
  † Apelt says (*Epochen der Geschichte der Menschheit*, bd. i., 1845, s.
223): "the remarkable law of the distances, which is usually known
under the name of Bode's law (or that of Titius), is the discovery of
Kepler, who, after many years of persevering industry, deduced it from
the observations of Tychq de Brahe."   See *Harmonices Mundi libri
quinque*, cap. 3.   Compare, also, Cournot's Additions to his French
translation of Sir John Herschel's *Astronomy*, 1834, § 434, p. 324, and
Fries, *Vorlesungen über die Sternkunde*, 1813, s. 325 (On the Law of
the Distances in the Secondary Planets).   The passages from Plato,
Pliny, Censorinus, and Achilles Tatius, in the Prolegomena to the
Aratus, are carefully collected in Fries, *Geschichte der Philosophie*, bd.
i., 1837, s. 146–150 ; in Martin, *Etudes sur le Timée*, t. ii., p. 38 ; and in
Brandis, *Geschichte der Griechisch-Römischen Philosophie*, th. ii., abth.
i., 1844, s. 364.

gation of the theory of gravitation, have immortalized the name of Kepler.* Cosmical considerations, even when based merely on feeble analogies and not on actual observations, riveted the attention more powerfully then, as they still frequently do, than the most important results of *calculating astronomy.*

After having described the important discoveries which in so small a cycle of years extended the knowledge of the regions of space, it still remains for me to revert to the advances in physical astronomy which characterize the latter half of this great century. The improvement in the construction of telescopes led to the discovery of Saturn's satellites. Huygens, on the 25th of March, 1655, forty-five years after the discovery of Jupiter's satellites, discovered the sixth of these bodies through an object-glass which he had himself polished. Owing to a prejudice, which he shared with other astronomers of his time, that the number of the secondary planetary bodies could not exceed that of the primary planets,† he did not seek to discover other satellites of Saturn. Dominicus Cassini discovered four of these bodies, the Sidera Lodivicea, viz., the seventh and outermost in 1671, which exhibits great alternation of light, the fifth in 1672, and the fourth and third in 1684, through Campani's object-glass, having a focal length of 100–136 feet; the two innermost, the first and second, were discovered more than a century later (1788 and 1789) by William Herschel, through his colossal telescope. The last-named of these satellites presents the remarkable phenomenon of accomplishing its revolution round the primary planet in less than one day.

Soon after Huygens's discovery of a satellite of Saturn, Childrey first observed the zodiacal light, between the years 1658 and 1661, although its relations in space were not determined until 1683 by Dominicus Cassini. The latter did not regard it as a portion of the sun's atmosphere, but believed, with Schubert, Laplace, and Poisson, that it was a detached revolving nebulous ring.‡ Next to the recognition of the existence of secondary planets, and of the free and concentrically divided rings of Saturn, the conjecture of the probable existence of the nebulous zodiacal light belongs incontestably to the grandest enlargement of our views regarding the planetary system, which had previously appeared so sim-

* Delambre, *Hist. de l'Astronomie Moderne*, t. i., p. 360.

† Arago, in the *Annuaire* for 1842, p. 560–564; also *Cosmos*, vol. i., p. 97.   ‡ Compare *Cosmos*, vol. i., p. 137–144.

ple. In our own time, the intersecting orbits of the small planets between Mars and Jupiter, the interior comets, which were first proved to be such by Encke, and the swarms of falling stars associated with definite days (since we can not regard these bodies in any other light than as such cosmical masses moving with planetary velocity), have enriched our views of the universe with a remarkable abundance of new objects.

During the age of Kepler and Galileo, our ideas were very considerably enlarged regarding the contents of the regions of space, or, in other words, the distribution of all created matter beyond the outermost circle of the planetary bodies, and beyond the orbit of any comet. In the same period in which (1572–1604) three new stars of the first magnitude suddenly appeared in Cassiopeia, Cygnus, and Ophiuchus, David Fabricius, pastor at Ostell, in East Friesland (the father of the discoverer of the sun's spots), in 1596, and Johann Bayer, at Augsburg, in 1603, observed in the neck of the constellation Cetus another star, which again disappeared, whose changing brightness was first recognized by Johann Phoçylides Holwarda, professor at Franeker (in 1638 and 1639), as we learn from a treatise of Arago, which has thrown much light on the history of astronomical discoveries.* The phenomenon was not singular in its occurrence, for, during the last half of the seventeenth century, variable stars were periodically observed in the head of Medusa, in Hydra, and in Cygnus. The manner in which accurate observations of the alternations of light in Algol are able to lead directly to a determination of the velocity of the light of this star, has been ably shown by the treatise to which I have alluded, and which was published in 1842.

The use of the telescope now excited astronomers to the

---

* *Annuaire du Bureau des Longitudes pour l'an* 1842, p. 312–353 (*Etoiles Changeantes ou Périodiques*). In the seventeenth century there were recognized, as variable stars, besides Mira Ceti (Holwarda, 1638), a Hydræ (Montanari, 1672), β Persei or Algol, and χ Cygni (Kirch, 1686). On what Galileo calls nebulæ, see his *Opere*, t. ii., p. 15, and Nelli, *Vita*, vol. ii., p. 208. Huygens, in the *Systema Saturninum*, refers most distinctly to the nebula in the sword of Orion, in saying of nebulæ generally, " Cui certe simile aliud nusquam apud reliquas fixas potui animadvertere. Nam ceteræ nebulosæ olim existimatæ atque ipsa via lactea, perspicillis inspectæ, nullas nebulas habere comperiuntur, neque aliud esse quam plurium stellarum congeries et frequentia." It is seen from this passage that the nebula in Andromeda, which was first described by Marius, had not been attentively considered by Huygens any more than by Galileo.

earnest observation of a class of phenomena, some of which could not even escape the naked eye. Simon Marius described in 1612 the nebula in Andromeda, and Huygens, in 1656, drew the figure of that in the stars of the sword of Orion. Both nebulæ might serve as types of a more or less advanced condensation of nebulous cosmical matter. Marius, when he compared the nebula in Andromeda to " a wax taper seen through a semi-transparent medium," indicated very forcibly the difference between nebulæ generally and the stellar masses and groups in the Pleïades and in Cancer, examined by Galileo. As early as the sixteenth century, Spanish and Portuguese sea-farers, without the aid of telescopic vision, had noticed with admiration the two Magellanic clouds of light revolving round the south pole, of which one, as we have observed, was known as " the white spot" or " white ox" of the Persian astronomer Abdurrahman Sufi, who lived in the middle of the tenth century. Galileo, in the *Nuncius Siderius*, uses the terms " *stellæ nebulosæ*" and " *nebulosæ*" to designate clusters of stars, which, as he expresses it, like *areolæ sparsim per æthera subfulgent*. As he did not bestow any especial attention on the nebula in Andromeda, which, although visible to the naked eye, had not hitherto revealed any star under the highest magnifying powers, he regarded all nebulous appearances, all his *nebulosæ*, and the Milky Way itself, as luminous masses formed of closely-compressed stars. He did not distinguish between the nebula and star, as Huygens did in the case of the nebulous spot of Orion. These are the feeble beginnings of the great works on *Nebulæ*, which have so honorably occupied the first astronomers of our own time in both hemispheres.

Although the seventeenth century owes its principal splendor at its beginning to the sudden enlargement afforded to the knowledge of the heavens, imparted by the labors of Galileo and Kepler, and at its close to the advance in mathematical science, due to Newton and Leibnitz, yet the greater number of the physical problems which occupy us in the present day likewise experienced beneficial consideration in the same century. In order not to depart from the character peculiarly appropriate to a history of the contemplation of the universe, I limit myself to a mere enumeration of the works which have exercised direct and special influence on general, or, in other words, on cosmical views of nature. With reference to the processes of light, heat, and magnetism, I would first name Huygens, Galileo, and Gilbert. While Huygens was occupied with the double refraction of light in crystals of Iceland spar, *i. e.*, with the separation of the pencils of light into two parts, he also discovered, in 1678, that kind of polarization of light which bears his name. The discovery of this isolated phenomenon, which was not published till 1690, and, consequently, only five years before the death of Huygens, was followed, after the lapse of more than a century, by the great discoveries of Malus, Arago, Fresnel, Brewster, and Biot.* Malus, in 1808, discovered polarization by reflection from polished surfaces, and Arago, in 1811, made the discovery of colored polarization. A world of wonder, composed of manifold modified waves of light, having new properties, was now revealed. A ray of light, which reaches our eyes, after traversing millions of miles, from the remotest regions of heaven, announces of itself, in Arago's polariscope, whether it is reflected or refracted, whether it emanates from a solid, or fluid, or gaseous body ; announcing even the degree of its intensity.† By pursuing this course, which leads us back through Huygens to the seventeenth century, we are instructed concerning the constitution of the solar body and its envelopes ; the reflected or the proper light of cometary tails and the zodiacal light ; the optical properties of our atmosphere ; and the position of the four neutral points of polarization‡ which Arago, Babinet, and Brewster discovered. Thus does man create new organs, which, when skillfully employed, reveal to him new views of the universe.

Next to polarization I should name the *interference* of light, the most striking of all optical phenomena, faint traces of which were also observed in the seventeenth century—by Grimaldi in 1665, and by Hooke, although without a proper understanding of its original and causal conditions.§ Modern times owe the discovery of these conditions, and the clear insight into the laws, according to which, (unpolarized) rays of light, emanating from one and the same source, but with a different length of path, destroy one another and produce darkness, to the successful penetration of Thomas Young. The laws of the in-

---

* On the important law discovered by Brewster, of the connection between the angle of complete polarization and the index of refraction, see *Philosophical Transactions of the Royal Society for the Year* 1815, p. 125–159.   † See *Cosmos*, vol. i., p. 39 and 52.

‡ Sir David Brewster, in Berghaus and Johnson's *Physical Atlas*, 1847, Part vii., p. 5 (*Polarization of the Atmosphere*).

§ On Grimaldi's and Hooke's attempt to explain the polarization of soap-bubbles by the interference of the rays of light, see Arago, in the *Annuaire* for 1831, p. 164 (Brewster's *Life of Newton*, p. 53).

terference of polarized light were discovered in 1816 by Arago and Fresnel. The theory of undulations advanced by Huygens and Hooke, and defended by Leonhard Euler, was at length established on a firm and secure basis.

Although the latter half of the seventeenth century acquired distinction from the attainment of a successful insight into the nature of double refraction, by which optical science was so much enlarged, its greatest splendor was derived from Newton's experimental researches, and Olaus Römer's discovery, in 1675, of the measurable *velocity of light*. Half a century afterward, in 1728, this discovery enabled Bradley to regard the variation he had observed in the apparent place of the stars as a conjoined consequence of the movement of the earth in its orbit, and of the propagation of light. Newton's splendid work on *Optics* did not appear in English till 1704, having been deferred, from personal considerations, till two years after Hooke's death ; but it would seem a well-attested fact that, even before the years 1666 and 1667,* he was in possession of the principal points of his optical researches, his theory of gravitation and differential calculus (method of fluxions).

In order not to sever the links which hold together the general primitive phenomena of matter in one common bond, I would here immediately, after my succinct notice of the optical discoveries of Huygens, Grimaldi, and Newton, pass to

---

* Brewster, *The Life of Sir Isaac Newton*, p. 17. The date of the year 1665 has been adopted for that of the invention of the method of fluxions, which, according to the official explanations of the Committee of the Royal Society of London, April 24, 1712, is " one and the same with the differential method, excepting the name and mode of notation." With reference to the whole unhappy contest on the subject of priority with Leibnitz, in which, strange to say, accusations against Newton's orthodoxy were even advanced, see Brewster, p. 189–218. The fact that all colors are contained in white light was already maintained by De la Chambre, in his work entitled *"La Lumière"* (Paris, 1657), and by Isaac Vossius (who was afterward a canon at Windsor), in a remarkable memoir entitled *" De Lucis Natura et Proprietate"* (Amstelod., 1662), for the knowledge of which I was indebted, two years ago, to M. Arago, at Paris. Brandis treats of this memoir in the new edition of Gehler's *Physikalische Wörterbuch*, bd. iv. (1827), s. 43, and Wilke notices it very fully in his *Gesch. der Optik*, th. i. (1838), s. 223, 228, and 317. Isaac Vossius, however, considered the fundamental substance of all colors (cap. 25, p. 60) to be sulphur, which forms, according to him, a component part of all bodies. In *Vossii Responsum ad Objecta, Joh. de Bruyn, Professoris Trajectini, et Petri Petiti*, 1663, it is said, p. 69, Nec lumen ullum est absque calore, nec calor ullus absque lumine. Lux sonus, anima (!) odor, vis magnetica, quamvis incorporea, sunt tamen aliquid. (De Lucis Nat., cap. 13, p. 29.)

the consideration of terrestrial magnetism and atmospheric temperature, as far as these sciences are included in the century which we have attempted to describe. The able and important work on magnetic and electric forces, the *Physiologia nova de Magnete*, by William Gilbert, to which I have frequently had occasion to allude,* appeared in the year 1600. This writer, whose sagacity of mind was so highly admired by Galileo, conjectured many things of which we have now acquired certain knowledge.† Gilbert regarded terrestrial magnetism and electricity as two emanations of a single fundamental force pervading all matter, and he therefore treated of both at once. Such obscure conjectures, based on analogies of the effect of the Heraclean magnetic stone on iron, and the attractive force exercised on dry straws by amber, when animated, as Pliny expresses it, with a soul by the agency of heat and friction, appertain to all ages and all races, to the Ionic natural philosophy no less than to the science of the Chinese physicists.‡ According to Gilbert's idea, the earth itself is a magnet, while he considered that the inflections of the lines of equal declination and inclination depend upon the distribution of mass, the configuration of continents, or the form and extent of the deep, intervening oceanic basins. It is difficult to connect the periodic variations which characterize the three principal forms of magnetic phenomena (the isoclinal, isogonic, and isodynamic lines) with this rigid system of the distribution of force and mass, unless we represent to ourselves the attractive force of the material particles modified by similar periodic changes of temperature in the interior of the terrestrial planet.

In Gilbert's theory, as in gravitation, the quantity of the material particles is merely estimated, without regard to the specific heterogeneity of substances. This circumstance gave his work, at the time of Galileo and Kepler, a character of cosmical greatness. The unexpected discovery of rotation-magnetism by Arago in 1825, has shown practically that every kind of matter is susceptible of magnetism ; and the most recent investigations of Faraday on dia-magnetic substances

---

* *Cosmos*, vol. i., p. 177, 179, and vol. ii., p. 278.

† Lord Bacon, whose comprehensive, and, generally speaking, free and methodical views, were unfortunately accompanied by very limited mathematical and physical knowledge, even for the age in which he lived, was very unjust to Gilbert. " Bacon showed his inferior aptitude for physical research in rejecting the Copernican doctrine which William Gilbert adopted" (Whewell, *Philosophy of the Inductive Sciences*, vol. ii., p. 378).   ‡ *Cosmos*, vol. i., p. 188.

have, under especial conditions of meridian or equatorial direction, and of solid, fluid, or gaseous inactive conditions of the bodies, confirmed this important result.   Gilbert had so clear an idea of the force imparted by telluric magnetism, that he ascribed the magnetic condition of iron rods on crosses of old church towers to this action of the Earth.*

The increased enterprise and activity of navigation to the higher latitudes, and the improvement of magnetic instruments, to which had been added, since 1576, the dipping needle (inclinatorium), constructed by Robert Norman, of Ratcliff, were the means, during the course of the seventeenth century, of extending the general knowledge of the periodical advance of a portion of the magnetic curves or lines of no variation.   The position of the magnetic equator, which was believed to be identical with the geographical equator, remained uninvestigated.   Observations of inclination were only carried on in a few of the capital cities of Western and Southern Europe.   Graham, it is true, attempted in London, in 1723, to measure, by the oscillations of a magnetic needle, the intensity of the magnetic terrestrial force, which varies both with space and time ; but, since Borda's fruitless attempt on his last voyage to the Canaries in 1776, Lemanon was the first who succeeded, in La Perouse's expedition in 1785, in comparing the intensity in different regions of the earth.

In the year 1683, Edmund Halley sketched his theory of four magnetic poles or points of convergence, and of the periodical movement of the magnetic line without declination, basing his theory on a large number of existing observations of declination of very unequal value, by Baffin, Hudson, James Hall, and Schouten.   In order to test this theory, and render it more perfect by the aid of new and more exact observations, the English government permitted him to make three voyages (1698–1702) in the Atlantic Ocean, in a vessel under his own command.   In one of these he reached 52° S. lat.   This expedition constituted an epoch in the history of telluric magnetism.   Its result was the construction of a general variation chart, on which the points at which navigators had found an equal amount of variation were connected together by curved

* The first observation of the kind was made (1590) on the tower of the church of the Augustines at Mantua.   Grimaldi and Gassendi were acquainted with similar instances, all occurring in geographical latitudes where the inclination of the magnetic needle is very considerable. On the first measurements of magnetic intensity by the oscillation of a needle, compare my *Relation Hist.*, t. i., p. 260–264, and *Cosmos*, vol. i., p. 186, 187.

lines. Never before, I believe, had any government fitted out a naval expedition for an object whose attainment promised such advantages to practical navigation, while, at the same time, it deserved to be regarded as peculiarly scientific and physico-mathematical.

As no phenomenon can be thoroughly investigated by a careful observer, without being considered in its relation to other phenomena, Halley, on his return from his voyage, hazarded the conjecture that the northern light was of a magnetic origin. I have remarked, in the general picture of nature, that Faraday's brilliant discovery (the evolution of light by magnetic force) has raised this hypothesis, enounced as early as in the year 1714, to empirical certainty.

But if the laws of terrestrial magnetism are to be thoroughly investigated—that is to say, if they are to be sought in the great cycle of the periodic movement in space of the three varieties of magnetic curves, it is by no means sufficient that the diurnal regular or disturbed course of the needle should be observed at the magnetic stations which, since 1828, have begun to cover a considerable portion of the earth's surface, both in northern and southern latitudes ;* but four times in every century an expedition of three ships should be sent out, to examine, as nearly as possible at the same time, the state of the magnetism of the Earth, so far as it can be investigated in those parts which are covered by the ocean. The magnetic equator, or the curve at which the inclination is null, must not merely be inferred from the geographical position of its nodes (the intersections with the geographical equator), but the course of the ship should be made continually to vary according to the observations of inclination, so as never to leave the track of the magnetic equator for the time being. Land expeditions should be combined with these voyages, in order, where masses of land can not be entirely traversed, to determine at what points of the coast-line the magnetic curves (especially those having no variation) enter. Special attention might also, perhaps, be deservedly directed to the movement and gradual changes in the oval configuration and almost concentric curves of variation of the two isolated closed systems in Eastern Asia, and in the South Pacific in the meridian of the Marquesas Group.† Since the memorable Antarctic expedition of Sir James Clark Ross (1839–1843), fitted out with admirable instruments, has thrown so much light over the polar regions of the southern hemisphere, and has determ-

* *Cosmos,* vol. i., p. 190–192.　　　　† *Cosmos,* vol. i., p. 182.

ined empirically the position of the magnetic south pole ; and since my honored friend, the great mathematician, Frederic Gauss, has succeeded in establishing the first general theory of terrestrial magnetism, we need not renounce the hope that the many requirements of science and navigation will lead to the realization of the plan I have already proposed.   May the year 1850 be marked as the first normal epoch in which the materials for a magnetic chart shall be collected ; and may permanent scientific institutions (academies) impose upon themselves the practice of reminding, every twenty-five or thirty years, governments favorable to the advance of navigation, of the importance of an undertaking whose great cosmical importance depends on its long-continued repetition.

The invention of instruments for measuring temperature (Galileo's thermoscopes of 1593 and 1602,* depending simultaneously on the changes in the temperature and the external pressure of the atmosphere) gave origin to the idea of determining the modifications of the atmosphere by a series of connected and successive observations.   We learn from the *Diario dell' Accademia del Cimento,* which exercised so happy an influence on the taste for experiments, conducted in a regular and systematic method during the brief term of its activity, that observations of the temperature were made with spirit thermometers similar to our own at a great number of stations, among others at Florence, in the Convent Degli Angeli, in the plains of Lombardy, on the mountains near Pistoja, and even in the elevated plain of Innspruck, as early as 1641, and five times daily.†   The Grand-duke Ferdinand II. employed the monks in many of the monasteries of his states to perform this task.‡   The temperature of mineral springs was also determined at that period, and thus gave occasion to many ques-

* On the oldest thermometers, see Nelli, *Vita e Commercio Letterario di Galilei* (Losanna, 1793), vol. i., p. 68–94 ; *Opere di Galilei* (Padovo, 1744), t. i., p. lv. ; Libri, *Histoire des Sciences Mathématiques en Italie,* t. iv. (1841), p. 185–197.   As evidences of first comparative observations on temperature, we may instance the letters of Gianfrancesco Sagredo and Benedetto Castelli in 1613, 1615, and 1633, given in Venturi, *Memorie e Lettere inedite di Galilei,* Part i., 1818, p. 20.

† Vincenzio Antinori, in the *Saggi di Naturali Esperienze, fatte nell' Accademia del Cimento,* 1841, p. 30–44.

‡ On the determination of the thermometric scale of the Accademia del Cimento, and on the meteorological observations continued for sixteen years by a pupil of Galileo, Father Raineri, see Libri, in the *Annales de Chimie et de Physique,* t. xlv., 1830, p. 354 ; and a more recent similar work by Schouw, in his *Tableau du Climat et de la Végétation de l'Italie,* 1839, p. 99–106.

tions regarding the temperature of the Earth.   As all natural
phenomena––all the changes to which terrestrial matter is
subject––are connected with modifications of heat, light, and
electricity, whether at rest or moving in currents, and as likewise the phenomena of temperature, acting by the force of
expansion, are most easily discernible by the sensuous perceptions, the invention and improvement of thermometers must
necessarily, as I have already elsewhere observed, indicate a
great epoch in the general progress of natural science.   The
range of the applicability of the thermometer, and the rational
deductions to be arrived at from its indications, are as immeasurable as the sphere of those natural forces which exercise
their dominion over the atmosphere, the solid portions of the
earth, and the superimposed strata of the ocean––alike over
inorganic substances, and the chemical and vital processes of
organic matter.

The action of radiating heat was likewise investigated, a
century before the important labors of Scheele, by the Florentine members of the Accademia del Cimento, by remarkable
experiments with concave mirrors, against which non-lumin-
ous heated bodies, and masses of ice weighing 500 lbs., actually and *apparently* radiated.*   Mariotte, at the close of
the seventeenth century, entered into investigations regarding
the relations of radiating heat in its passage through glass
plates.   It has seemed necessary to allude to these isolated
experiments, since in more recent times the doctrine of the
radiation of heat has thrown great light on the cooling of the
ground, the formation of dew, and many general climatic
modifications, and has led, moreover, through Melloni's admirable ·sagacity, to the contrasting diathermism of rock salt
and alum.

To the investigations on the changes in the temperature of
the atmosphere, depending on the geographical latitude, the
seasons of the year, and the elevation of the spot, were soon
added other inquiries into the variation of pressure and the
quantity of vapor in the atmosphere, and the often-observed
periodic results, known as the *law of rotation* of the winds.
Galileo's correct views respecting the pressure of the atmosphere led Torricelli, a year after the death of his great teacher,
to the construction of the barometer.   It would appear that
the fact that the column of mercury in the Torricellian column
stood higher at the base of a tower or hill than at its summit,

* Antinori, *Saggi dell' Accad. del Cim.*, 1841, p. 114, and in the *Ag-
giunte* at the end of the book, p. lxxvi.

was first observed at Pisa by Claudio Beriguardi ;* and five years later in France, at the suggestion of Pascal, by Perrier, the brother-in-law of the latter, when he ascended the Puy de Dôme, which is nearly one thousand feet higher than Vesuvius. The idea of employing barometers for measuring elevations now presented itself readily ; it may, perhaps, have been suggested to Pascal in a letter of Descartes.† It is not necessary to enter into any especial explanation of the influence exercised on the enlargement of physical geography and meteorology by the barometer when used as a hypsometrical instrument in determining the local relations of the Earth's surface, and as a meteorological instrument in ascertaining the influence of atmospheric currents. The theory of the atmospheric currents already referred to was established on a solid foundation before the close of the seventeenth century. Bacon had the merit, in 1664, in his celebrated work entitled *Historia Naturalis et Experimentalis de Ventis*,‡ of considering the direction of the winds in their dependence on thermometric and hydrometric relations; but, unmathematically denying the correctness of the Copernican system, he conjectured the possibility " that our atmosphere may daily turn round the earth like the heavens, and thus occasion the tropical east wind."

Hooke's comprehensive genius here also diffused order and light.§ He recognized the influence of the rotation of the Earth, and the existence of the upper and lower currents of warm and cold air, which pass from the equator to the poles, and return from the poles to the equator. Galileo, in his last *Dialogo*, had indeed also regarded the trade winds as the consequence of the rotation of the Earth ; but he ascribed the detention of the particles of air within the tropics (when compared with the velocity of the Earth's rotation) to a vaporless purity of the air in the tropical regions.‖ Hooke's more cor-

---

* Antinori, p. 29.

† *Ren. Cartesii Epistolæ* (Amstelod., 1682), Part iii., ep. 67.

‡ Bacon's *Works*, by Shaw, 1733, vol. iii., p. 441. (See *Cosmos*, vol. i., p. 315.)

§ Hooke's *Posthumous Works*, p. 364. (Compare my *Relat. Historique*, t. i., p. 199.) Hooke, however, like Galileo, unhappily assumed a difference in the velocity of the rotation of the Earth and of the atmosphere. See *Posth. Works*, p. 88 and 363.

‖ Although, according to Galileo's views, the detention of the particles of air is one of the causes of the trade winds, yet his hypothesis ought not to be confounded, as has recently been done, with that of Hooke and Hadley. Galileo, in the *Dialogo quarto* (*Opere*, t. iv., p. 311), makes Salviati say, " Dicevamo pur' ora che' l'aria, come corpo tenue, e fluido, e non saldamente congiunto alla terra, pareva che non

rect view was taken up by Halley late in the eighteenth
century, and was then more fully and satisfactorily explained
with reference to the action of the velocity of rotation peculiar to each parallel of latitude. Halley, prompted by his
long sojourn in the torrid zone, had even earlier (1686) published an admirable empirical work on the geographical extension of trade winds and monsoons. It is surprising that
he should not have noticed, in his magnetic expeditions, the
law of rotation of the winds, which is so important for the
whole of meteorology, since its general features had been recognized by Bacon and Johann Christian Sturm, of Hippol-
stein (according to Brewster, the actual discoverer of the
differential thermometer*).

In the brilliant epoch characterized by the foundation of
mathematical natural philosophy, experiments were not wanting for determining the connection existing between the humidity of the atmosphere, and the changes in the temperature and the direction of the winds. The Accademia del
Cimento had the felicitous idea of determining the quantity
of vapor by evaporation and precipitation. The oldest Florentine hygrometer was accordingly a condensation-hygrometer—an apparatus in which the quantity of the discharged

avesse necessità d'obbedire al suo moto, se non in quanto l' asprezza
della superficie terrestre ne rapisce, e seco porta una parte a se contigua,
che di non molto intervallo sopravanza le maggiori altezze delle mon-
tagne; la qual pozzion d'aria tanto meno dovrà esser renitente alla
conversion terrestre, quanto che ella è ripiena di vapori, fumi, ed esala-
zioni, materie tutte participanti delle qualità terrene : e per conseguen-
za atte nate per lor natura (?) a i medesimi movimenti. Ma dove, man-
cassero le cause del moto, cioè dova la superficie del globo avesse grandi
spazii piani, e meno vi fusse della mistione de i vapori terreni, quivi ces-
serebbe in parte la causa, per la quale l' aria ambiente dovesse total-
mente obbedire al rapimento della conversion terrestre ; sì che in tali
uoghi, mentre che la terra si volge verso Oriente, si dovrebbe sentir con-
tinuamente un vento, che si ferisse, spirando da Levante verso Ponente;
e tale spiramento dovrebbe farsi più sensibile, dove la vertigine del
globo fusse più veloce: il che sarebbe ne i luoghi più remoti da i Poli,
e vicini al cerchio massimo della diurna conversione. L'esperienza ap-
plaude molto a questo filosofico discorso, poichè ne gli ampi mari sotto-
posti alla Zona torrida, dove anco l'evaporazioni terrestri mancano (?)
si sente una perpetua aura muovere da Oriente. . . . ."

* Brewster, in the *Edinburgh Journal of Science*, vol. ii., 1825, p. 145.
Sturm has described the Differential Thermometer in a little work, entitled *Collegium Experimentale Curiosum* (Nuremberg, 1676), p. 49.
On the Baconian law of the rotation of the wind, which was first extended to both zones, and recognized in its ultimate connection with
the causes of all atmospheric currents by Dove, see the detailed treatise
of Muncke, in the new edition of Gehler's *Physikal. Wörterbuch*, bd.
x., s. 2003–2019 and 2030–2035.

precipitated water was determined by weight.* In addition
to the condensation-hygrometer, which, by the aid of the ideas
of Le Roy in our own times, has gradually led to the exact
psychrometrical methods of Dalton, Daniell, and August, we
have (in accordance with the examples set by Leonardo da
Vinci†) the absorption-hygrometer, composed of substances
taken from the animal and vegetable kingdoms, made by San-
tori (1625), Torricelli (1646), and Molineux. Catgut and the
spikes of grasses were employed almost simultaneously. Instruments of this kind, which were based on the absorption by
organic substances of the aqueous vapor contained in the atmosphere, were furnished with indicators or pointers, and small
counter-weights, very similar in their construction to the hair
and whalebone hygrometers of Saussure and De Luc. The
instruments of the seventeenth century were, however, deficient in the fixed points of dryness and humidity so necessary
to the comparison and comprehension of the results, and which
were at length determined by Regnault (setting aside the susceptibility acquired by time in the hygrometrical substances
employed). Pictet found the hair of a Guanche mummy
from Teneriffe, which was perhaps a thousand years old, sufficiently susceptible in a Saussure's hygrometer.‡

The electric process was recognized by William Gilbert as
the action of a proper natural force allied to the magnetic
force. The book in which this view is first expressed, and in
which the words *electric force*, electric emanations, and electric attraction are first used, is the work of which I have already frequently spoken,§ and which appeared in the year

* Antinori, p. 45, and even in the *Saggi*, p. 17–19.
† Venturi, *Essai sur les Ouvrages Physico-mathématiques de Leonard
de Vinci*, 1797, p. 28.
‡ *Bibliothèque Universelle de Genève*, t. xxvii., 1824, p. 120.
§ Gilbert, *De Magnete*, lib. ii., cap. 2–4, p. 46–71. With respect to
the interpretation of the nomenclature employed, he already said,
Electrica quæ attrahit eadem ratione ut electrum; versorium non mag-
neticum ex quovis metallo, inserviens electricis experimentis. In the
text itself we find as follows: Magneticè ut ita dicam, vel electricè
attrahere (vim illam electricam nobis placet appellare . . . .) (p. 52);
effluvia electrica, attractiones electricæ. We do not find either the abstract expression *electricitas* or the barbarous word *magnetismus* introduced in the eighteenth century. On the derivation of ἤλεκτρον, "the
attractor and the attracting stone," from ἕλξις and ἕλκειν, already indicated in the Timæus of Plato, p. 80, c., and the probable transition
through a harder ἔλεκτρον, see Buttmann, *Mythologus*, bd. ii. (1829),
s. 357. Among the theoretical propositions put forward by Gilbert
(which are not always expressed with equal clearness), I give the following: "Cum duo sint corporum genera, quæ manifestis sensibus

1600, under the title of " Physiology of Magnets and of the Earth as a great Magnet (de magno magnete tellure)." " The property," says Gilbert, " of attracting light substances, when rubbed, be their nature what it may, is not peculiar to amber, which is a condensed earthy juice cast up by the waves of the sea, and in which flying insects, ants, and worms lie entombed as in eternal sepulchers (æternis sepulchris). The force of attraction belongs to a whole class of very different substances, as glass, sulphur, sealing wax, and all resinous substances, rock crystal, and all precious stones, alum, and rock salt." Gilbert measured the strength of the excited electricity by means of a small needle, not made of iron, which moved freely on a pivot (*versorium electricum*), and perfectly similar to the apparatus used by Haüy and Brewster in testing the electricity excited in minerals by heat and friction. " Friction," says Gilbert further, " is productive of a stronger effect in dry than in humid air; and rubbing with silk cloths is most advantageous. The globe is held together as by an electric force (?) Globus telluris per se electrice congregatur et cohæret ; for the tendency of the electric action is to produce the cohesive accumulation of matter (motus electricus est motus coacervationis materiæ)." In these obscure axioms we trace the recognition of *terrestrial electricity*—the expression of a force—which, like magnetism, appertains as such to matter. As yet we meet with no allusions to repulsion, or the difference between insulators and conductors.

Otto von Guericke, the ingenious inventor of the air pump, was the first who observed any thing more than mere phenomena of attraction. In his experiments with a rubbed piece of sulphur, he recognized the phenomena of repulsion, which

nostris motionibus corpora allicere videntur, Electrica et Magnetica; Electrica naturalibus ab humore effluviis ; Magnetica formalibus efficientiis seu potius primariis vigoribus, incitationes faciunt. Facile est hominibus ingenio acutis, absque experimentis et usu rerum labi, et errare. Substantiæ proprietates aut familiaritates, sunt generales nimis, nec tamen veræ designatæ causæ, atque, ut ita dicam, verba quædam sonant, re ipsâ nihil in specie ostendunt. Neque ista succini credita attractio, a singulari aliquâ proprietate substantiæ, aut familiaritate assurgit ; cum in pluribus aliis corporibus eundem effectum, majori industria invenimus, et omnia etiam corpora cujusmodicunque proprietatis, ab omnibus illiis alliciuntur." (*De Magnete,* p. 50, 51, 60, and 65.) Gilbert's principal labors appear to fall between the years from 1590 to 1600. Whewell justly assigns him an important place among those whom he terms " practical reformers of the physical sciences." Gilbert was surgeon to Queen Elizabeth and James I., and died in 1603. After his death there appeared a second work, entitled " *De Mundo nostro Sublunari Philosophia Nova.*"

subsequently led to the establishment of the laws of the sphere of action, and of the distribution of electricity. He heard the first sound, and saw the first light in artificially-produced electricity. In an experiment instituted by Newton in 1675, the first traces of the electric charge in a rubbed plate of glass were seen.* We have here only sought the earliest germs of electric knowledge, which, in its great and singularly-retarded development, has not only become one of the most important branches of meteorology, but has also thrown much light on the internal action of terrestrial forces, since magnetism has been recognized as one of the simplest forms under which electricity is manifested.

Although Wall in 1708, Stephen Gray in 1734, and Nollet conjectured the identity of friction-electricity and of lightning, it was first proved with empirical certainty in the middle of the eighteenth century by the successful efforts of the celebrated Benjamin Franklin. From this period the electric process passed from the domain of speculative physics into that of cosmical contemplation—from the recesses of the study to the freedom of nature. The doctrine of electricity, like that of optics and of magnetism, experienced long periods of extremely tardy development, until in these three sciences the labors of Franklin and Volta, of Thomas Young and Malus, of Œrsted and of Faraday, roused their cotemporaries to an admirable degree of activity. Such are the alternations of slumber and of suddenly-awakened activity that appertain to the progress of human knowledge.

But if, as we have already shown, the relations of temperature, the alternations in the pressure of the atmosphere, and the quantity of the vapor contained in it, were made the object of direct investigation by means of the invention of appropriate, although still very imperfect physical instruments, and by the acute penetration of Galileo, Torricelli, and the members of the Accademia del Cimento, all that refers to the chemical composition of the atmosphere remained, on the other hand, shrouded in obscurity. The foundations of pneumatic chemistry were, it is true, laid by Johann Baptist von Helmont and Jean Rey in the first half of the seventeenth century, and by Hooke, Mayow, Boyle, and the dogmatizing Becher in the closing part of the same century ; but, however striking may have been the correct apprehension of detached and important phenomena, the insight into their connection was still wanting. The old belief in the elementary simplic-

* Brewster, *Life of Newton*, p. 307.

ity of the air, which acts on combustion, on the oxydation of metals, and on respiration, constituted a most powerful impediment.

The inflammable or light-extinguishing gases occurring in caverns and mines (the *spiritus letales* of Pliny), and the escape of these gases in the form of vesicles in morasses and mineral springs, had already attracted the attention of Basilius Valentinus, a Benedictine monk of Erfurt (probably at the close of the fifteenth century), and of Libavius, an admirer of Paracelsus, in 1612. Men drew comparisons between that which was accidentally observed in alchemistical laboratories, and that which was found prepared in the great laboratories of nature, especially in the interior of the Earth. The working of mines in strata, rich in ores (especially those containing iron pyrites, which become heated by oxydation and contact-electricity), led to conjectures of the chemical relation existing between metals, acids, and the external air having access to them. Even Paracelsus, whose visionary fancies belong to the period of the first discovery of America, had remarked the evolution of gas when iron was dissolved in sulphuric acid. Van Helmont, who first employed the term *gas*, distinguished it from atmospheric air, and also, by its non-condensibility, from vapors. According to him, the clouds are vapors, and become converted into gas, when the sky is very clear, " by means of cold and the influence of the stars." Gas can only become water after it has been again converted into vapor. Such were the views entertained in the first half of the seventeenth century regarding the meteorological process. Van Helmont was not acquainted with the simple method of taking up and separating his *gas sylvestre* (the name under which he comprehended all uninflammable gases which do not maintain combustion and respiration, and differ from pure atmospheric air) ; but he caused a light to burn in a vessel under water, and observed that, when the flame was extinguished, the water entered, and the *volume of air* diminished. Van Helmont likewise endeavored to show by determinations of weight (which we find already given by Cardanus) that all the solid portions of plants are formed from water.

The alchemistic opinions of the Middle Ages regarding the composition of metals, and the loss of their brilliancy by combustion in the open air (incineration, calcination), led to a desire of investigating the conditions by which this process was attended, and the changes experienced by the calcined metals, and by the air in contact with them. Cardanus, as early as

in 1553, had noticed the increase of weight that accompanies
the oxydation of lead, and, perfectly in accordance with the
idea of the myth of Phlogiston, had attributed it to the escape
of a "celestial fiery matter," causing levity ; and it was not
until eighty years afterward that Jean Rey, a remarkably
skillful experimenter at Bergerac, who had investigated with
the greatest care the increase of weight during the calcination
of lead, tin, and antimony, arrived at the important conclusion that this increase of weight must be ascribed to the access of the air to the metallic calx. "Je responds et soutiens
glorieusement," he says, "que ce surcroît de poids vient de
l'air qui dans le vase a été espessi."*

Men had now discovered the path which was to lead them
to the chemistry of the present day, and through it to the
knowledge of a great cosmical phenomenon, viz., the connection between the oxygen of the atmosphere and vegetable life
The combination of ideas, however, which presented itself to
the minds of distinguished men, was strangely complicated in
its nature.   Toward the close of the seventeenth century a
belief arose in the existence of nitrous particles (*spiritus nitro-
aëreus pabulum nitrosum*), which, contained in the air, and
identical with those which are fixed in saltpetre, were supposed to possess the necessary requirements for combustion ;
an opinion which, obscurely expressed by Hooke in his *Micro-
graphia* (1671), is found more fully developed by Mayow in
1669, and by Willis in 1671.   "It was maintained that the
extinction of flame in a closed space is not owing to the over-
saturation of the air with vapors emanating from the burning
body, but is the consequence of the entire absorption of the
spiritus nitro-aëreus contained in the nitrogenous air." The
sudden increase of the glowing heat when fusing saltpetre
(emitting oxygen) is strewed upon coals, and the formation of

---

* Rey, strictly speaking, only mentions the access of air to the oxyds;
he did not know that the oxyds themselves (which were then called
the earthy metals) are only combinations of metals and air.   According to him, the air makes "the metallic calx heavier, as sand increases
in weight when water hangs about it."   The calx is susceptible of being saturated with air.   "L'air *espaissi* s'attache à la chaux, ainsi le
poids augmente du commencement jusqu'à la fin: mais quand tout en
est affublé, elle n'en sçauroit prendre d'avantage.   Ne continuez plus
votre calcination soubs cet espoir, vous perdriez vostre peine."   Rey's
work thus contains the first approach to the better explanation of a
phenomenon, whose more complete understanding subsequently exercised a favorable influence in reforming the whole of chemistry.   See
Kopp, *Gesch. der Chemie*, th. iii., s. 131–133. (Compare, also, in the same
work, th. i., s. 116–127, and th. iii., s. 119–138, as well as s. 175–195.)

saltpetre on clay walls in contact with the atmosphere, appear to have contributed jointly to the adoption of this view. The nitrous particles of the air influence, according to Mayow, the respiration of animals, the result of which is to generate animal heat, and to deprive the blood of its dark color ; and, while they control all the processes of combustion and the calcination of metals, they play nearly the same part in the antiphlogistic chemistry as oxygen. The cautious and doubting Robert Boyle was well aware that the presence of a certain constituent of atmospheric air was necessary to combustion, but he remained uncertain with regard to its nitrous nature.

Oxygen was to Hooke and Mayow an ideal object—a delusion of the intellect. The acute chemist and vegetable physiologist Hales first saw oxygen evolved in the form of a gas when, in 1727, he was engaged at Mennige in calcining a large quantity of lead under a very powerful heat. He observed the escape of the gas, but he did not examine its nature, or notice the vivid burning of the flame. Hales had no idea of the importance of the substance he had prepared. The vivid evolution of light in bodies burning in oxygen, and its properties, were, as many persons maintain, discovered independently—by Priestley in 1772–1774, by Scheele in 1774–1775, and by Lavoisier and Trudaine in 1775.*

The dawn of pneumatic chemistry has been touched upon in these pages with respect to its historical relations, because, like the feeble beginning of electrical science, it prepared the way for those grand views regarding the constitution of the atmosphere and its meteorological changes which were manifested in the following century. The idea of specifically distinct gases was never perfectly clear to those who, in the seventeenth century, produced these gases. The difference between atmospheric air and the irrespirable light-extinguishing or inflammable gases was now again exclusively ascribed to the admixture of certain vapors. Black and Cavendish first showed, in 1766, that carbonic acid (fixed air) and hydrogen (combustible air) are specifically different aëriform fluids. So long did the ancient belief of the elementary simplicity of the atmosphere check all progress of knowledge. The final knowledge of the chemical composition of the atmosphere, acquired by means of the delicate discrimination of its quantitative re-

* Priestley's last complaint of that which " Lavoisier is considered to have appropriated to himself," is put forth in his little memoir entitled " *The Doctrine of Phlogiston Established*," 1800, p. 43.

lations by the beautiful researches of Boussingault and Dumas, is one of the brilliant points of modern meteorology.

The extension of physical and chemical knowledge, which we have here briefly sketched, could not fail to exercise an influence on the earliest development of geognosy. A great number of the geognostic questions, with the solution of which our own age has been occupied, were put forth by a man of the most comprehensive acquirements, the great Danish anatomist, Nicolaus Steno (Stenson), in the service of the Grand-duke of Tuscany, Ferdinand II.; by another physician, Martin Lister, an Englishman, and by Robert Hooke, the "worthy rival" of Newton.* Of Steno's services in the geognosy of position I have treated more circumstantially in another work.† Leonardo da Vinci, toward the close of the fifteenth century (probably when he was planning the canals in Lombardy which intersect the alluvial and tertiary formations), Fracastoro in 1517, on the occasion of the accidental exposure of rocky strata, containing fossil fishes, at Monte Bolca, near Verona, and Bernard Palissy, in his investigations regarding fountains in 1563, had indeed recognized the existence of traces of an earlier oceanic animal world. Leonardo, as if with a presentiment of a more philosophical classification of animal forms, terms conchylia "*animali che hanno l'ossa di fuora.*" Steno, in his work on the substances contained in rocks (De Solido intra Solidum naturalitér Contento), distinguishes (1669) between (primitive?) rocky strata which have become solidified before the creation of plants and animals, and therefore contain no organic remains, and sedimentary strata (turbidi maris sedimenta sibi invicem imposita) which alternate with one another, and cover the first-named strata. All fossiliferous strata were originally deposited in horizontal beds. This inclination (or fall) has been occasioned partly by the eruption of subterranean vapors, generated by central heat (ignis in medio terræ), and partly by the giving way of the feebly-supported lower strata.‡ The valleys are the result of this falling in."

Steno's theory of the formation of valleys is that of De Luc, while Leonardo da Vinci, like Cuvier, regards the valleys as

* Sir John Herschel, *Discourse on the Study of Natural Philosophy*, p. 116.

† Humboldt, *Essai Géognostique sur le Gisement des Roches dans les deux Hémispheres*, 1823, p. 38.

‡ Steno, *De Solido intra Solidum naturaliter Contento*, 1669, p. 2, 17, 28, 63, and 69 (fig. 20–25).

the former beds of streams.* In the geognostic character of the soil of Tuscany, Steno recognized convulsions which must, in his opinion, be ascribed to six great natural epochs (Sex sunt distinctæ Etruriæ facies ex præsenti facie Etruriæ collectæ). The sea had broken in at six successive periods, and, after continuing to cover the interior of the land for a long time, had retired within its ancient limits. All petrifactions were not, however, according to his opinion, referable to the sea ; and he distinguished between pelagic and fresh-water formations. Scilla, in 1670, gave drawings of the petrifactions of Calabria and Malta; and among the latter, our great anatomist and zoologist, Johannes Müller, has recognized the oldest drawing of the teeth of the gigantic Hydrarchus of Alabama (the *Zeuglodon cetoides* of Owen), a mammal of the great order of the Cetacea.† The crown of these teeth is formed similarly to those of seals.

Lister, as early as 1678, made the important assertion that each kind of rock is characterized by its own fossils, and that " the species of Murex, Tellina, and Trochus, which occur in the stone quarries of Northamptonshire, are indeed similar to those existing in the present seas, but yet, when more closely examined, they are found to differ from them." They are, he says, specifically different.‡ Strictly conclusive proofs of the truth of these grand conjectures could not, however, be advanced in the then imperfect condition of descriptive morphology. We here indicate the early dawn and speedy extinction of light prior to the noble palæontological researches of Cuvier and Alexander Brongniart, which have given a new form to the geognosy of sedimentary formations.§ Lister, whose at-

* Venturi, *Essai sur les Ouvrages Physico-mathématiques de Leonard de Vinci*, 1797, § 5, No. 124.

† Agostino Scilla, *La vana Speculazione disingannata dal Senso*, Nap., 1670, tab. xii., fig. 1. Compare Joh. Müller, *Bericht über die von Herrn Koch, in Alabama Gesammelten Fossilen Knochenreste seines Hydrachus* (the Basilosaurus of Harlan, 1835 ; the Zeuglodon of Owen, 1839 ; the Squalodon of Grateloup, 1840 ; the Dorudon of Gibbes, 1845), read in the Royal Academy of Sciences at Berlin, April—June, 1847. These valuable fossil remains of an ancient world, which were collected in the State of Alabama (in Washington county, near Clarksville), have become, by the munificence of our king, the property of the Zoological Museum at Berlin since 1847. Besides the remains found in Alabama and South Carolina, parts of the Hydrarchus have been found in Europe, at Leognan near Bordeaux, near Linz on the Danube, and, in 1670, in Malta.

‡ Martin Lister, in the *Philosophical Transactions*, vol. vi., 1671, No. lxxvi., p. 2283.

§ See a luminous exposition of the earlier progress of palæontological

tention had been drawn to the regular succession of strata in England, first felt the want of geognostic maps. Although these phenomena, and their dependence on ancient inundations (either single or repeated), riveted the attention of men, and, mingling belief and knowledge together, gave origin in England to the so-called systems of Ray, Woodward, Burnet, and Whiston ; yet, owing to the total want of mineralogical distinction between the constituents of compound minerals, all that relates to crystalline and massive rocks of eruption remained unexplored. Notwithstanding the opinions held with respect to a central heat in the Earth, earthquakes, hot springs, and volcanic eruptions were not regarded as the consequence of the reaction of the planet against its external crust, but were attributed to trifling local causes, as, for instance, the spontaneous combustion of beds of iron pyrites. The unscientific experiments of Lemery (1700) unhappily exercised a long-continued influence on volcanic theories, although the latter might certainly have been raised to more general views by the richly-imaginative *Protogæa* of Leibnitz (1680).

The *Protogæa*, occasionally even more imaginative than the many metrical attempts of the same author which have lately been made known,[*] teaches " the scorification of the cavernous, glowing, once self-luminous crust of the Earth, the gradual cooling of the radiating surface enveloped in vapors, the precipitation and condensation of the gradually-cooled, vaporous atmosphere into water, the sinking of the level of the sea by the penetration of water into the internal cavities of the earth, and, finally, the breaking in of these caves, which occasions the fall, or horizontal inclination of these strata." The physical portion of this wild and fanciful view presents some features which will not appear to merit entire rejection by the adherents of our modern geognosy, notwithstanding its more perfect development in all its branches. Among these better traits we must reckon the movement and heat in the interior of the globe, and the cooling occasioned by radiation from the surface ; the existence of an atmosphere of vapor ; the pressure exercised by these vapors on the Earth's strata during their consolidation ; and the two-fold origin of the mass-

studies, in Whewell's *History of the Inductive Sciences*, 1837, vol. iii., p. 507–545.

[*] Leibnizens, *Geschichtliche Aufsätze und Gedichte*, edited by Pertz, 1847, in the *Gesammelte Werke : Geschichte*, bd. iv. On the first sketch of the *Protogæa* of 1691, and on its subsequent revisions, see Tellkampf, *Jahresbericht der Bürgerschule zu Hannover*, 1847, s. 1–32.

es by fusion and solidification, or by precipitation from the
waters. The typical character and mineralogical differences
of rocks, or, in other words, the associations of certain mostly
crystallized substances recurring in the most remote regions,
are as little made a subject of consideration in the *Protogæa*
as in Hooke's geognostic views. Even in the last-named
writer, physical speculations on the action of subterranean
forces in earthquakes, in the sudden upheaval of the sea's
bottom and of littoral districts, and in the origin of islands and
mountains, hold a prominent place. The nature of the organic remains of a former world even led him to conjecture that
the temperate zone must originally have enjoyed the heat of a
tropical climate.

It still remains for us to speak of the greatest of all geognostic phenomena—the mathematical figure of the Earth—in
which we distinctly trace a reflection of the primitive world
in the condition of fluidity of the rotating mass, and its solidification into our terrestrial spheroid. The main outlines of
the figure of the Earth were sketched as early as the close of
the seventeenth century, although the relation between the
polar and equatorial axes was not ascertained with numerical
exactness. Picard's measurement of a degree, made in 1670
with instruments which he had himself improved, is so much
the more important, since it was the means of inducing Newton to resume with renewed zeal his theory of gravitation
(which he discovered as early as 1666, but had subsequently
neglected), by offering to that profound and successful investigator the means of proving how the attraction of the Earth
maintained the Moon in its orbit, while urged on its course
by the centrifugal force. The fact of the compression of the
poles of Jupiter, which was much earlier recognized,* had, as
it is supposed, induced Newton to reflect on the causes of a
form which deviated so considerably from sphericity. The
experiments on the actual length of the seconds pendulum by
Richer at Cayenne in 1673, and by Varin on the western
coast of Africa, had been preceded by others of less decisive
character, prosecuted in London, Lyons, and Bologna at a
difference of 7° of latitude.†

The decrease of gravity from the poles to the equator, which
even Picard had long denied, was now generally admitted.
Newton recognized the polar compression, and the spheroidal
form of the earth as a consequence of its rotation ; and he

* *Cosmos*, vol. i., p. 164.
† Delambre, *Hist. de l'Astronomie Mod.*, t. ii., p. 601.

even ventured to determine numerically the amount of this compression, on the assumption of the homogeneous nature of the mass. It remained for the comparative measurements of degrees in the eighteenth and nineteenth centuries, at the equator, near the north pole, and in the temperate zones of both the southern and northern hemispheres, to determine exactly the mean amount of this compression, and by that means to ascertain the true figure of the Earth. The existence of this compression announces, as has already been observed in the " Picture of Nature,"* that which may be named the most ancient of all geognostic events—the condition of general fluidity of a planet, and its earlier and progressive solidification.

We began our description of the great epoch of Galileo, Kepler, Newton, and Leibnitz with the discoveries in the regions of space by means of the newly-invented telescope, and we now close it with the figure of the Earth, as it was then recognized from theoretical conclusions. " Newton was enabled to give an explanation of the system of the universe because he succeeded in discovering the force† from whose action the laws of Kepler necessarily result, and which most correspond with these phenomena, since these laws corresponded to and predicted them." The discovery of such a force, the existence of which Newton has developed in his immortal work, the *Principia* (which comprise the general sciences of nature), was almost simultaneous with the opening of the new paths to greater mathematical discoveries by means of the invention of the infinitesimal calculus. Intellectual labor shows itself in all its exalted grandeur where, instead of requiring external material means, it derives its light exclusively from the sources opened to pure abstraction by the mathematical development of thought. There dwells an irresistible charm, venerated by all antiquity, in the contemplation of mathematical truths— in the everlasting revelations of time and space, as they reveal

* *Cosmos*, vol. i., p. 163. The dispute regarding priority as to the knowledge of the Earth's compression, in reference to a memoir read by Huygens in 1669 before the Paris Academy, was first cleared up by Delambre in his *Hist. de l'Astr. Mod.*, t. i., p. lii., and t. ii., p. 558. Richer's return to Europe occurred indeed in 1673, but his work was not printed until 1679; and as Huygens left Paris in 1682, he did not write the *Additamentum* to the Memoir of 1669, the publication of which was very late, until he had already before his eyes the results of Richer's Pendulum Experiments, and of Newton's great work, *Philosophiæ Naturalis Principia Mathematica*.

† Bessel, in *Schumacher's Jahrbuch für* 1843, s. 32.

themselves in tones, numbers, and lines.* The improvement
of an intellectual instrument of research—analysis—has powerfully accelerated the reciprocal fructification of ideas, which
is no less important than the rich abundance of their creations.
It has opened to the physical contemplation of the universe
new spheres of immeasurable extent in the terrestrial and celestial regions of space, revealed both in the periodic fluctuations of the ocean and in the varying perturbations of the
planets.

RETROSPECT OF THE EPOCHS THAT HAVE BEEN SUCCESSIVELY
CONSIDERED.—INFLUENCE OF EXTERNAL OCCURRENCES ON THE
DEVELOPMENT OF THE RECOGNITION OF THE UNIVERSE AS ONE
WHOLE.—MULTIPLICITY AND INTIMATE CONNECTION OF THE SCIEN-
TIFIC EFFORTS OF RECENT TIMES.—THE HISTORY OF THE PHYSIC-
AL SCIENCES BECOMES GRADUALLY ASSOCIATED WITH THE HISTO-
RY OF THE COSMOS.

I APPROACH the termination of my bold and difficult undertaking. Upward of two thousand years have been passed in
review before us, from the early stages of civilization among
the nations who dwelt around the basin of the Mediterranean
and the fruitful river valleys of Western Asia, to the beginning of the last century, to a period, therefore, at which general views and feelings were already beginning to blend with
those of our own age. I have endeavored, in seven sharply-
defined sections, forming, as it were, a series of as many separate pictures, to present a history of the physical contemplation of the universe, or, in other words, the history of the
gradual development of the knowledge of the universe as a
whole. To what extent success may have attended the attempt to apprehend the mass of accumulated matter, to seize
on the character of the principal epochs, and to indicate the
paths on which ideas and civilization have been advanced, can
not be determined by him who, with a just mistrust of his remaining powers, is alone conscious that the image of so great
an undertaking has been present to his mind in clear though
general outlines.

At the commencement of our consideration of the period
of the Arabs, and in beginning to describe the powerful influence exercised by the admixture of a foreign element in
European civilization, I indicated the limits beyond which
the history of the Cosmos coincides with that of the physical

* Wilhelm von Humboldt, *Gesammelte Werke*, bd. i., s. 11.

sciences. According to my idea, the historical recognition of the gradual extension of natural science in the two spheres of terrestrial and celestial knowledge (geography and astronomy) is associated with certain periods and certain active intellectual events, which impart a peculiar character and coloring to those epochs. Such, for instance, were the undertakings which led Europeans into the Euxine, and permitted them to conjecture the existence of another sea-shore beyond the Phasis; the expeditions to tropical lands rich in gold and incense; the passage through the Western Straits, or the opening of that great maritime route on which were discovered, at long intervals of time, Cerne and the Hesperides, the northern tin and amber lands, the volcanic islands of the Azores, and the New Continent of Columbus, south of the ancient settlement of the Scandinavians. To the consideration of the movements which emanated from the basin of the Mediterranean, and the most northern part of the neighboring Arabian Gulf, and of the expeditions on the Euxine and to Ophir, succeed, in my historical delineation, the campaigns of the Macedonian conqueror, and his attempts to fuse together the west and the east; the influence exercised by Indian maritime trade and by the Alexandrian Institute under the Ptolemies; the universal dominion of the Romans under the Cæsars; and, lastly, the taste evinced by the Arabs for the study of nature and of natural forces, especially with reference to astronomy, mathematics, and practical chemistry, a taste that exercised so important and beneficial an influence. According to my view, the series of events which suddenly enlarged the sphere of ideas, excited a taste for the investigation of physical laws, and animated the efforts of men to arrive at the ultimate comprehension of the universe as a whole, terminated with the acquisition of an entire hemisphere which had till then lain concealed, and which constituted the greatest geographical discovery ever made. Since this period, as we have already remarked, the human mind has brought forth great and noble fruits without the incitement of external occurrences, and, as the effect of its own inherent power, developed simultaneously in all directions.

Among the instruments which man formed for himself, like new organs, as it were, to heighten his powers of sensuous perception, there was one which exercised an influence similar to that of some great and sudden event. By the power of penetrating space possessed by the telescope, considerable portions of the heavens were almost at once explored, the number of known heavenly bodies was increased, and attempts

made to determine their forms and orbits.   Mankind now
first attained to the possession of the " celestial sphere" of the
Cosmos.   Sufficient foundation for a seventh section of the
history of the contemplation of the universe seemed to be afforded by the importance of the acquisition of this celestial
knowledge, and of the unity of the efforts called forth by the
use of the telescope.   If we compare another great invention,
and one of recent date, the voltaic pile, with the discovery of
this optical instrument, and reflect on the influence which it
has exercised on the ingenious electro-chemical theory; on
the production of the metals; of the earths and alkalies; and
on the long-desired discovery of electro-magnetism, we are
brought to the consideration of a series of phenomena called
forth at will, and which, by many different paths, lead to a
profound knowledge of the rule of natural forces, but which
constitute rather a section in the history of physical science
than a direct portion of the history of cosmical contemplation.
It is this multiplied connection between the various departments of modern knowledge that imparts such difficulty to the
description and limitation of its separate branches.   We have
very recently seen that electro-magnetism, acting on the direction of the polarized ray of light, produces modifications
like chemical mixtures.   Where, by the intellectual labors
of the age, all knowledge appears to be progressing, it is as
dangerous to attempt to describe the intellectual process, and
to depict that which is constantly advancing as already at the
goal of its efforts, as it is difficult, with the consciousness of
one's own deficiencies, to decide on the relative importance
of the meritorious efforts of the living and of the recently departed.

In the historical considerations I have almost every where,
in describing the early germs of natural knowledge, designated
the degree of development to which it has attained in recent
times.   The third and last portion of my work will, for the
better elucidation of the general picture of nature, set forth
those results of observation on which the present condition of
scientific opinions is principally based.   Much that, according to other views than mine, regarding the composition of a
book of nature, may have appeared wanting, will there find
its place.   Excited by the brilliant manifestation of new discoveries, and nourishing hopes, the fallacy of which often continues long undetected, each age dreams that it has approxi-
mated closely to the culminating point of the recognition and
comprehension of nature.   I doubt whether, on serious reflec-

tion, such a belief will tend to heighten the enjoyment of the present. A more animating conviction, and one more consonant with the great destiny of our race, is, that the conquests already achieved constitute only a very inconsiderable portion of those to which free humanity will attain in future ages by the progress of mental activity and general cultivation. Every acquisition won by investigation is merely a step to the attainment of higher things in the eventful course of human affairs.

That which has especially favored the progress of knowledge in the nineteenth century, and imparted to the age its principal character, is the general and beneficial endeavor not to limit our attention to that which has been recently acquired, but to test strictly, by measure and weight, all earlier acquisitions ; to separate certain knowledge from mere conjectures founded on analogy, and thus to subject every portion of knowledge, whether it be physical astronomy, the study of terrestrial natural forces, geology, or archæology, to the same strict method of criticism. The generalization of this course has, most especially, contributed to show, on each occasion, the limits of the separate sciences, and to discover the weakness of certain studies in which unfounded opinions take the place of certain facts, and symbolical myths manifest themselves under ancient semblances as grave theories. Vagueness of language, and the transference of the nomenclature of one science to another, have led to erroneous views and delusive analogies. The advance of zoology was long endangered, from the belief that, in the lower classes of animals, all vital actions were attached to organs similarly formed to those of the higher classes. The knowledge of the history of the development of plants in the so-called Cryptogamic Cormophytes (mosses and liverworts, ferns, and lycopodiaceæ), or in the still lower Thallophytes (algæ, lichens, and fungi), has been still more obscured by the supposed general discovery of analogies with the sexual propagation of the animal kingdom.[*]

If art may be said to dwell within the magic circle of the imagination, the extension of knowledge, on the other hand, especially depends on contact with the external world, and this becomes more manifold and close in proportion with the increase of general intercourse. The creation of new organs (instruments of observation) increases the intellectual and not

[*] Schleiden, *Grundzüge der wissenschaftlichen Botanik*, th. i., 1845, s. 152, th. ii., s. 76 ; Kunth, *Lehrbuch der Botanik*, th. i., 1847, s. 91–100, und 505.

unfrequently the physical powers of man. More rapid than
light, the closed electric current conveys thought and will to
the remotest distance. Forces, whose silent operation in elementary nature, and in the delicate cells of organic tissues,
still escape our senses, will, when recognized, employed, and
awakened to higher activity, at some future time enter within
the sphere of the endless chain of means which enable man to
subject to his control separate domains of nature, and to ap-
proximate to a more animated recognition of the Universe as
a Whole.

END OF VOL. II.

Library of Congress-Cataloging-in-Publication Data

Humboldt, Alexander von. 1769–1859.
    [Kosmos. English]
    Cosmos : a sketch of a physical description of the universe /
by Alexander von Humboldt ; translated by E. C. Otté.
       p.   cm. — (Foundations of natural history)
Previously published: New York : Harper & Brothers, 1858.
Includes index.
ISBN 0-8018-5502-0 (vol. 1, pbk. : alk. paper)
ISBN 0-8018-5503-9 (vol. 2, pbk. : alk. paper)
    1. Cosmology.   I. Series
QB981.H8613   1997
508—dc20   96-36421
                                             CIP